LANDSCAPE SURVEYING

SECOND EDITION

LANDSCAPE SURVEYING

SECOND EDITION | HARRY L. FIELD

DELMAR
CENGAGE Learning™

Australia • Brazil • Japan • Korea • Mexico • Singapore • Spain • United Kingdom • United States

DELMAR
CENGAGE Learning™

Landscape Surveying, Second Edition
Harry L. Field

Vice President, Editorial: Dave Garza
Director of Learning Solutions: Matthew Kane
Senior Acquisitions Editor: Sherry Dickinson
Managing Editor: Marah Bellegarde
Senior Product Manager: Christina Gifford
Editorial Assistant: Scott Royael
Vice President, Marketing: Jennifer Baker
Marketing Director: Debbie Yarnell
Marketing Manager: Erin Brennan
Marketing Coordinator: Erin DeAngelo
Production Director: Wendy Troeger
Production Manager: Mark Bernard
Senior Content Project Manager: Katie Wachtl
Senior Art Director: Dave Arsenault

Cover and Design Credits

Level: © Olivier Le Queinec/www.Shutterstock.com

Drafting Tools: © Steven Bower/www.Shutterstock.com

Digital Surveying Equipment: © Dmitry Kalinovsky/www.Shutterstock.com

Tiered Lawn: © OneSmallSquare/www.Shutterstock.com

GPS Surveying: © Henryk Sadura/www.Shutterstock.com

Brick Walkway: © Yobidaba/www.Shutterstock.com

Spring Gardens: © 2009fotofriends/www.Shutterstock.com

Image Credits

All figures are © Cengage Learning 2012

Library of Congress Control Number: 2011932581

ISBN-13: 978-1-111-31060-8

ISBN-10: 1-111-31060-2

Delmar
5 Maxwell Drive
Clifton Park, NY 12065-2919
USA

Cengage Learning is a leading provider of customized learning solutions with office locations around the globe, including Singapore, the United Kingdom, Australia, Mexico, Brazil, and Japan. Locate your local office at: **international.cengage.com/region**

Cengage Learning products are represented in Canada by Nelson Education, Ltd.

To learn more about Delmar, visit **www.cengage.com/delmar**

Purchase any of our products at your local college store or at our preferred online store **www.cengagebrain.com**

Printed in the United States of America
1 2 3 4 5 6 7 15 14 13 12 11

PREFACE

Surveying is an old and honorable profession. Surveyors must be both artists and scientists. Surveyors must be artists because they must be able to interpret the characteristics of a site and determine the best way to complete the job. They must be scientists because they must know the principles of surveying and be able to produce accurate results.

A characteristic of a profession is a set of fundamental principles that governs the profession and directs the behavior of the members. Surveyors must adhere to two sets of principles: professional and practical. A surveyor must pay close attention to professional ethics and the principles that form the foundation of their profession. They must also understand practical principles such as the necessity of accurate data, or that a benchmark must be easily identifiable and have a stable elevation. All of these practical principles, and more, are included in this text.

Surveying also includes principles based on complicated methods and higher levels of mathematics such as trigonometry, geometry, and calculus. Some of these include establishing control points, calculating areas, balancing traverses, and laying out curves. Combining mathematical-based methods with expensive new technology has led to the perception that surveying must be accomplished by a professional surveyor or engineer. In many instances this is true, but individual knowledgeable in the principles of surveying, using mechanical or electronic equipment, can complete many surveying jobs.

The purpose of this text is to provide an explanation of the basic principles of surveying and a practical application of these principles. This is accomplished by explaining the principles in nonengineering terms, using many illustrations, and providing practical examples. The principles of surveying have not changed, although new methods of data collection have been adopted. The nineteenth-century surveyor used astronomical sights for locating positions, and mechanical instruments to measure distance, elevation, and angles. The modern surveyor locates sites by GPS and uses electronic equipment to record information into an electronic field book. New technology has made surveying a less arduous task, and the amount of human resources required to complete a survey has declined dramatically, but electronics have not replaced humans. Most of this electronic equipment is very expensive and is not economically feasible for any but the professional surveyor. Many surveying jobs around the home and small business and in professions such as agriculture, construction, and landscape architecture do not require a professional surveyor.

This text is intended for the individual or teacher who has a desire to learn or teach the practical applications of surveying. It is designed to be used as a class textbook, but it is also intended to be an excellent resource for the individual who wants to learn on his or her own how to measure distance; complete differential, profile, and topographic surveys; and measure horizontal angles without the requirement of a degree in engineering.

NEW TO THIS EDITION

The entire text has been rewritten to bring better clarity to the material presented. Updates have been made to Chapter 10, incorporating the newest trends in technology used for surveying. A new chapter on determining area has been included.

AVAILABLE SUPPLEMENTS

Instructor On-line Resource containing instructor's guide with solutions to end of chapter activities, as well as tips for instructors to assist their students in successfully completing those assignments.

ACKNOWLEDGMENTS

The author wishes to express his thanks to his family for their understanding and patience throughout the writing of this book. He also wishes to thank his colleagues in the Biosystems and Agricultural Engineering Department at Oklahoma State University for being willing to answer questions and provide wise counsel during this project.

The author and Delmar Learning also wish to thank the following reviewers for their time and content expertise:

Justin Snyder, Alamance Community College, Graham, NC

Internet Disclaimer

The author and Delmar Learning affirm that the Web site URLs referenced herein were accurate at the time of printing. However, due to the fluid nature of the Internet, we cannot guarantee their accuracy for the life of the edition.

Harry Field, Ed.D.
January 2011

ABOUT THE AUTHOR

Harry L. Field, Ed.D., is an associate professor at Oklahoma State University in the Biosystems and Agricultural Engineering Department. He received his B.S. and M.S. degrees in Agricultural Mechanization and Agriculture Education, respectively, from Kansas State University and his Ed.D. in Administration, Curriculum, and Instruction from the University of Nebraska–Lincoln. He provides in-service training for secondary teachers as well as 4-H leaders and also produces and writes educational materials. He is a member of the National Association of Colleges and Teachers in Agriculture (NACTA) and the National Agricultural Mechanics Career Development Event Committee (NAMCDE). Dr. Field has also been awarded both State and National Honorary FFA degrees.

CONTENTS

2 Equipment 18

8 Topographic Survey 119

9 Traverse Survey 146

CHAPTER 1

Principles of Land Measurement and Surveying

Objectives

After reading this chapter, the student should be able to do the following:

- Define the term surveying.
- Explain the history of surveying.
- Understand the terms common to surveying.
- Know the two common categories of surveys.
- List the common types of surveys.
- Explain the necessity of selecting the appropriate level of accuracy and precision.
- Understand the difference between random and systematic errors.

Terms To Know

surveying
Public Land Survey System (PLSS)
metes and bounds
oblate spheroid
ellipse
ellipsoid
level
spirit level
level surface
plane surveying
geodetic surveying
vertical line
plane
vertical plane
horizontal line
horizontal plane
distance
horizontal distance
slope distance
angle
vertex
horizontal angle
vertical angle
horizontal zero
zenith zero
benchmark
United States Geological Survey (USGS)
elevation
National Geodetic Vertical Datum of 1929
backsight
foresight
intermediate foresight
true foresight
turning point
geodetic surveys
plane surveys
degrees-minutes-seconds
sexagesimal system
decimal degrees
differential surveying
profile
profile surveying
topographic surveying
property surveys
construction survey
traverse
accuracy
precision
random error
double reading
three-wire leveling
systematic errors

INTRODUCTION

Land is all around us. We walk on it, build houses and commercial buildings on it, drive on it, fence it, dig trenches in it, and farm it. Life as we know it would not exist without land. Its presence permeates our lives to the point that we seldom think about it. Yet in order to plan or analyze the use of land, measurements must be made.

Many of our uses of the land require us to measure, mark, or locate points on, above, or below the surface. We often do this without thinking about the principles we are using. Property is located and marked before a fence is built. A carpenter carefully marks the corners of a building before starting construction. Engineers and planners may spend months deciding on the location of a road and staking it out. The slope and other features of an area must be measured before a pond is built or a drainageway constructed. The property corners of a parcel of land are located and marked during the transfer of ownership.

The principles of surveying are used in all of the examples mentioned above. The complexity of surveying can range from taking a few minutes and two sticks to lay out a 90° corner, to spending several days with thousands of dollars worth of equipment establishing a road or power line right-of-way to establishing survey control monuments, the most complex survey.

This chapter will define surveying and some of the essential terms in surveying. Many of these terms will be used throughout the text. This chapter should be an important reference for understanding the more complicated principles presented in later chapters.

WHAT IS SURVEYING?

Surveying *is the art and science of measuring and locating points and angles on, above and below the surface of the earth.* In this definition, the terms "art" and "science" are used because good surveying is both. Webster defines art as "skill acquired by experience, study, or observation" and "the conscious use of skill and creative imagination esp. in the production of aesthetic objects."

An example of the art of surveying is being able to reconnoiter a site and determine the "best" instruments and methods to use to collect the desired data. "Best" is a very subjective term. What is best for one individual, crew, or site may not be the best if the conditions change, such as moving to a different site, completing a different type of survey, or working for a different government agency. The best instruments and methods produce the required data with the least consumption of resources. This includes determining what data needs to be collected, the most appropriate surveying method to use, the best location of the instruments, etc.

For professional surveyors and commercial construction operations, time is money. An experienced surveyor can look at a site and determine the best method for collecting the necessary data. This ability is an art because it cannot be learned from a textbook or in a classroom. This "art" is developed through natural abilities and from experience.

In many cases, standards and procedures have been developed to provide guidance in this area. For example, if the purpose of a survey is to establish the legal description of a parcel of land, then standard procedure requires an instrument that measures angles with known and verified accuracy be used. Distances must also be measured within the stated limits and a procedure called "balancing the traverse" should be used.

Surveying is also a science. Webster defines science as *"knowledge covering general truths or the operation of general laws esp. as obtained and tested through scientific method."* This definition is appropriate for many aspects of surveying. The principles and practices that have been developed to collect accurate data provide an example of the use of science in surveying. Accurate data is valuable in surveying, while inaccurate data can be costly because it may result in design errors. A rule to remember is that bad data is worse than no data.

Procedures have been developed and tested over time that, if followed, control errors. These include procedures to set up and use the instruments, record data, and to complete the calculations.

HISTORY OF SURVEYING

Evidence of measuring land and marking boundaries is almost as old as civilization. Hunter-gatherer societies had no need to measure or mark land because they were nomadic and did not have a concept of land ownership. As cultures evolved from "hunting-gathering" to farming, they developed the desire to mark out and claim ownership of the land.

The basis of modern techniques can be traced to ancient Egypt. In Egypt, the most fertile farmland was located along the Nile River and each year the river would flood, completely covering this prime farmland. The flood would erase many if not all of the boundary markers. The Egyptians developed methods to replace the field corners after each flood so each family could identify their area. Certain individuals identified as surveyors were responsible for this task. History indicates that the primary concern was not the assurance that each family farmed the same land every year, but that the correct taxes were collected.

Evidence exists that many early cultures built water collection devices, irrigation canals, large buildings, and complex road systems. For years, archeologists were not able to determine how these feats were accomplished. There is now sufficient evidence for one archeologist to suggest, based on the pieces of pottery he has found, that at least one early South American culture developed a clay bowl level which was used as a surveying instrument, Figure 1-1.

A complete bowl has not been discovered, so the illustration is this author's best guess as to what this instrument actually looked like based on the description of the pottery pieces that have been found. The bowl had two holes opposite each other, the same vertical distance from the bowl rim. It is surmised that it was filled with water up to a mark, and if it were mounted on a tripod or suspended, it could be tilted. The bowl would be positioned at the bottom of a slope and tilted. Water levels at different marks on the opposite side would represent different slopes. The user could sight through the holes to establish the slope.

Figure 1-1 Clay bowl level.

Having a tradition of land ownership, the first Europeans in North America set about establishing an accurate legal system of boundaries and land ownership. The first land descriptions were modeled after the system that was used in Europe. The King of England, working from primitive maps, gave large grants of land and charters for land to many different individuals and groups. These land grants and charters were described with very rudimentary boundary descriptions. Some of the boundary descriptions overlapped, and to make matters worse, some individuals interpreted the grant boundary lines to give them the best land and exclude land that they did not want. This resulted in many boundary conflicts.

By the Revolutionary War, many boundaries were in dispute and lawyers spent a large portion of their time settling these conflicts. The number of land disputes expanded dramatically after the Revolutionary War. The Continental Congress increased the number of land boundary problems because it did not have sufficient funds to pay soldiers after the war, so they granted them land instead of wages. The land grants did not cause the problem; the land boundary problems arose because a good plan for identifying the land did not exist. The soldiers were told to identify their claim by staking the corners and filing a written description of the corner marker locations at a local land office established for that purpose. This type of land identification is called **metes and bounds**.

Individuals naturally set their stakes to claim the best land with no regard to direction and location of other stakes. This caused a sudden rash of land disputes. These land disputes added fuel to the desire for a better system. Consequently, the Continental Congress formed a committee charged with the task of determining a better method for measuring and identifying land.

The result was the adoption of the **Public Land Survey System** (PLSS) commonly called the **Rectangular System of Land Description** (RSLD). The adoption of the PLSS required that all land must be surveyed. During these surveys, every section corner, and in most cases half section corners, were located and marked with a permanent marker. Most of these markers still exist and continue to be used as starting points for modern surveys. The PLSS is explained in more depth in Chapter 3.

SURVEYING TERMS

To understand the methods and techniques of any subject, it is important to understand the terminology used. This is especially true for surveying because the meaning of some terms is different from common usage. A good understanding of the following terms and definitions is essential for understanding surveying.

Oblate Spheroid

Oblate Spheroid is a term that has been used to describe the shape of the Earth. The earth is not a sphere. The distance between the poles is less than the diameter at the equator. An oblate spheroid is a solid obtained by rotating an ellipse on its shortest axis, Figure 1-2.

Because of the earth's relief, a rotated ellipsoid does not represent the surface. It is used to describe mean sea level, which is used as the reference for true elevations. Different mathematical models are used to describe the earth's shape. The scientific discipline that measures and models the earth is called Geodesy, also called **geodetics**. Some electronic instruments provide the option for the user to select the geodetic model of the earth's shape that they want to use, for example NAD83.

Level

The term **level** has at least three meanings in surveying. It is often used to describe the relationship of objects with each other and the relationship between objects and the horizon. In surveying, an object or line is considered level when it is perpendicular to a vertical line.

When the term is used to describe the relative position of two or more objects, objects that appear to be at the same elevation are called level. In this application the alignment of the objects is based on visual methods, not measurements.

Level is often defined as being parallel with the horizon. Being parallel with the horizon is only true when the horizon is a level surface. A level surface is defined in Figure 1-7. The normal relief of the earth's surface usually prevents the horizon from being a level surface. The exception is mean sea level.

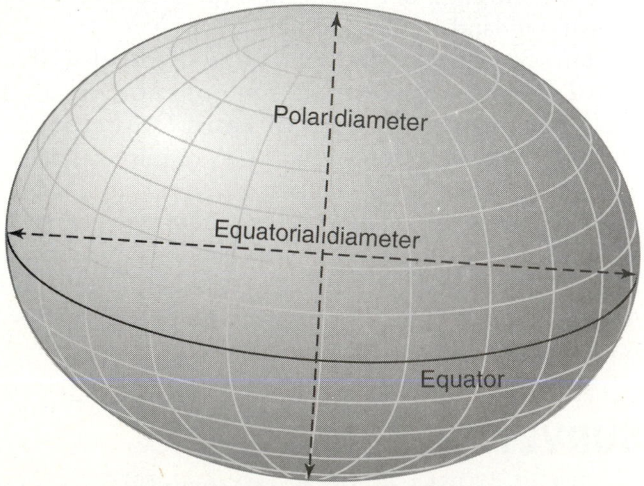

Figure 1-2 Oblate spheroid ellipsoid.

Level can also be defined as being perpendicular to a vertical line. The earth's gravitational force will cause a weighted line to point towards the center of the earth. Any line perpendicular to a vertical line would be level.

The condition of being level is usually determined by a tool called a **spirit level**. An air bubble in a small container of liquid moves as the container is tilted, Figure 1-3.

When the air bubble is in the middle, the tool is level. The container is usually a tube or cylinder, Figure 1-3, or a disk, Figure 1-4.

A spirit level is usually incorporated within a frame or as part of another tool. An example is a carpenter's level, Figure 1-5, or a survey level, Figure 1-6. Levels are discussed in more detail in Chapter 2.

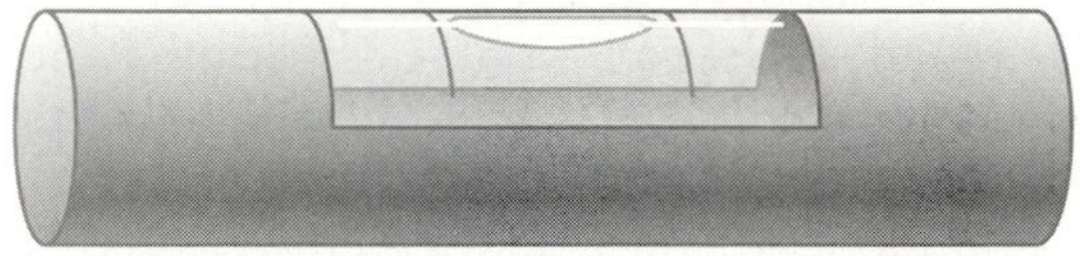

Figure 1-3 Tubular level.

Figure 1-4 Disk or bull's-eye spirit level.

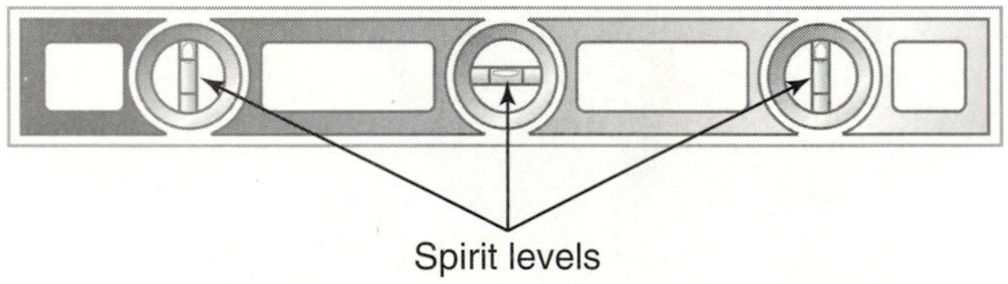

Figure 1-5 Carpenter's level.

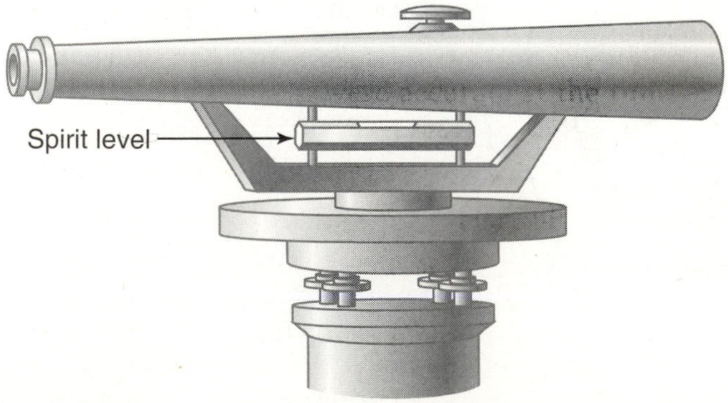

Figure 1-6 Surveying level.

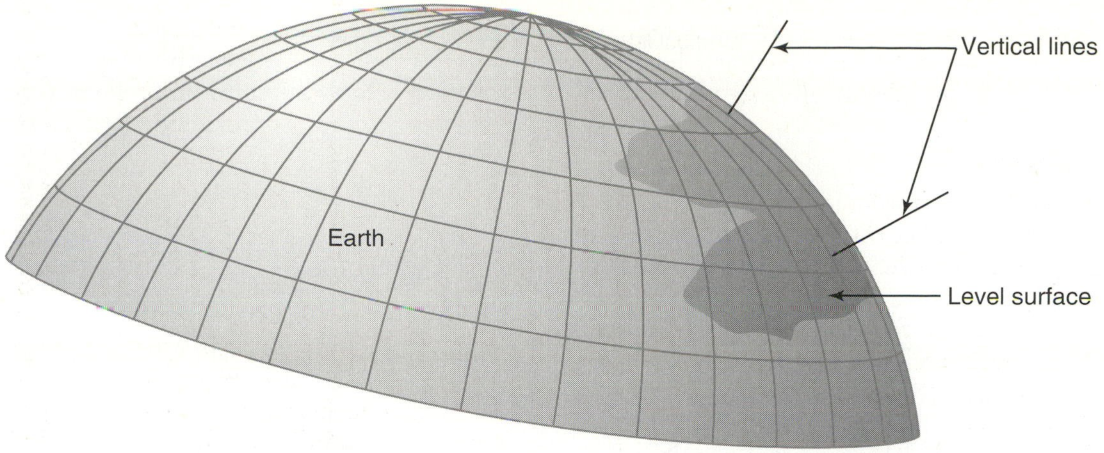

Figure 1-7 Level surface.

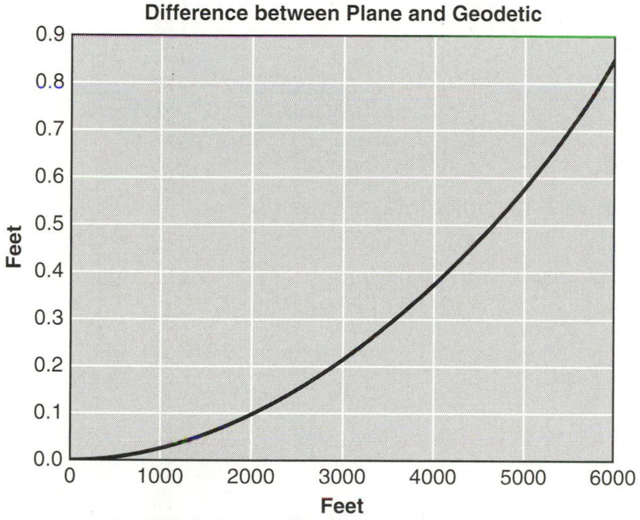

Figure 1-8 Difference between a plane and a level surface.

Level Surface

A **level surface** is a continuous surface that is at all points perpendicular to the direction of gravity. A level surface is not a flat surface or a plane. A level surface is a surface that follows the theoretical surface of the earth. Mean sea level is the level surface used as a reference for true elevations. A large body of still water, such as a lake, best illustrates a level surface, Figure 1-7.

For a small distance, the difference between a level surface and a horizontal plane is indistinguishable and not a concern for many surveys. Figure 1-8 is an approximation of the differences between a horizontal plane and a level surface at the equator. The difference will be greater for the same horizontal distance as the latitude increases.

Difference between Plane and Level Surface

The difference between a horizontal plane and a level surface is the primary difference between two types of surveying, **plane surveying** and **geodetic surveying**. Plane surveying assumes the earth is a flat plane. As long as the area is not very large, this is an acceptable assumption. Geodetic surveying does not use this assumption. Instead, geodetic surveying adjusts all elevations for the curvature of the earth. The decision to use plane or geodetic surveying must include the allowable error for the survey. Figure 1-8 shows that for a distance of 2000 feet the difference between a plane and a level surface is only 0.1 feet, an acceptable error for a low-precision survey. An error of 0.1 feet would not be acceptable for a high precision survey.

Vertical Lines

A **vertical line**, also called a plumb line, is a line that follows the direction of gravity, Figure 1-9. At any point on the earth's surface, a string with an attached weight will point towards the gravitational center of the earth, forming a vertical line. Vertical lines are perpendicular to the earth's level surface established by mean sea level.

A carpenter's level can also be used to establish a vertical line because it has spirit levels that are mounted perpendicular to the long dimension of the frame. These spirit levels are used to establish if objects are "plumb" vertically.

Vertical Plane

A plane is defined as a flat surface. Therefore, a vertical plane is a flat surface that is vertical, Figure 1-10. A vertical plane will incorporate a vertical line. For

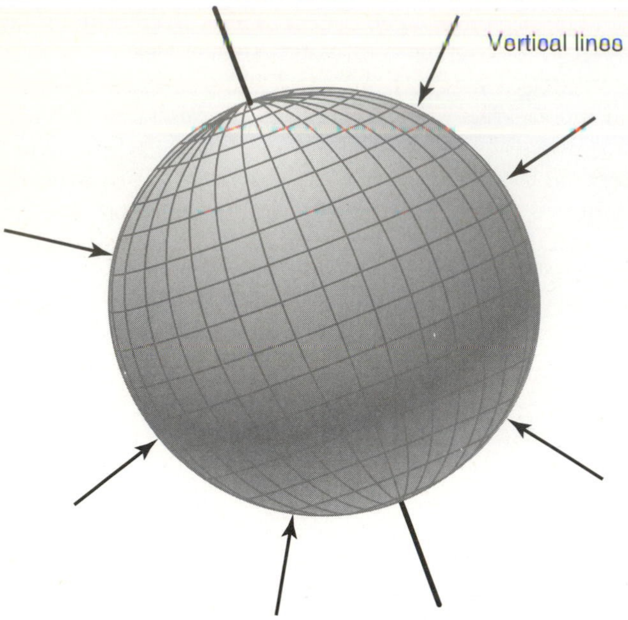

Figure 1-9 Vertical lines.

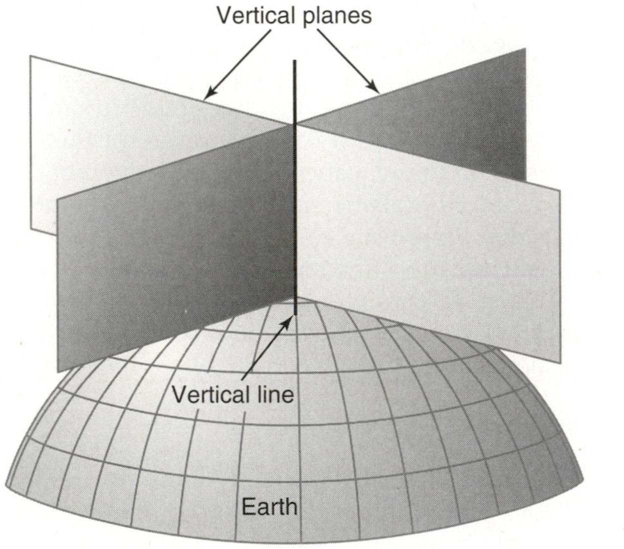

Figure 1-10 Vertical plane.

any vertical line, there are an infinite number of vertical planes. The walls of a building are usually vertical planes.

Horizontal Lines

A **horizontal line** is formed when a line is established perpendicular to a vertical line, or when a line is established parallel with a level surface, Figure 1-11. At any one point, there are an unlimited number of horizontal lines.

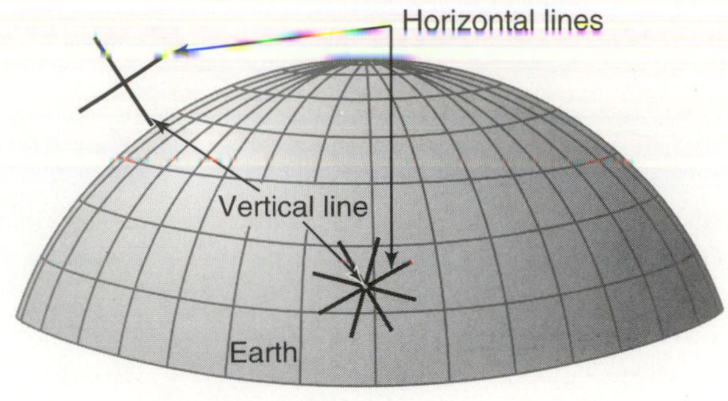

Figure 1-11 Horizontal lines.

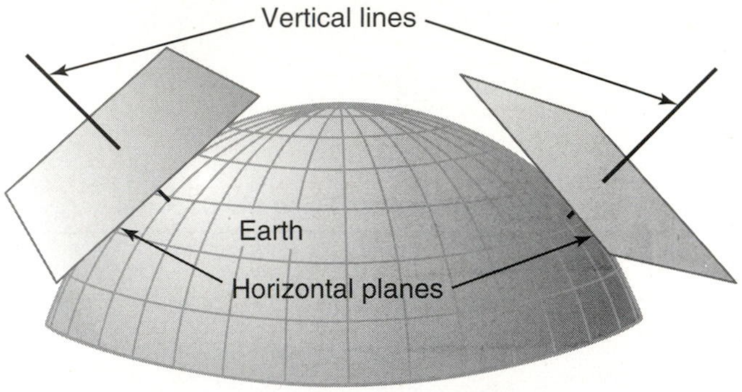

Figure 1-12 Horizontal planes.

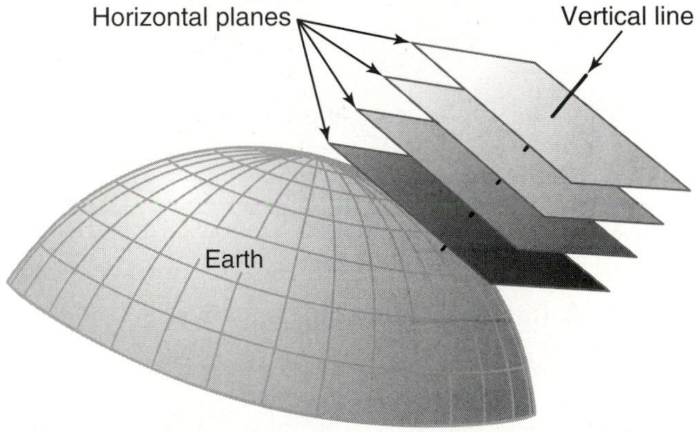

Figure 1-13 Multiple horizontal planes.

Horizontal Plane

A **horizontal plane** is a plane that is perpendicular to a vertical line, Figure 1-12.

A vertical line can have an infinite number of horizontal planes. Each elevation on a vertical line is a different horizontal plane, Figure 1-13. This is an important concept because every time a survey instrument is set up, the height of the instrument will be different. If the height is different for every setup of the instrument, then the line-of-sight through the instrument will form a horizontal plane at a different elevation every time the instrument is set up.

Horizontal Distance

A **distance** is the amount of separation between two points, lines, surfaces, or objects measured along the shortest path joining them, Figure 1-14. A **horizontal distance** is a distance measured on a horizontal line or plane.

If a distance is measured by laying surveyor's tape on the surface of the earth, the distance measured is not a horizontal distance. This is called **slope distance,** or sometimes it is called surface distance. If the chain is held level and pulled tight with the correct amount of tension, the distance measured is a horizontal distance.

Angle

An **angle** is formed by the intersection of two lines. Angles are a measure of the rate of divergence of two lines. An angle has three parts: a baseline, **vertex**, and second line. The vertex is the junction, or point of divergence, of the two lines, Figure 1-15.

To distinguish the two lines surveyors call the baseline the backsight and the second line the foresight, Figure 1-16.

The instrument is placed over the vertex to lay out or measure the angle. Surveying measures and lays out both horizontal and vertical angles.

Horizontal Angle

A **horizontal angle** is an angle measured on a horizontal plane, Figure 1-17. A horizontal angle is the angle between two vertical planes.

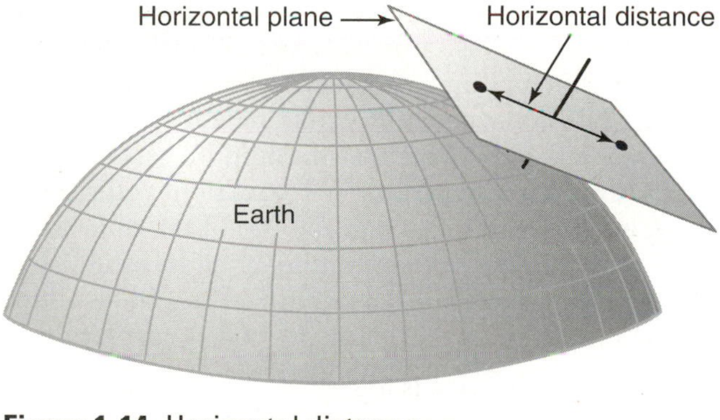

Figure 1-14 Horizontal distance.

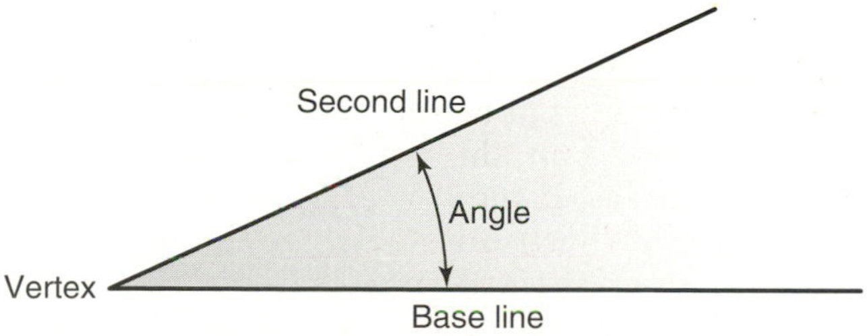

Figure 1-15 Parts of an angle.

Vertical Angle

A **vertical angle** is an angle measured on a vertical plane. When measuring vertical angles in surveying, two different baselines can be used, **horizontal zero** or **zenith zero**. When horizontal zero is used, angles measured upward from a horizontal line or horizontal plane are referred to as plus angles (+), and angles measured downward from a horizontal line or plane are referred to as minus angles (−). The (+) and (−) signs are used to indicate the direction the angle was turned, not mathematical positive or negative, Figure 1-18.

When zenith zero angles are used, zero degrees is vertically overhead and 180 degrees is vertically down, Figure 1-19.

Benchmark

A **benchmark** (BM) is a point of known or assumed elevation. To be considered a benchmark, the point must be identified by a permanent or semi-permanent structure that will not be affected by frost heave, traffic vibrations, or environmental changes. Surveying standards have very specific guidelines on the appropriate structure for benchmarks. In the United States, the National Geodetic Survey (NGS)

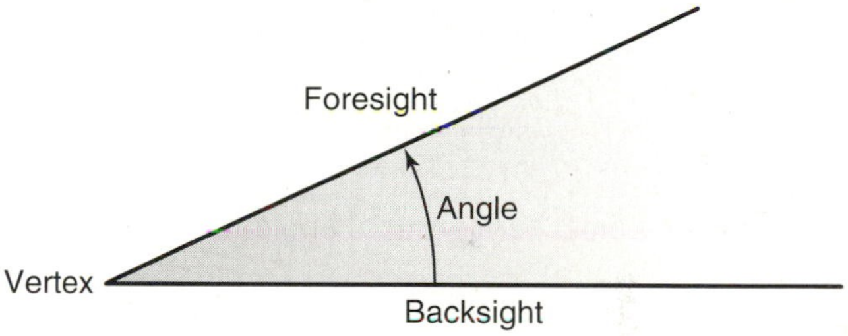

Figure 1-16 Angle nomenclature when surveying.

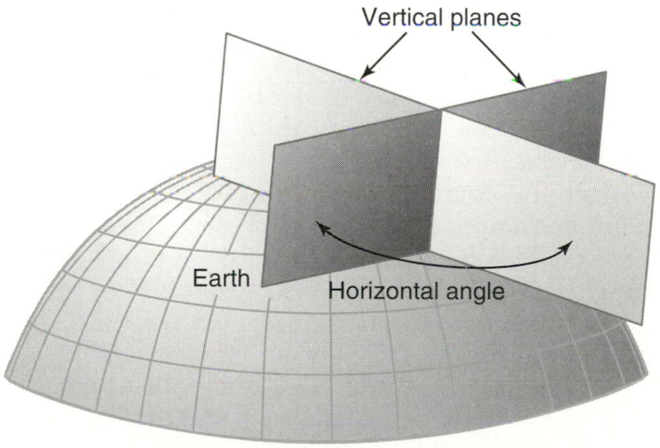

Figure 1-17 Horizontal angle.

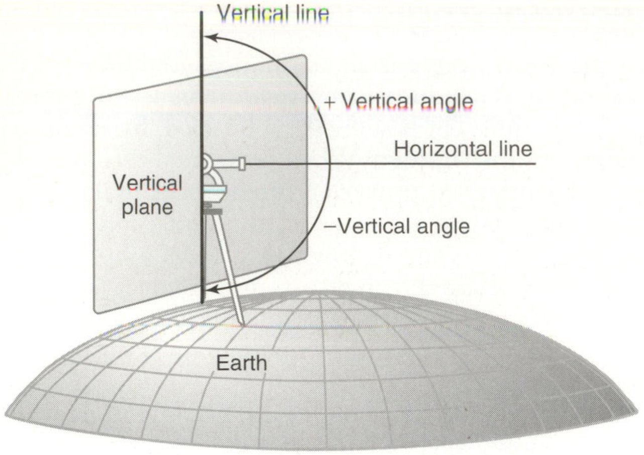

Figure 1-18 Horizontal zero vertical angle.

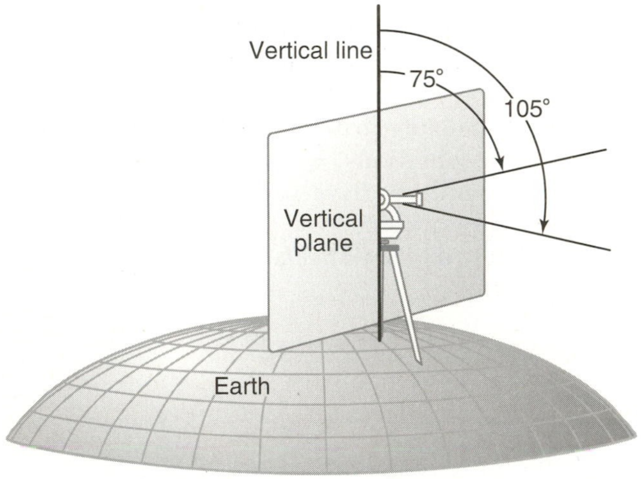

Figure 1-19 Zenith zero vertical angle.

division of National Oceanic and Atmospheric Administration (NOAA) maintains the records for all of the benchmarks established by government agencies. Datasheets for these benchmarks are available at the NGS web site, http://www.ngs.noaa.gov/cgi-bin/datasheet.prl?Type=DATASHEETS. Benchmarks established by other agencies, companies, etc. may not be available.

It is also common practice to establish a permanent or temporary benchmark at a construction project. This becomes the reference point during construction. If they meet the standards for a benchmark, they may be left in place and used for additional projects.

Benchmarks are reference points that form the basis for many surveys and construction projects. Great care must be taken to insure benchmarks are not disturbed.

Elevation

Elevation (EL) is the distance above or below a reference level surface. For surveying, the reference level surface is mean sea level. Because the earth is not a true sphere, mean sea level is a man-made reference surface. One of the first reference level surfaces used in the United States was the **National Geodetic Vertical Datum of 1929**. It was established by connecting 26 tidal benchmarks along the Atlantic, Gulf of Mexico, and Pacific Coasts. The 1929 datum was updated in the 1980s.

For many construction surveys the true elevation is not important; in these situations a local benchmark can be established and used as the reference point. This type of benchmark may use any assumed elevation. When an assumed elevation is used for the benchmark, it is common practice to use an elevation of 100.00 feet or any other even positive value to insure that elevations are not negative.

Difference in Elevation

The difference in elevation is the vertical distance between two level surfaces or planes, Figure 1-20.

In a situation where the elevation of each of the level surfaces (1, 2, and 3) in Figure 1-20 is known, the difference in elevation can be calculated between the earth (1) and surface 2 or between the earth and surface 3. It can also be determined between surface 2 and 3.

Backsight

The term backsight has two uses in surveying. When shooting elevations, a **backsight** (BS) is a rod reading taken on a point of known or assumed elevation. The elevation of the point would be known if the station was a true benchmark. It could be assumed if it was a temporary benchmark. A backsight rod reading is the vertical distance from the top of the object that the rod is resting on to the line-of-sight established by the instrument. The backsight rod reading is added to the elevation at the point the rod is resting on to determine the height of the instrument.

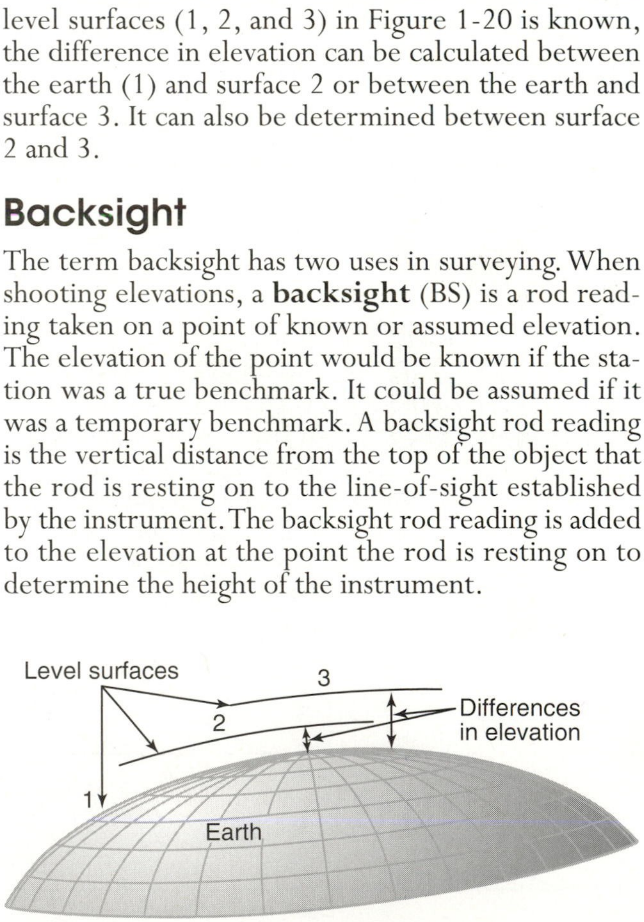

Figure 1-20 Difference in elevation.

The term backsight is also used when turning angles. A backsight angle is a sighting on a station that is used to zero set the angle scale on the instrument.

Foresight

The term foresight also has two uses. When shooting elevations, the **foresight** is a rod reading taken on a point of unknown elevation. It is used to determine the elevation of the point where the measurement is taken. In surveying two different types of foresights are used, **intermediate foresight** (IFS), and **true foresight** (FS).

An intermediate foresight is a rod reading on a point that *will not* be used as a turning point or benchmark. Extra care and procedures must be used when recording these rod readings because the checks for error will not catch a mistake in these readings. This is discussed in more detail in Chapter 6.

A true foresight is a rod reading on an unknown point that *will* be used for a turning point or for a benchmark. The person on the instrument must also be careful when recording these readings. The error checks will catch a mistake in the rod reading of a true foresight, which prevents the use of data that has errors. The problem is that the errors cannot be corrected. If the error is excessive, the data has no value and the survey must be repeated.

When turning angles the term foresight refers to the station that the instrument is turned to when measuring the angle. Additional information on turning angles can be found in Chapter 7.

Turning Point

A **turning point** (TP) is a station that is used as a temporary benchmark. The purpose of the turning point is to provide a point of known elevation that can be used to reestablish the height of the instrument after it has been moved, Figure 1-21. The turning point should be a stake or other durable structure that has a stable elevation. Elevations of the turning points are recorded in the notes, but should not be used for design work unless the top of the structure being used is flush with the ground surface.

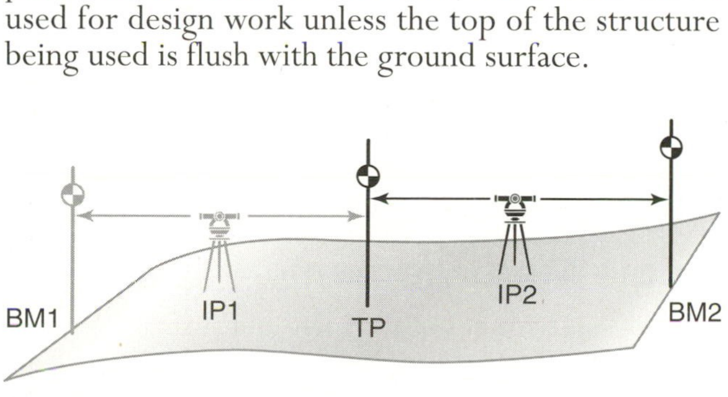

Figure 1-21 Turning point.

In Figure 1-21, the instrument would be set up at instrument position one (IP1); a backsight would be taken on benchmark one (BM1) and a foresight on the turning point (TP). Adding the value of the backsight to the elevation of BM1 results in the height of the instrument. Subtracting the value of the foresight from the height of the instrument results in the elevation of the turning point. The TP becomes a new station that can be used as a reference elevation. Even though turning points are intended to be temporary, the rod must be placed on a structure that will remain stable until the survey is completed. Standing the rod on the ground is not an acceptable turning point. After the foresight on the TP is recorded, the instrument would be moved to IP2, and the process repeated until the survey is completed. There is no limit on the number of turning points that can be used to complete a survey. Each turning point increases the complexity of the notes and increases the amount of time to complete the survey. With practice and planning the surveyor should be able to keep the number of turning points to a minimum.

TYPES OF SURVEYS AND METHODS

Professional surveyors complete many different types of surveys. Eight of the common ones are:

- Distance measurement
- Angle measurement
- Differential
- Profile
- Topographic
- Traverse
- Construction
- Property

Several factors are used to select the best type of survey. These include:

- The area of the survey
- The complexity of the terrain
- On land or water
- Time available
- Resources available
- The intended use of the data

The most important is the intended use of the data. For example, if the purpose of the survey was to collect information to design a sidewalk, a profile survey would be required. If the desire were to level a pad for a building, then a topographic survey would be conducted.

More than one type of survey may be required to complete a job. A topographic survey for a building site may require a differential survey to determine the elevation of the site. It may also require a property survey to insure the zoning setbacks are not being violated. Each job must be carefully analyzed to determine the type(s) of surveys needed to collect the required data. This is a critical part of the survey if the desire is to collect the data with the least amount of expenditure of resources.

All eight of these surveys can be completed using the geodetic method or the plane method. The primary difference between geodetic and plane surveying is the approach used to account for the curvature of the earth.

Geodetic Surveying

Geodetic surveying uses a level surface, Figure 1-7, as the reference for all elevations. Because mechanical and electronic surveying instruments use a horizontal line-of-sight as a reference plane, an equation is used to determine the level surface elevation from a line-of-sight instrument rod reading. Geodetic surveys are used when a high degree of accuracy is needed or when the survey will cover long distances and large areas, Figure 1-22.

Geodetic surveys are highly technical and time-consuming. Each rod reading must be adjusted by calculating the amount of curvature of the earth that occurs between the position of the instrument and the position of the rod.

Data recorded with instruments that are designed to use the Global Positioning System (GPS) do not need to be adjusted because GPS uses a mathematical model of the earth to determine the level surface elevations and positions. This system is explained in Chapter 10.

Plane Surveying

Plane surveys assume the earth is flat and that all rod readings are measurements from a flat surface (plane). We know that the surface of the earth is a level surface, not a flat surface. Therefore, in plane surveying every rod reading has a small amount of error. For general construction and for all surveys that use short distances and small areas, this error can be ignored, Figure 1-8.

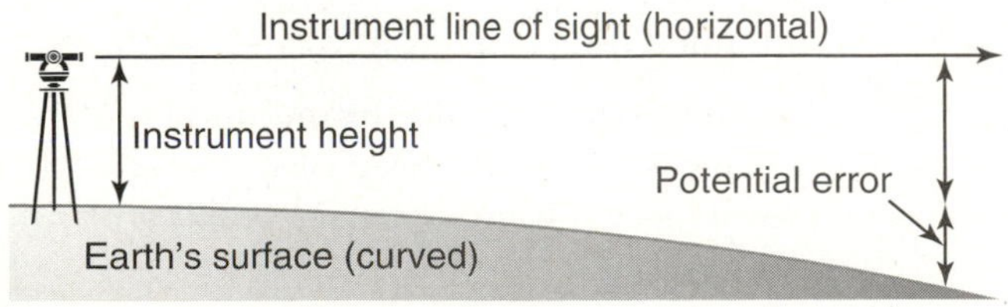

Figure 1-22 Geodetic surveying.

The eight common types of surveys will be briefly discussed in the following sections. Additional information on the first six is included in subsequent chapters.

Construction and property surveys are beyond the scope of this text. Information on these two types of surveys can be found in texts written for civil engineering. These texts will also include more information on the six types of surveys included in the subsequent chapters of this text.

Distance Measurement

Distance measuring is determining the distance between two or more points. Many activities involving land require measuring distances. Several different instruments and methods are used. The best method to use will depend on the intended use of the data, the accuracy and precision required, and the resources available. For example, chaining is one of the most accurate methods, but it requires several people and it is a slow process. Odometer wheels are not as accurate, but they only require one person and the measurements can be taken as fast as a person can walk. Common distance measuring methods are:

- Pacing
- Chaining
- Stadia
- Odometer wheel
- Rangefinder
- Electronic distance measuring (EDM)
- Laser

A distance can be measured as either a slope (surface) distance or true horizontal distance. The methods and equipment used to measure distance are determined by which type of distance is required. For applications such as determining the number of lengths of water pipe that are needed and similar low-precision jobs, the difference between slope distance and horizontal distance can be ignored for slopes up to five percent. At 5% slope, the amount of error in one hundred feet is 0.38014 feet, or 4-1/2 inches. In situations that require a higher degree of precision the difference between slope distance and horizontal distance must be accounted for.

It is important to know when horizontal distance is required and when slope distance is sufficient. Additional information on methods and principles of measuring distance is included in Chapter 4.

Angle Measurement

Angles can be measured or laid out directly or indirectly. The method used must be decided before selecting equipment because different instruments and techniques are used for each method.

Direct Measurements of Angles

Direct measurements of angles require the use of a survey instrument. Survey instruments can output angle measurements in **degrees-minutes-seconds** (DMS) or **decimal degrees** (DD). DMS is the **Sexagesimal System** of measuring angles. In the sexagesimal system there are 60 minutes in each degree and 60 seconds in each minute. DD is a method of recording angles that expresses parts of a degree as a decimal.

Indirect Angle Measurements

Angles can also be measured or laid out without using an instrument. Common indirect methods of measuring angles are:

- Chord
- 3-4-5
- Tape-sine

Additional information on measuring angles can be found in Chapter 7.

Differential Surveys

Differential surveying is used to determine or establish the difference in elevation between two or more points. A common use of differential surveys is to establish a new benchmark. The process of determining if concrete forms are at the correct elevation, and are level, is also a use of differential surveying. Differential surveys are covered in more depth in Chapter 5.

Profile Surveys

A **profile** is a side view of an object. **Profile surveying** is used to establish a side view of the earth's surface. Once the data is collected, it is usually plotted, and the plot can then be used to make decisions about such design considerations as the slope, depths of cuts, volumes of cuts, and fills. A profile survey is usually done before constructing routes, such as roadways, sidewalks, pipelines, drains, and other utilities. Figure 1-23 is an example of a profile plot.

More information on conducting profile surveys can be found in Chapter 6.

Topographic Surveys

Topographic surveying is used to collect the data required to draw a topographic map. A topographic map is a three-dimensional drawing of the earth's surface. Figure 1-24 is an example of a three-dimensional topographic data.

Topographic maps are very useful for planning and preliminary design work. On topographic maps, each horizontal line connects points of constant elevation, Figure 1-25. The two traditional methods used to collect the required data for topographic maps are the grid method, and the angle and distance method. These methods are still used for small areas, but for large areas most topographical data is collected by GPS equipment.

Chapter 8 includes additional information on conducting topographical surveys and drawing topographic maps.

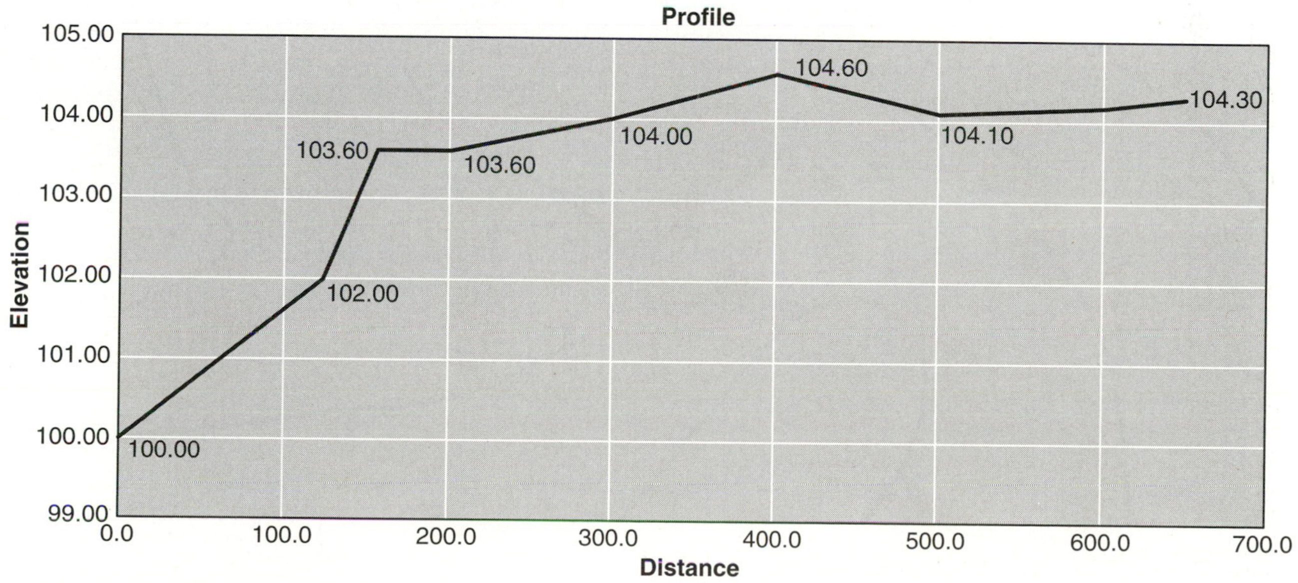

Figure 1-23 Plot of profile survey.

Property Surveys

Property surveys are used either to establish property lines when subdividing a parcel of land or to check existing property lines. A property survey should be completed before starting any construction or before buying property. Property lines can become confused as land transfers through owners. Fences may not be built on the property lines. Corner markers may be destroyed or moved. For these reasons, it is not safe to rely on the memories of landowners or existing fences or other property markers. If the area has been surveyed using the Public Land Survey System (PLSS), the property survey will start with the PLSS description, Chapter 3. For parcels less than a section or areas with an irregular shape, accurately measured angles and distances are used to define the property lines.

Property surveys may not resolve boundary disputes. It is natural for an individual to think that with modern equipment and techniques it would be possible to locate a property line very accurately. This is not necessarily true. Many factors may cause inaccuracies. The surveyor may be working with old, incomplete, or inaccurate information. If two different surveyors attempt to locate a property line but use different starting points, they may not agree on the location of the property line. For this reason, many states have passed fence laws that determine property lines. The wording and criteria for fence laws vary from state to state, but the intent is the same. A fence that meets the design requirement and that has existed for a prescribed number of years becomes the legal property line even if a property survey shows the fence is not on the property line. The "fence property line" can be moved if both property owners agree to accept the survey and move the fence.

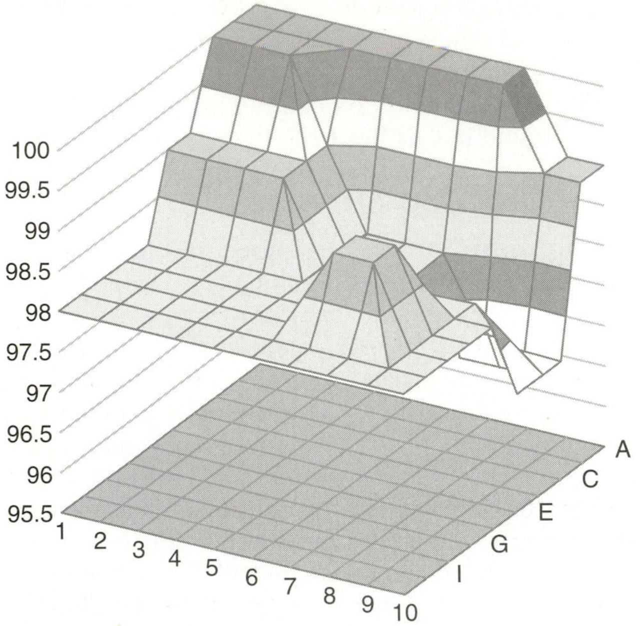

Figure 1-24 Plot of topographic data.

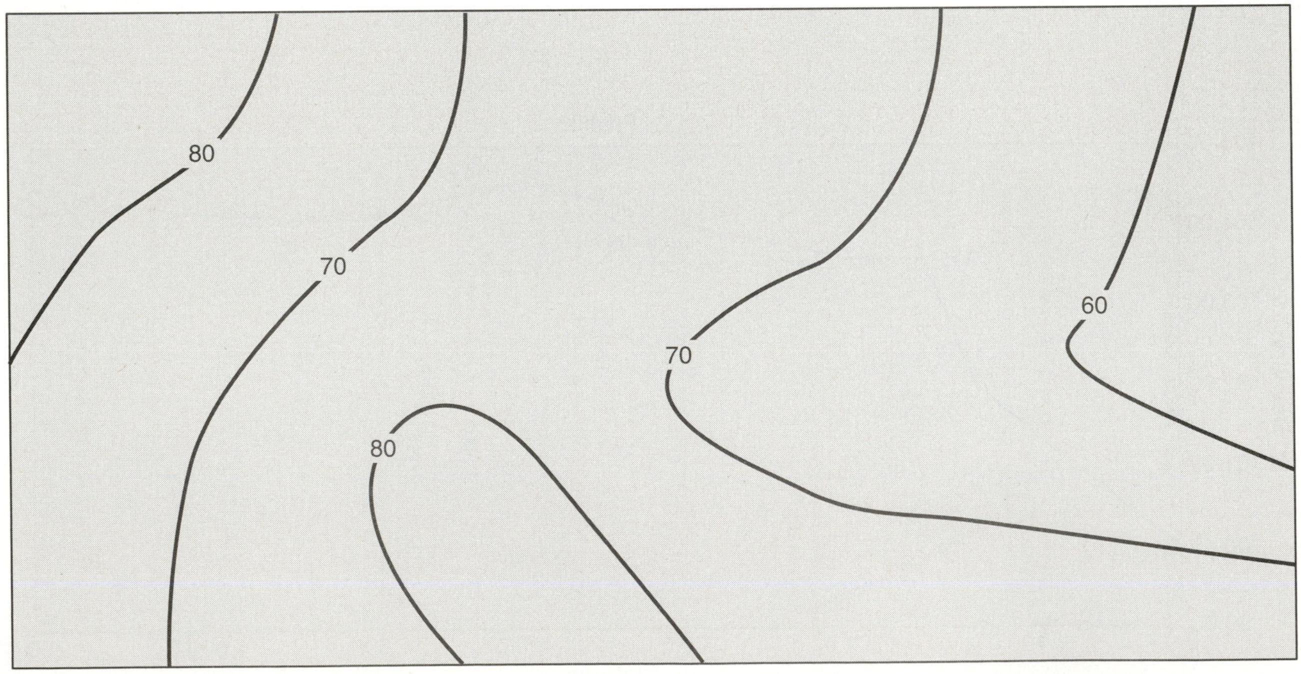

Figure 1-25 Example topographic map.

Construction Surveys

A **construction survey** is used to either lay out a proposed construction project, or to evaluate a project after it is built. Construction surveys may be relatively simple or very complex, depending on the size and the type of project.

Traverse

A **traverse** is a series of connected straight lines. These lines may establish a closed polygon. A traverse survey requires very careful measurements of distances and angles. Figure 1-26 is an example of a traverse survey that was conducted using a transit and a tape/chain.

Once data is collected, mathematical procedures can be used to balance the traverse. A traverse must be balanced because there is only one possible combination of angles and distances that will form a closed polygon. If errors occur in measuring angles or distances, the data will not produce a polygon. Some error always occurs; therefore, the traverse must be balanced. The balancing procedure may be an arbitrary adjustment of angles and distances or a mathematical process that calculates the best fit of angles and distances to form a polygon. Figure 1-27 is a plot of the measured angles and distances recorded for a five-sided area.

Notice that because of errors in measuring the distances and angles the plot of the data in Figure 1-27 does not form a closed polygon. When the traverse is balanced and replotted the figure will close. This is illustrated in Figure 1-28.

Additional information concerning traverses is included in Chapter 9.

DATA USE

Measurements are collected for many different purposes. The intended use of data must be considered when determining the equipment and methods to use. For example, if the purpose of a job is to determine the lengths of water pipe required to lay a water line, measurements to the nearest 10 feet are probably sufficient. Instruments and methods that measure angles to the nearest minute and distances to the nearest thousands (0.001) of a foot would be required to locate the legal boundaries for a piece of property. The surveyor must use appropriate equipment and methods to collect the desired information. Two additional characteristics of surveying data are accuracy and precision.

Accuracy and Precision

Accuracy and precision are two important concepts to keep in mind when collecting any measurements. The level of accuracy and precision needed to determine how many rolls of barbed wire are required to build a fence is different from the level of accuracy and precision needed to construct a building. Some standards and common practices have been developed

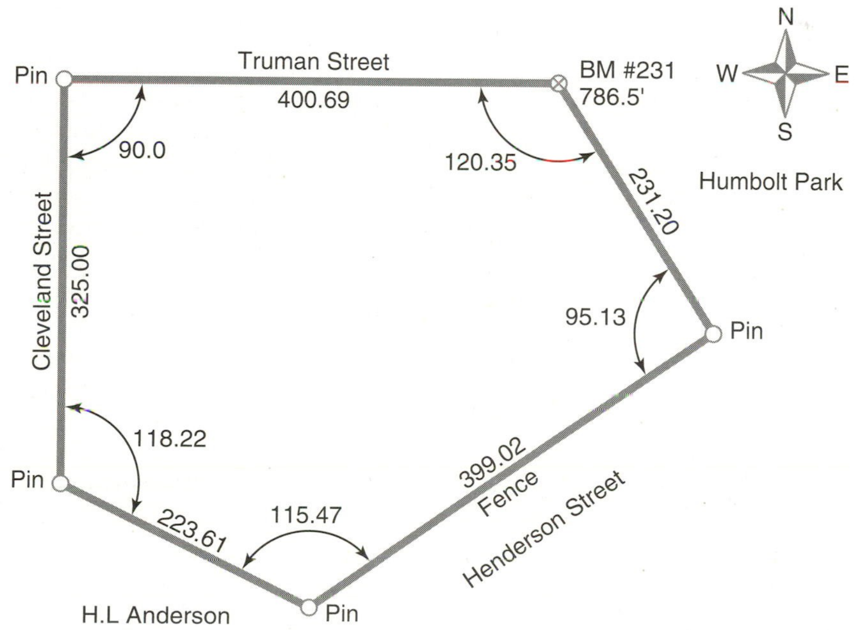

Figure 1-26 Transit and tape traverse.

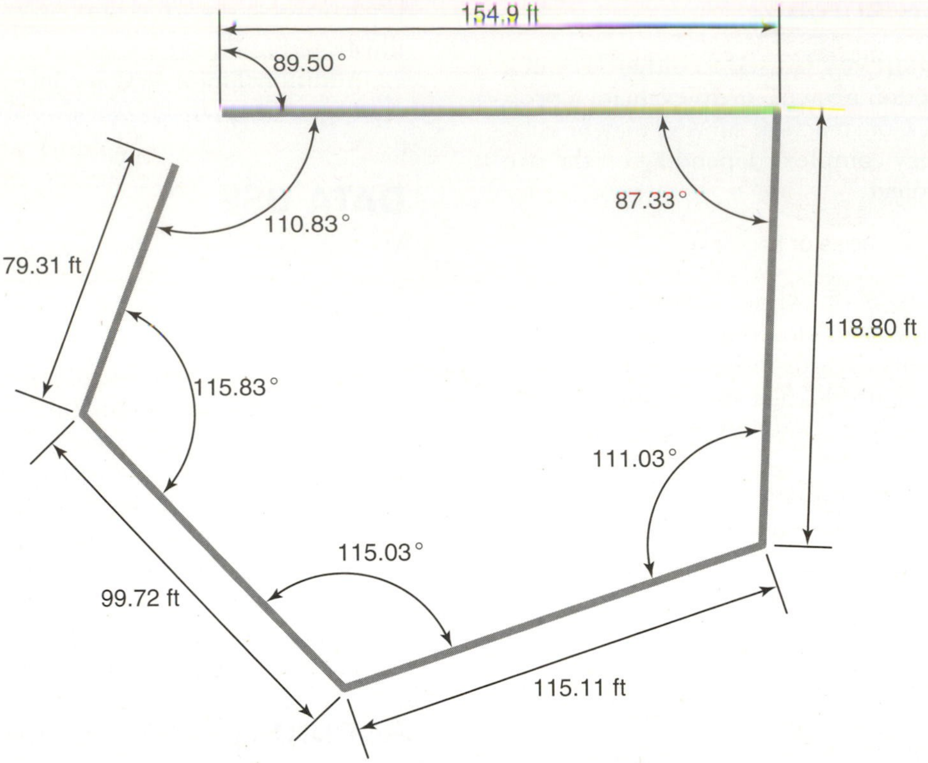

Figure 1-27 Plot of measured traverse.

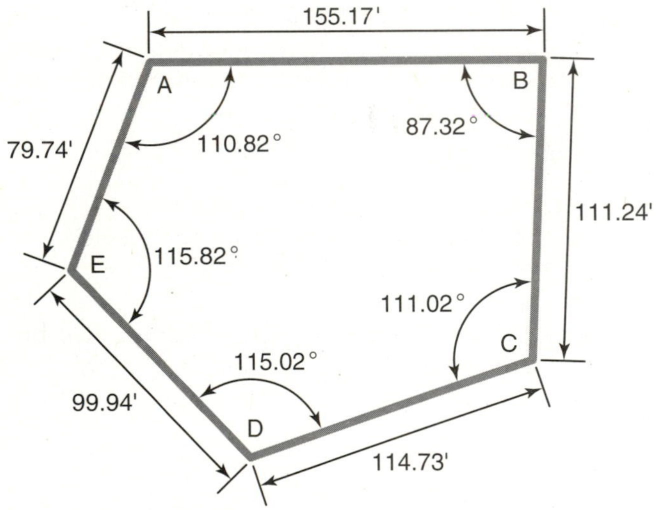

Figure 1-28 Plot of balanced traverse.

to provide a guide in this area but in many cases, decisions are based on experience.

The principles of accuracy and precision can be illustrated graphically by using the distribution of shots at a target, Figure 1-29. Four different situations can exist. They can be neither precise nor accurate, precise but not accurate, accurate but not precise, and both accurate and precise.

Mathematically, **accuracy** is defined as the number of significant digits included in a measurement. The instrument determines the maximum number of significant digits in a measurement and the surveyor must insure that additional digits are not included during computations.

All measuring devices are accurate to only plus or minus one of the smallest units. To illustrate this point, consider a one-foot ruler graduated in eighths of an inch. The accuracy of all measurements made with the ruler is accurate to 1/8 (or 0.125) inch. If a higher or lower level of accuracy is required, a measuring tool that has the capability to measure at the required level must be used.

Precision is defined as the unit of measure. The smaller the increment used, the more precise the measurement. A measurement in units of inches is more precise than a measurement with units of feet. Ounces are more precise than pounds. Some distance measuring instruments used for surveying have a precision of 0.001 feet. These devices are much more precise than a ruler that has the smallest measurement of 1/8 of an inch.

The required level of accuracy and precision must be determined before any measurements are taken

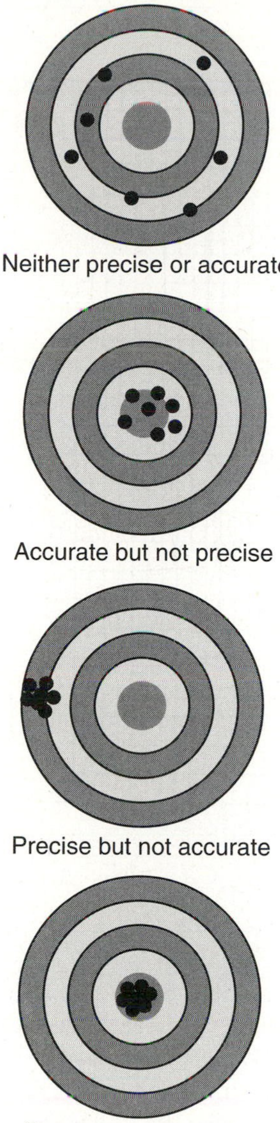

Figure 1-29 Accuracy and precision.

because the level of accuracy and precision required will influence the best method and instruments to use.

Field Notes

Surveying and land measurement require the collection and recording of information. Methods of recording information that provide a systematic, accurate way of recording data have been developed. The purpose of these methods is to improve the ease of using the information and to reduce the chance of errors. *Incorrect information is worse than no information* because it can lead to wrong decisions. Wrong decisions can cause very expensive mistakes. Field data can also be considered public documents and used as evidence in court.

FIELD BOOKS

Surveyors have traditionally used a specific type of small book called a field book to record data. Field books are made up of very high quality paper that is weather-resistant. The information in field books must meet four requirements. It must be:

- Clear
- Concise
- Complete
- Correct

To insure the data meets these four c's, specific guidelines have been adopted for organizing information in the field book. These guidelines are discussed in Chapter 2. Figure 1-30 is a page from a typical field book. It shows that the lines on the page are unique.

The uses of a field book will vary with the project. On large projects, professional surveyors will use a new book for every survey. The original book is kept in a safe and copies of the information are made for daily use. If a surveyor is surveying small projects, he or she may use one book for several different projects. The customer and the complexity of the job dictate the use of the field book. In either case, the original information should be stored in a safe and copies made for daily use. Modern electronic surveying instruments with their digital recording of data have reduced the reliance on field books for recording data, but the need for storing the data in a secure location is still important.

ERROR AND ERROR MANAGEMENT

Surveying data must be *accurate to be useful.* To insure the data is accurate the surveyor must understand the types of errors that can occur and learn how to manage them.

Common errors for each type of survey will be included in the appropriate section. In general, errors can be divided into two categories—random and systematic.

Random Error

A **random error** is an error that does not occur with a predictable pattern. Random errors are very hard to manage because the cause or when they will occur is unknown. In surveying, random errors are managed by following established procedures. Procedures such as the note check have been adopted because years of experience have shown it is a successful method for finding mathematical errors in a data table. Other examples of random errors include

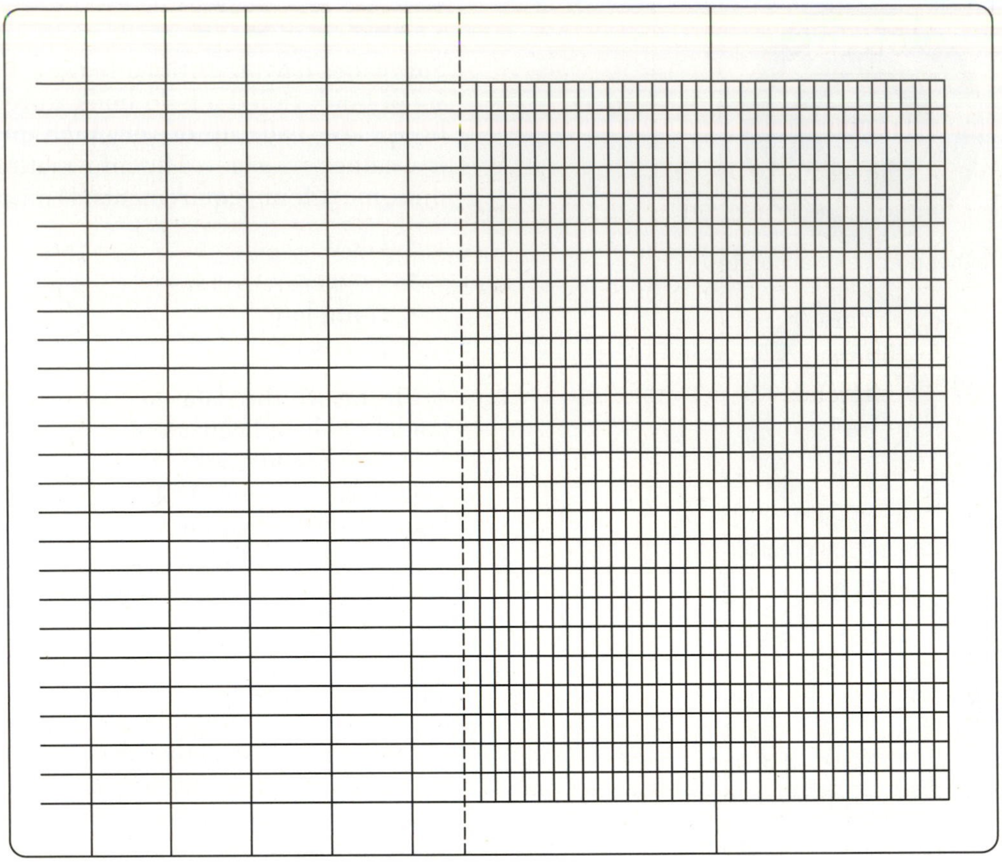

Figure 1-30 A page from a typical field book.

an instrument reader who switches two numbers when recording a rod reading or an individual who reads the rod incorrectly.

For rod reading errors procedures such as **double reading** and **three-wire leveling** were developed. When double reading, the instrument reader calls out the rod reading and the note keeper writes it down. Then the instrument person rotates the instrument away from the rod and back again and reads the rod a second time. Double reading provides dual protection against the possible errors of reading the rod wrong and recording the data incorrectly.

Three-wire reading utilizes all three horizontal crosshairs in the instrument, if they are available. In this method all three crosshairs are read and recorded in the field book and then averaged. The averaged value is used as the measurement.

Systematic Errors

Systematic errors are predictable. They are often associated with the limitations of, or problems associated with, measuring instruments. Systematic errors are easier to manage because once the error for an instrument or condition is known, a correction factor can be calculated and the appropriate adjustments can be made to correct the data.

An example of a systematic error is temperature correction for a surveyor's tape. The amount of error is predictable because the expansion and contraction of the steel is predictable. An equation is used to determine the amount of correction needed. This number is added to or subtracted from each measurement to determine the correct distance.

Summary

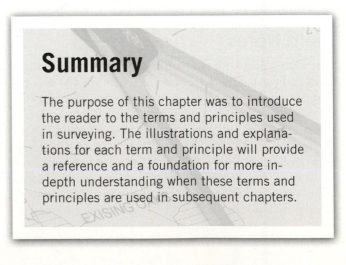

The purpose of this chapter was to introduce the reader to the terms and principles used in surveying. The illustrations and explanations for each term and principle will provide a reference and a foundation for more in-depth understanding when these terms and principles are used in subsequent chapters.

Student Activities

Answer the following questions.

1. Explain the difference between a plane and a level surface.
2. How is horizontal distance different from surface distance?
3. What are the three parts of an angle?
4. What is the possible range of vertical angles when horizontal zero is used?
5. Describe a benchmark.
6. Explain the difference between precision and accuracy.
7. What is the difference between random and systematic errors?

CHAPTER 2

Equipment

Objectives

After reading this chapter, the reader should be able to:

- List the three categories of equipment.
- Identify the three common types of chains.
- Read a first foot, extended foot, and fully graduated chain.
- Measure distance using a chain.
- Measure distance using an odometer.
- Measure distance using an optical rangefinder.
- Use a hand level.
- Level a four-legged level.
- Identify the parts of an automatic level.
- Identify the parts of a dumpy level.
- Identify the parts of a transit.
- Read Vernier scales.
- List the information that should be recorded in a field book.
- Keep notes in a field book.
- Read a rod.

Terms To Know

Gunter's chain
graduations
cut chain
add chain
fully graduated chain
odometer wheel
rangefinder
inclinometer
electronic distance measuring
hand levels
slope
Abney level
dumpy level
parallax
automatic level
laser level
transits
vernier
least count
electronic transit
construction transit
transit level
theodolite
total station
surveying pin
rod
target
rocking the rod
rod level
range pole
plumb bob
clockwise
counterclockwise

INTRODUCTION

Surveying is heavily dependent on the use of instruments and equipment. Proper selection, use, and care of this equipment will greatly influence the quality of data collected and the amount of resources that will be required to collect the data. The instruments and equipment used may be as simple as two tree branches or as complex as a GPS unit which costs tens of thousands of dollars. This chapter will discuss the common equipment used for land measurement and surveying.

One of the most important principles to remember is that surveying instruments are precision instruments, and they are easily damaged by rough use or improper care. Instruments can be easily damaged by being dropped, stored improperly, or the having the movements "forced" without being unlocked. A damaged instrument may function, but the data it produces will be wrong, and the operator may not know the data is incorrect. Instruments used by professionals must be checked on a regular basis by trained individuals to insure they are producing accurate results.

CATEGORIES OF EQUIPMENT

Surveying equipment can be divided into five categories: distance measuring, measuring angles and elevations, accessories, combination instruments that measure both angles and distances, and GPS.

Distance Measuring

When measuring distance one method can be used that does not require the use of an instrument. This is called pacing. Pacing is explained in Chapter 4. Equipment used for measuring distance include:

- Pacing
- Tapes and chains
- Odometer wheel
- Levels (stadia)
- Rangefinders
- EDM's
- Lasers

Instruments for Measuring Angles and Elevations

Instruments used for measuring angles and elevations include:

- Hand level
- Dumpy/farm level
- Automatic level
- Laser level
- Transit
- Theodolite
- Construction transit

Accessories

Many accessories are used when measuring distance angles and elevations. Some of these include:

- Pins
- Field books
- Rods and targets
- Range poles
- Stakes/flags
- Plumb bobs
- Paint

Professional surveyors use combination instruments that can measure distance and angle at the same time. The most popular instrument in this category is the total station. A total station is an electronic transit with a built-in EDM.

GPS instruments are in a category of their own because they are a space-based system. All other surveying equipment is land based. Global positioning systems and instruments are discussed in Chapter 10.

Surveying equipment, instruments and accessories are available with different features and capabilities. The following sections will discuss some of these.

EQUIPMENT FOR MEASURING DISTANCE

Distance measuring equipment ranges in technology from fiber tapes to EDM equipment that uses microwaves or lasers to GPS units that use satellites in space. When preparing to measure a distance, one question that must always be answered is, "What equipment should I use?" The best equipment for measuring distance is based on the use of the data and additional factors such as the topography, the skill of the surveyor, and even what equipment is available. If use of the data dictates a high degree of accuracy is required, then chaining or an EDM should be used. If a low level of accuracy is acceptable, then an odometer wheel or even pacing can be used. If the route can be walked a chain could be used. If it is not possible to walk the route, then a line-of-sight instrument like an EDM must be used. Regardless of the equipment used, the surveyor must insure it is in proper working order and that it is used correctly.

Tapes and Chains

The **Gunter's chain** was the standard chain for many years. A Gunter's chain is 66 feet long, comprised of 100 links, with each length being 7.92 inches. The links were made of heavy wire and connected by rings. The handle was threaded and was used to adjust the length of the chain to compensate for wear, Figure 2-1.

The Gunter's chain was primarily used for land surveying because of its relationship to a mile, 80 chains = 1 mile.

In the United States, the Gunter's chain was replaced by the engineer's chain. Engineer's chains are composed of links also, but they were designed to be 100 feet long and each link was one foot. The modern 100-foot steel tape eventually replaced engineer's chains. Although the modern surveyor's chain looks like a steel tape, the term chain is still often used to distinguish them from cloth tapes. Steel tapes (chains) are manufactured from a special steel alloy that does not stretch with tension and gives them a predictable expansion and contraction rate from temperature change. Cloth tapes tend to stretch when tension is applied and therefore their error is unpredictable. The designed accuracy of a steel surveyor's chain is only true if it is at standard temperature, usually 72°F, and has a standard tension applied, usually 15 pounds. For surveys with a high level of accuracy, the measured distances must be adjusted for the expansion or contraction of the tape. This is called the temperature correction.

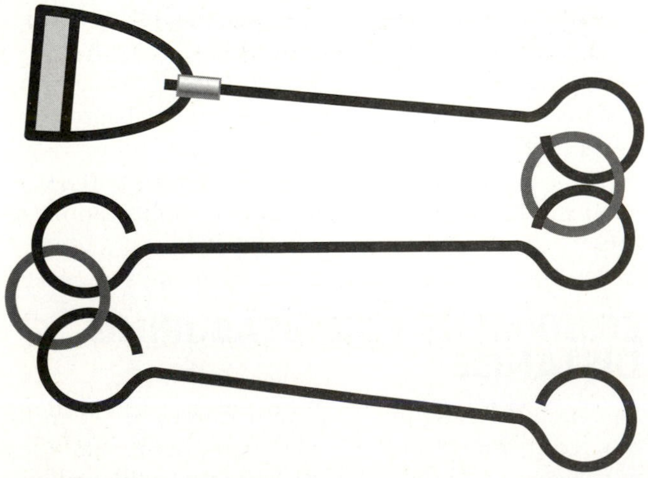

Figure 2-1 Handle and two links of early surveyor's chain.

The biggest disadvantage of traditional steel chains is that they rust very easily. Modern technology has produced a combination of steel and plastic that combines the accuracy of steel with the durability of plastic.

Surveying chains are available with a variety of graduations. A **graduation** refers to the subdivisions of the whole unit. For example, an inch is a graduation of a foot and 1/2 of an inch is a graduation of an inch. Some chains are fully graduated, like a carpenter's tape, but most only have graduations every foot. These chains include a graduated foot on each end for measuring partial feet.

An individual must be careful when using any tapes or chains because several different graduations are used. Good quality nonmetallic tapes are available for the non-professional to use and can be purchased in feet and inches or decimal feet. A carpenter's tape is often used for measuring short distances. It will be in feet, inches, and fractional graduations for each inch. Fractions of 1/2, 1/4, 1/8, and 1/16 of an inch are common. Surveyor's chain uses decimal feet and is available with units of tenths (0.1) hundredths (0.01) and thousandths (0.001) of a foot. Three common styles of chains are used, fully graduated, cut chain, and add chain.

Fully Graduated Chain

The fully graduated chain has one or two sets of graduations between each foot mark. The chain in Figure 2-2 has nine lines, ten spaces, between each foot. The precision is 1/10 of a foot, or 0.1 feet. The zero mark at the rear of the chain is held on one point, and the distance is read at the head end of the chain. The correct reading for the chain in Figure 2-2 is 99.4 feet. When the distance is less than a whole chain and less than a whole foot, the distance can be read directly from the chain.

Cut Chain

When using a cut chain and the distance is less than a full chain and less than a whole foot, the head person holds the tape on the nearest foot mark and the rear person takes out the slack and reads the partial foot. The error occurs because the distance being

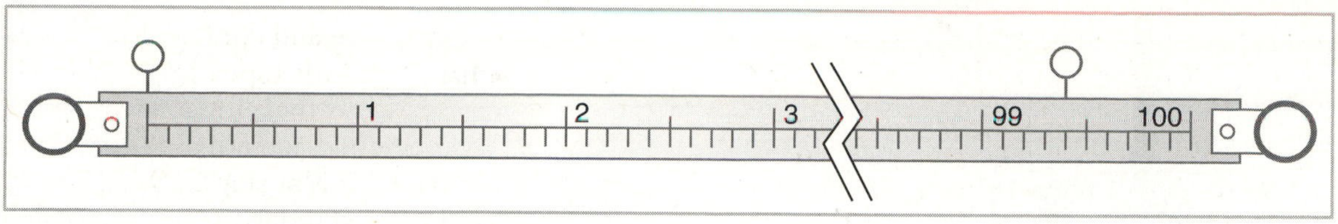

Figure 2-2 Fully graduated chain, reading is 99.4 feet.

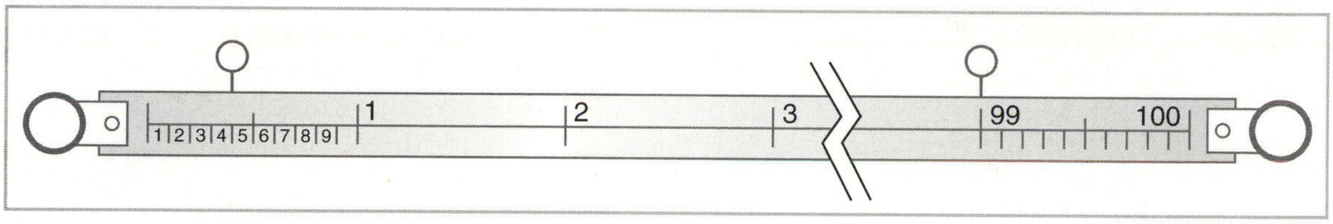

Figure 2-3 Cut chain, reading is 98.6 feet.

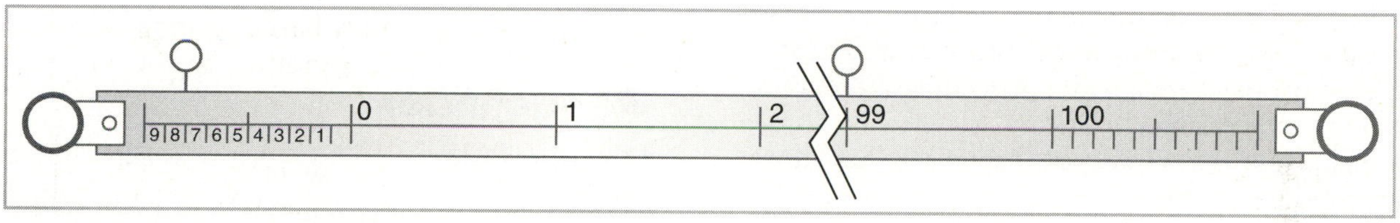

Figure 2-4 Add chain, reading 99.8 feet.

measured is less than the foot mark at the head of the chain. In Figure 2-3, the head person holds the chain on the 99-foot mark and the rear person has the chain on the 0.4-foot mark. The correct reading is 98.6 feet because the distance is 0.4 feet less than 99 feet. Because the distance from zero to one foot is used to measure the partial foot, the partial foot must be subtracted from the measurement at the head of the chain.

Add Chain

The **add chain** is similar to a first foot chain, except that an additional foot has been added to the zero end for the graduated foot. It is also possible that the 100-foot end will also have an extended foot. The head of the chain is positioned at the nearest whole foot and the chain is read at the rear position. The correct reading for the chain in Figure 2-4 is $99 + 0.8 = 99.8$ feet.

Add chains have their own chance for error. The user must be careful to use the correct zero-foot and 100-foot marks. The initial reading must be made using the zero-foot mark on the chain not the end of the extended foot. If the end of the extended foot is used as zero feet, the measurement will have an error of one foot. For chain with an extended foot on both ends, the error could be 2 feet for every full chain that is measured.

In addition to the three styles of chains, manufacturers also place the zero-foot mark at different points on the chain. For some styles of chains the zero mark is at the end of the chain. For others, it will be several inches in from the end of the chain. It is very important to inspect the chain and determine which style of tape is being used before recording measurements.

Using Surveyor's Chains

Surveyors' chains must be used carefully. These chains break easily if a loop is pulled tight. The surveyor chain should always be wiped clean and oiled after each use. Chains can be thrown and rolled by hand, but a chain holder, as shown in Figure 2-5, is very handy and reduces the chance of breaking the chain.

When winding up a chain on a holder, the holder handle should be held in the left hand, and the 100 foot end is attached with the numbers up. The crank handle should be carefully turned as the individual watches the chain to insure loops do not form in the chain.

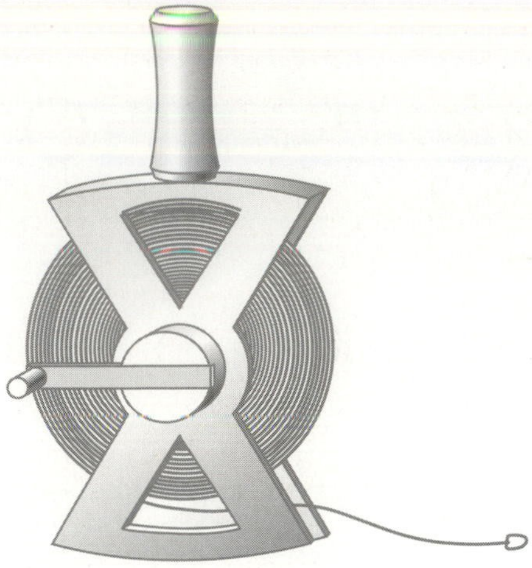

Figure 2-5 Chain on a holder.

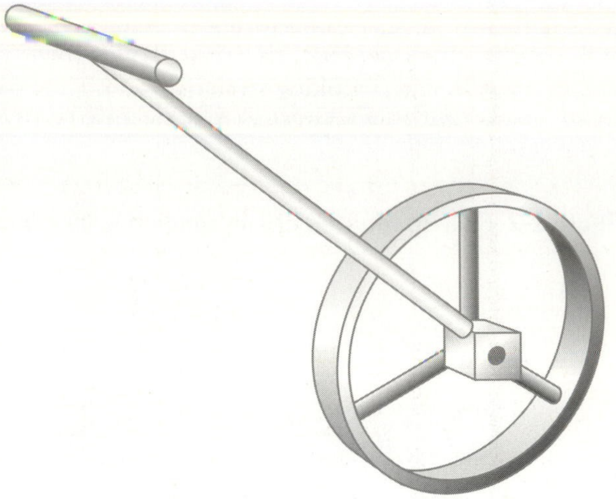

Figure 2-6 Odometer wheel.

Until the invention of electronic measuring devices, chaining was the most accurate method of measuring distance. A reasonable attention to detail will produce accurate measurements to 0.01 foot. With the appropriate chain and procedures, accuracy of 0.001 foot is possible.

Chains are still very useful, especially for short distances and where there is a clear path for the route. Electronic distance measuring has replaced chains for many surveying jobs such as measuring long distances, where the route cannot be easily walked or when the slope changes dramatically and frequently.

Odometer Wheel

As the name implies an **odometer wheel** is a combination of an odometer and a wheel, Figure 2-6. An odometer is a mechanical revolution counter. The size of the wheel influences the instrument error. Wheel diameters range from a few inches to several feet. Small wheels are intended to be used on hard smooth surfaces. When measuring on rough ground or in tall vegetation, a large diameter wheel should be used. Odometer wheels are available with several different levels of precision and measurement systems. Odometer wheels used for measuring distance are manufactured with dials that read in whole feet, feet and inches, decimal feet, or metric units.

To measure a distance with an odometer wheel the odometer is set to zero. Then the wheel is placed at the starting point, and the operator "walks the distance." A common error when using odometer wheels is to hold the handle directly above the wheel when setting it to zero and then tilting the handle down to the walking position. This practice will cause an error in the readings. Because the odometer wheel is rolled across the surface, all measurements are surface distance. Odometer wheels are very useful for low-precision measurements such as to determine the amount of fencing materials or pipe needed, and for estimating areas for application of chemicals.

Rangefinder

A **rangefinder** is a device that is usually handheld and can be used for determining distances. Rangefinders can be divided into two categories, optical and electronic.

Optical Rangefinder

An optical rangefinder is a low-tech, inexpensive instrument that uses the principles of right triangles to determine distances, Figure 2-7. Optical rangefinders

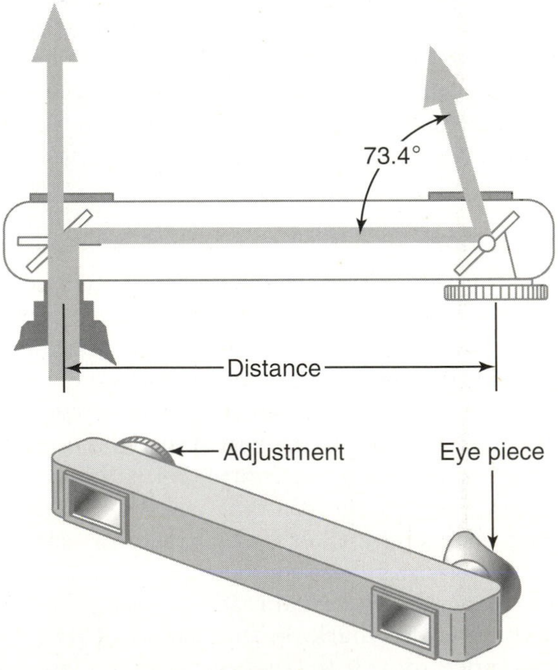

Figure 2-7 Optical rangefinder.

use a splitting mirror and a tilting mirror. When you look through the eyepiece, the view is split into two images. One image is looking straight ahead. The other image is a shadow image produced by the mirror offset to the side.

These instruments are commonly used for hunting and golfing, but they can also be used for land measurement when a low level of precision is acceptable. They are limited to measuring from the user's position to an object. They can be difficult to use if the background is very complex, such as a group of trees.

Using an Optical Rangefinder The instrument is used by focusing the instrument on a well-defined object. If the instrument is not set at the correct distance, a shadow image will be visible. The distance is determined by rotating the adjustment wheel until the shadow image is superimposed on the original image. When the images are superimposed, the distance can be read from the scale on the instrument.

The principle of optical rangefinders is based on trigonometric function. If you know the length of one side and one angle of a right triangle, then you can calculate the length of any remaining sides. The rangefinder does this using a mechanical linkage and a calibrated wheel.

For the example in Figure 2-7, if the distance between the splitting mirror and the tilting mirror is one foot, then the distance from the rangefinder to the object is:

$$\text{Tan}\,\theta = \frac{\text{opp}}{\text{adj}}$$

$$\text{adj} = \frac{\text{opp}}{\text{Tan}\,\theta}$$

$$= \frac{1.0\text{ ft.}}{\text{Tan } 0.19} = 301\text{ ft.}$$

The primary limitation of optical rangefinders is measuring range, accuracy, and precision. For example, the useable range may only be between 20 and 400 yards, or 50 and 1000 yards. The accuracy may be as low is 1/100 and the precision as large as one yard. An accuracy of 1/100 for a rangefinder with a precision of one yard means that the maximum error for any measurement is one yard for every 100 yards measured. Their primary advantage is that they are totally mechanical. They do not require any batteries to operate.

Electronic Rangefinder

Electronic rangefinders use an electronic impulse, usually microwave or laser, to measure distance. The devices range from inexpensive handheld instruments with a precision of one yard to tripod mounted devices with a precision less than an inch. Simple handheld rangefinders that have a low level of precision can be purchased for two hundred dollars or less. The type used for surveying can cost several thousands of dollars and are usually called electronic distance measuring (EDM) devices.

Electronic Distance Measuring

The term **electronic distance measuring** is used to describe a category of distance measuring devices that use microwaves or other forms of energy beams to determine the distance between itself and an object, between itself and a reflector, or between a master and a slave instrument. These instruments measure distance by rearranging the equation for velocity. In the standard equation, velocity equals distance divided by time.

$$V = \frac{D}{T} \qquad D = VT$$

Rearranged, distance equals velocity multiplied by time. EDM's determine distance by sending out a signal of known velocity and measuring the time lapse for the signal to return. The instrument must know the number of complete cycles and the point on a partial cycle when the signal returns to the instrument. If a single beam is used, the instrument can determine the point in a cycle when the beam returns, but not the number of complete cycles. Therefore, instruments use multiple beams with different wavelengths. Each wavelength will reach the sending unit at a different point in the cycle. Comparing the return point on the cycle, for beams with different frequencies, provides the CPU program with the information it needs to determine the number of cycles and the point on the partial cycle that occurred between when the signal was sent and when it was received, Figure 2-8.

Laser Rangefinder

A laser emits a visible or invisible beam that is very coherent: it diverges at a slow rate. This means that the diameter of the beam stays relatively consistent for a long distance. Some laser rangefinders use a reflector at the second station to reflect the signal back to the instrument, but others are reflectorless. Reflectorless lasers are point-and-shoot instruments. The object being measured is first centered in the viewfinder, or located with a visible dot of light. Then the rangefinder displays either the distance or the operator activates the laser for the distance. The accuracy and precision of the measurements vary because some instruments are intended to be used for

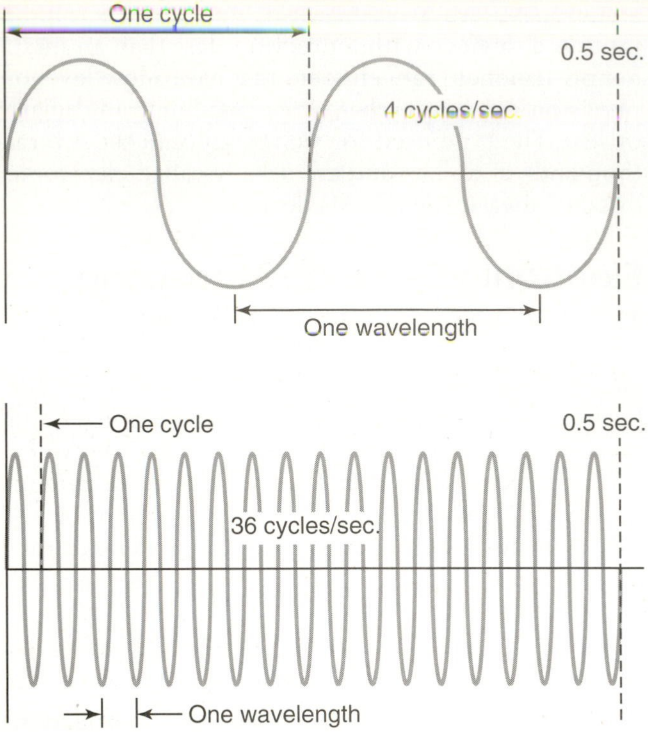

Figure 2-8 Comparison of wavelengths.

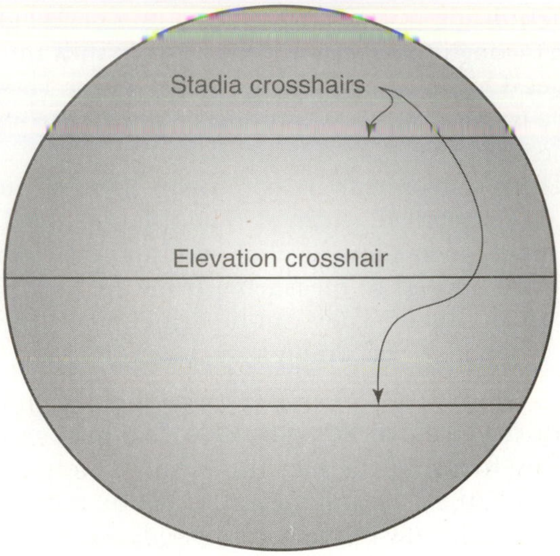

Figure 2-9 View of stadia crosshairs.

estimating flooring or roof areas while others are designed for professional surveyors.

Manufacturers provide a variety of additional features on laser rangefinders. One manufacturer includes a compass that is visible in the same screen as the distance. With this instrument a person can stand in one spot and measure the bearing and distance to any object within the range of the instrument. Another feature that is available is a built-in inclinometer. The addition of an **inclinometer** allows the user to determine vertical angles. With this ability, the rangefinder can be used to determine the height of objects. The combination of an EDM and an electronic transit results in an instrument called a total station. These instruments are described in the section on surveying instruments.

Advantages and Disadvantages of EDM's

Electronic distance measuring has several advantages:

- The operator does not need to walk the distance.
- Measurements can be taken across obstacles such as water, trees, and rough terrain. The only requirement is that a signal can reach the second point and return in a straight line.
- Instruments can be purchased which have the capability of downloading the data into a computer, data logger, or electronic field book, thereby eliminating the errors associated with manually recording data.
- Once leveled, EDM instruments can produce a reading every few seconds.
- They require fewer individuals to take measurements.
- They will measure short or long distances.

Disadvantages of EDM's include:

- The accuracy is influenced by air density.
- Handheld units are difficult to hold on the target at long distances.
- They use microprocessors: no electricity = no measurement.
- High quality instruments are expensive.

Distance by Stadia

Measuring distance by stadia requires the use of a surveying instrument with stadia crosshairs. Stadia crosshairs are additional horizontal crosshairs mounted an equal distance above and below the horizontal crosshair, Figure 2-9.

The distance between the stadia crosshairs is designed so that for most instruments when the difference between the top and bottom cross hair rod readings is one foot the instruments is 100 feet from the rod. This method is covered in more detail in Chapter 4.

EQUIPMENT FOR MEASURING ANGLES AND ELEVATIONS

A variety of instruments are available for measuring angles and elevations. It is important to select the best instrument for the job. Lacking personal experience, the best source of information for a specific instrument is the owner's manual or a manufacturer's catalogue. This section will discuss the common categories of surveying instruments.

Hand Levels

As the name implies, a **hand level** is a level that is held in the operator's hands. Hand levels are the simplest style of level used in surveying. They use a spirit level and a single crosshair. This style of hand level is used to insure that chains are level when measuring horizontal distance with plumb bobs, and estimating slope and changes in elevation. The common magnification is from zero to 5x. The more sophisticated hand levels will include stadia hairs for measuring horizontal distance.

The hand level illustrated in Figure 2-10 uses an external spirit level for controlling the vertical angle of the instrument. With this type, you must be able to hold the instrument horizontally while looking through the lens. Alternative designs include an internal level and a prism which provides a split view. When looking through the eyepiece you can see on one side the bubble for the spirit level and on the other half the crosshair. It is easier to hold this style of instrument in a horizontal position.

Hand levels are primarily used for estimating changes in elevation and estimating **slope**. Slope is the rate of change in elevation. Measuring slope with hand levels is accomplished by standing at the bottom of the slope, and while holding the level in a horizontal position, making note of a landmark where the line-of-sight strikes the ground. Using the distance to this point and the user's eye height above the ground, the slope can be calculated.

$$\% \text{ Slope} = \frac{\text{Rise}}{\text{Run}} \times 100$$

Rise is the eye height of the user, and the run is the distance from the observation point to where the line-of-sight strikes the ground. The distance can be measure by pacing. This results in a rough estimate of slope because of the low precision of measuring distance by pacing and because the measured distance is the slope distance, not horizontal distance, Figure 2-11.

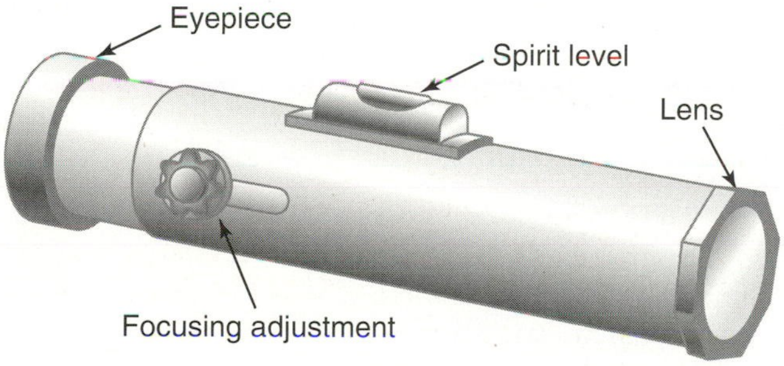

Figure 2-10 Hand level.

Abney Level

Abney levels are more sophisticated hand levels. They usually have a direct reading scale for vertical angles and slope, stadia hairs, and better magnification and optics. The precision of the slope calculation is better than a hand level because measuring distance by stadia can have a precision of 1/10 of a foot, and the distance measured using stadia is horizontal distance, Figure 2-12. The use of stadia for measuring distance is explained in Chapter 4.

Most Abney Levels have adjustments for both focusing and magnification. When used with a rod and target they provide sufficient accuracy for preliminary surveying.

Using an Abney Level

Abney levels use a split viewing area. When the user looks through the eyepiece, half of the area is used to view the spirit bubble and the remaining area is for viewing the target.

To measure vertical angles or slope, the tilting lock is loosened, and the instrument is aligned on a target the same height as the user's eye. While

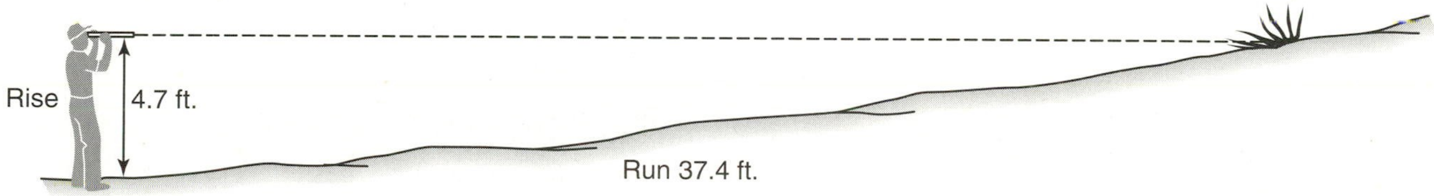

$$\% \text{ slope} = \frac{4.7}{37.4} \times 100 = 12.6\%$$

Figure 2-11 Example of estimating slope.

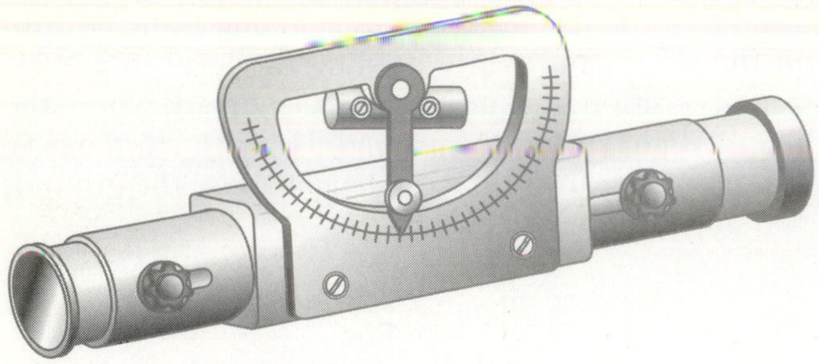

Figure 2-12 Abney level.

holding the instrument on the target, the level is tilted until the bubble is centered. Once the tilting lock is secured, the angle or slope can be read from the scale.

The accuracy of both hand levels and Abney levels is improved if they are used in conjunction with a stick or rod of known height. For example, if the centerline of the level is held on the 5.0-foot mark on a stick, the instrument height is five feet. Any difference in the rod reading at the unknown station from five feet is a difference in elevation. The use of a stick will also make it easier to hold the level steady.

Dumpy Level

The **dumpy level** is one of the simplest types of surveying levels mounted on a tripod. The use of a tripod improves the accuracy of the instrument and provides a reference for horizontal angles. A dumpy level consists of a telescope with a spirit level mounted in parallel with the line-of-sight of the telescope. The telescope will have at least one horizontal crosshair, mounted inline with the line-of-sight, and usually will have a vertical crosshair and two stadia crosshairs, Figure 2-13. See Figure 2-9 for an illustration of the crosshairs.

The telescope and spirit level are mounted on a mechanism (leveling plate) that rotates in a 360° horizontal circle. The entire mechanism is mounted on a plate that can be attached to a tripod and leveled.

Accuracy is insufficient for precise surveying, but is acceptable for general work such as leveling forms for concrete, determining slope, and shooting profiles for drainage work. A dumpy level will also contain a horizontal scale for measuring horizontal angles. The precision of the angle scale will vary with the manufacturer and model of instrument. They will usually have a precision of at least one degree, but a precision of 10 minutes is possible.

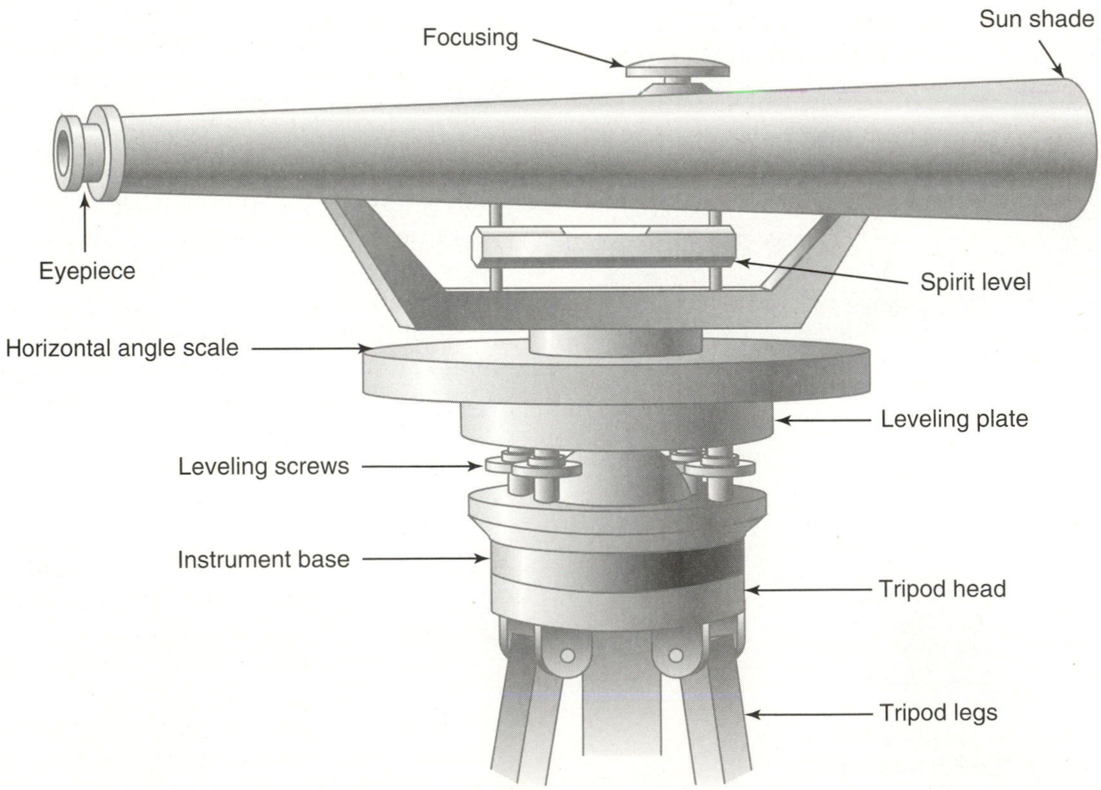

Figure 2-13 Dumpy level.

Using the Dumpy Level

The first step in using the level is to set up the tripod. The legs of the tripod should be spread until the tripod is stable and the head of the tripod is a good viewing height for the operator. The points of the legs should be planted firmly in the ground. During this process, the head of the tripod should be as close to horizontal as possible. This will reduce the amount of time that will be required to level the instrument after it is mounted.

After the tripod is set, the instrument is mounted on the head. All instruments are attached by either a large diameter thread, 3.5 inches with eight threads per inch, or a smaller diameter threaded bolt. During attachment, the base of the instrument should be loosened so one hand can grasp the instrument at all times as it is threaded onto the head of the tripod. This prevents dropping the instrument.

Once the instrument is mounted on the tripod, the next step is to level the instrument (set the telescope horizontal). Turning the four leveling screws in the correct sequence levels the instrument. This process requires some practice before an individual can level an instrument quickly because dumpy levels use a four screw (four leg) leveling mechanism.

Leveling a Four Legged Instrument

The first step in leveling a dumpy instrument is aligning the telescope over any two of the adjusting screws, Illustration A Figure 2-14. The next step is to adjust the leveling screws until the bubble of the spirit level is between the lines, Illustration B Figure 2-14. The direction the leveling screws should be turned depends on which way the instrument must be tilted to make it level.

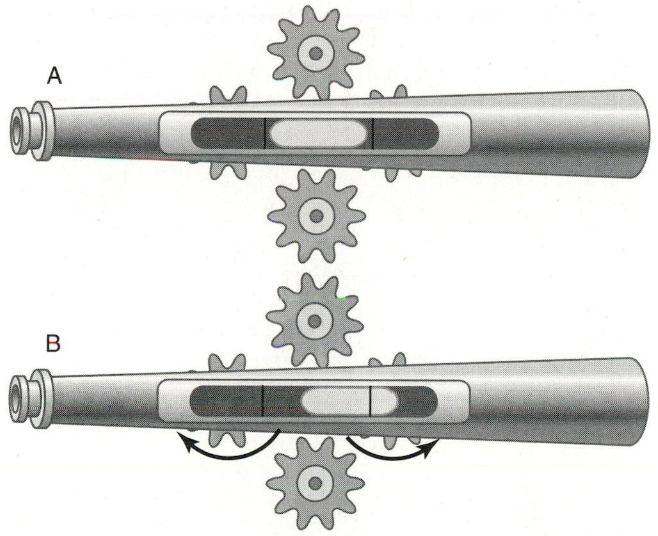

Note: The spirit level has been moved on top of the telescope for illustrative purposes.

Figure 2-14 Leveling a four legged instrument.

Note: The bubble will move in the same direction as the left thumb moves when the left leveling screw is turned. The two screws are turned in opposite directions until the bubble is between the lines on the spirit level vial. In Illustration B of Figure 2-14, the right hand screw is turned **counterclockwise** and the left-hand screw is turned **clockwise** because the right side of the telescope needs to be lowered to level the instrument. The opposite rotation of the leveling screws would be used if the opposite movement were desired.

During this process it is important to turn both screws at the same rate. Usually one hand of the operator is dominant and this hand will tend to turn that screw faster. When this occurs the adjustment can either become loose, in which case the telescope will teeter-totter on the base, or the adjusting screws will become too tight to turn. Forcing the adjusting screws too tight can warp the adjusting mechanism and permanently damage the instrument.

The second step is to rotate the telescope 90 degrees and repeat the leveling process. These two steps may need to be completed several times before the instrument remains level as it is rotated. The instrument is level when the bubble stays within the lines on the spirit level as the telescope is rotated 360 degrees.

Adjusting for Parallax

Parallax is a condition that occurs when the line-of-sight of the eye is not aligned with the line-of-sight of the telescope. This condition can occur anytime an instrument with a telescope is not adjusted correctly. This will cause an error every time the rod is read. Parallax is removed by adjusting the eyepiece until the crosshairs are their darkest.

Note: On some instruments the eyepiece is adjusted by rotating the eyepiece and on others it is adjusted by sliding the eyepiece in and out. If more than one individual is reading the instrument, the eyepiece must be adjusted each time a different person reads the instrument.

Once the instrument is leveled and the eyepiece adjusted for parallax, the focusing knob is used to bring the rod and target into focus.

When using any surveying instrument, never force any movement. If a part is designed to move, but will not, do not try to force it. Instead, check to be certain the locks have been released. A slight pressure is all that should be required to move any part of the instrument.

Automatic Level

The term "automatic" does not mean the automatic instrument levels itself. The instrument is designed to automatically compensate for small movements in the instrument and keep the line-of-sight level. Once the instrument is nearly level, an internal compensator completes the leveling process and maintains the line-of-sight in a horizontal position. The compensator also prevents the instrument from being knocked out of level by slight bumps. The effect of the wind is also minimized because the instrument can compensate for slight movement in the tripod. The use of the compensation mechanism allows the instrument to be leveled with three leveling screws instead of four. It is also common for automatic levels to use a bull's-eye spirit level instead of a tube level. The use of three leveling screws and a bull's-eye level results in a faster setup time, Figure 2-15.

Automatic levels are available in several different models. Some are more accurate and more precise than dumpy levels, but all are less accurate and precise than transits and total stations.

Using an Automatic Level An automatic instrument is leveled by aligning the telescope over one of the leveling screws and adjusting the other two screws until the instrument is level. The pentaprism is a mirror that allows the operator to see the bull's-eye level from the side. Before taking a reading the eyepiece must be adjusted until the crosshairs are the darkest lines to adjust for parallax.

The collimator is used to align the instrument on the rod or range pole. To use the collimator the operator must back away from the instrument about an arm's length and then align their eye with the collimator. Then they can superimpose the image in the collimator on the rod or range pole. The rod or range pole should be within the field of view when the operator looks through the eyepiece.

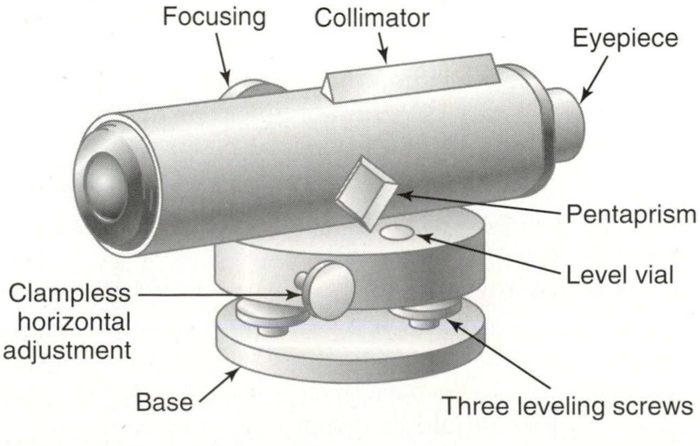

Figure 2-15 Automatic level.

Laser Levels

A **laser level** is an instrument that uses a beam of laser light to establish the line-of-sight, reference line. The different types can be divided into four categories: (1) single beam invisible, (2) single beam visible, (3) circular beam visible and (4) circular beam invisible. Circular beam lasers can also be categorized as rotating or non-rotating. A single beam laser will produce a single dot or a short line. A circular beam laser produces a 360-degree beam.

An invisible beam laser requires the use of a detector. The detector is moved up and down the face of the rod until it is centered on the laser beam. The pointer on the detector is used to read the height of the beam at the rod. Some use an electronic tone that beeps at a different rate when the detector is above and below the laser beam, and becomes a solid tone when the detector is on line with the laser beam. Others will use flashing lights in the same manner. Some use a display screen with a triangle that indicates the direction the detector should move to be aligned with the laser. Instruments may use a combination of these methods. The detectors for some laser levels also indicate if the laser is out of level. This feature reduces the chance of errors caused by an instrument that has been knocked out of level.

One distinct advantage of laser levels is that they can be operated by a single person. The laser level is mounted on a tripod and leveled. Once turned on, the laser does not require any supervision. The surveyor can walk around the area and record rod readings anywhere within the range of the beam, Figure 2-16.

Another advantage of this system is that multiple detectors can be used with a single laser. This allows more than one person to record data simultaneously.

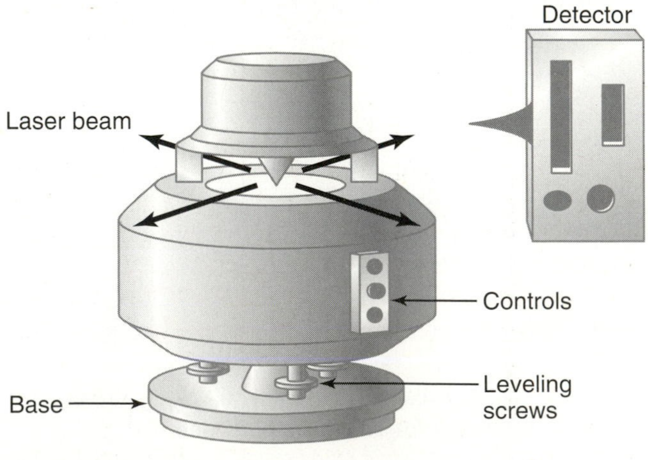

Figure 2-16 Laser level.

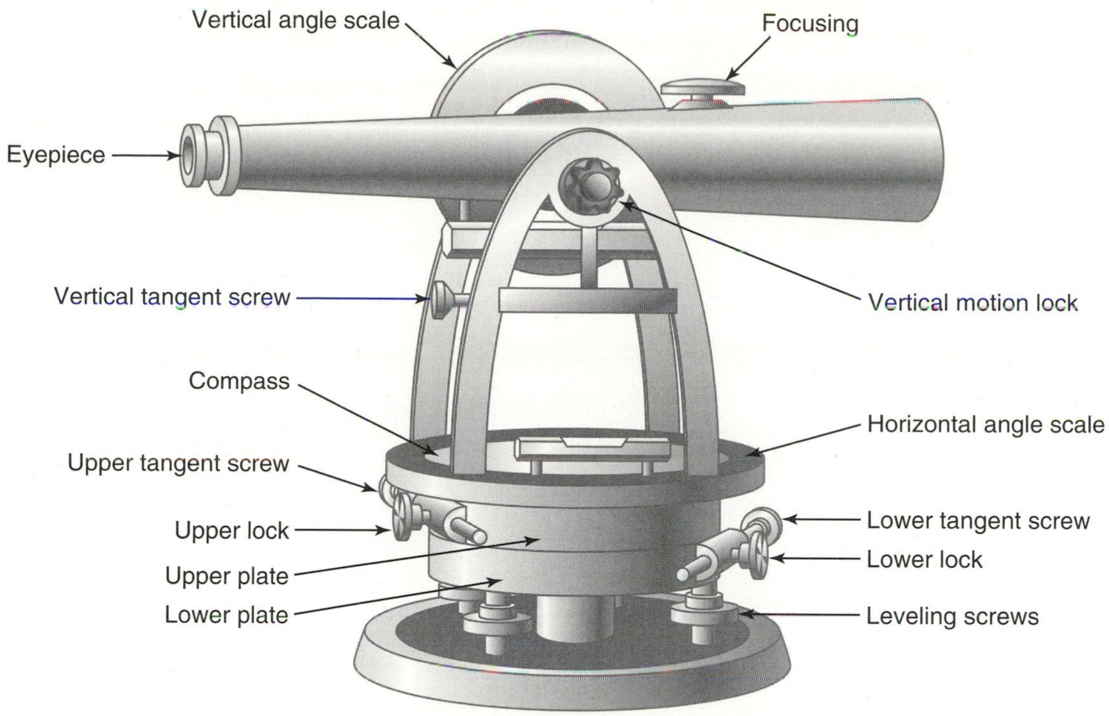

Figure 2-17 Transit.

Popular options for laser levels include an angle capability and self-leveling. Laser levels with the ability to be set at an angle are very useful for establishing a slope. This is a handy feature when installing drains, grading land, and working with grades in two different planes. The self leveling feature makes the instrument easier to use because the operator just attaches it to the tripod and turns it on. Built-in motors level the instrument.

Transits

Transits are the most complicated and precise mechanical surveying instrument. Consequently, they are useful for a wide variety of surveying jobs. Features of transits include a telescope that can be rotated 360 degrees horizontally and vertically, horizontal and vertical Vernier angle scales with a precision of minutes and usually seconds, higher power telescopes, and a magnetic compass, Figure 2-17.

Transits are capable of measuring both horizontal and vertical angles with a high degree of precision. Even though total stations and GPS instruments have replaced transits as the primary surveying instrument for professional surveyors, they are still very useful when a high degree of accuracy and precision are required.

Using a Transit

The process of setting up a transit is identical to a dumpy level. Before any readings can be taken it must be leveled using the four leveling screws, the eyepiece must be adjusted for parallax, and the telescope must be focused on the rod. The numerous adjustments increase the capabilities of the transit, but they also make it harder to set up and increase the opportunities for operator mistakes.

One common mistake is failing to set the telescope at zero vertical degrees when using it for measuring elevations. Transits use Vernier scales to read the horizontal and vertical angles. Reading a Vernier scale quickly and accurately requires practice.

Vernier Scales

Transits and other instruments use a Vernier on the angle scales to provide another level of precision. A Vernier scale can be single, Figure 2-18 or double, Figure 2-19. The single scale is used when the instrument is read in only one direction.

When a double Vernier scale is used the angles can be read in both the clockwise and counter clockwise directions. More information on turning angles and reading double Verniers can be found in Chapter 7.

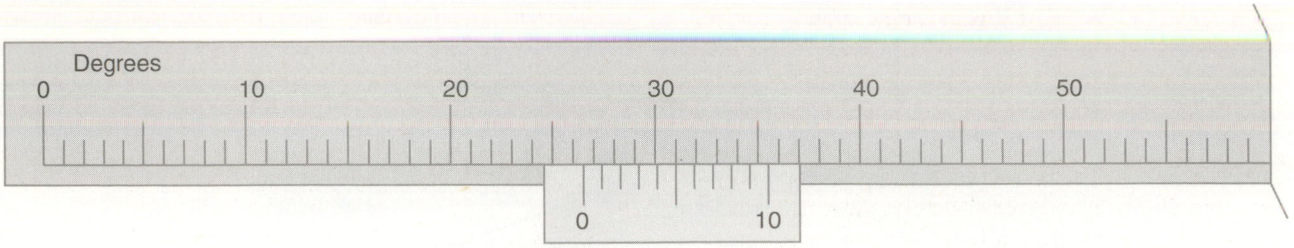

Figure 2-18 Single Vernier, the reading is 26.5 degrees.

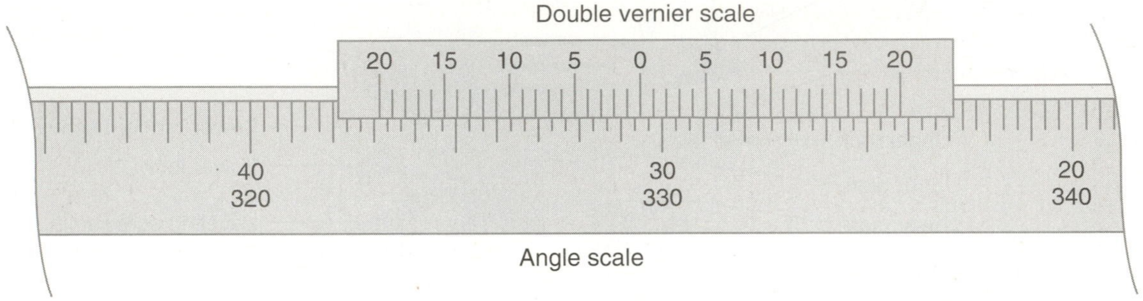

Figure 2-19 Double Vernier scale, clockwise reading is 30° 32' and the counterclockwise reading is 329° 28'.

Verniers are a mechanical means of increasing the physical size of the last unit on the main scale to provide an additional level of precision.

When reading a Vernier the first step is to determine the smallest possible reading, commonly called **least count**. This is accomplished by determining the smallest reading on the angle scale and then determining the precision of the Vernier. For example, in Figure 2-18 the precision of the main scale is one degree and the Vernier is divided into nine lines, ten spaces. The precision, least count, of the Vernier scale is 1/10, 0.1, of a degree.

The precision of the main scale for the double Vernier, Figure 2-19, is 20 minutes, 1 degree divided by 3. Note the Vernier scale ranges from zero to 20 without any additional subdivisions. The least count for this Vernier is one minute.

The lines on Vernier scales are very fine and the scale is usually small. Some type of handheld magnification is useful. Additional information on reading Vernier scales is available in Chapter 7.

Other Instruments

Manufacturers of surveying equipment often develop instruments based on new technology or for specialized uses. Examples of these are electronic transits, transit levels, theodolites, and total stations.

Electronic Transits

The **electronic transit** has the same capabilities as a mechanical transit. The differences are digital readouts for angle readings and modernized adjustments. This reduces reading errors and speeds up the process of collecting data. Some electronic transits also provide mounts for an EDM and connections for downloading the data in an electronic format.

Construction Transit

The term **construction transit** and **transit level** refers to a group of instruments that have characteristics of both transits and levels. They consist of a dumpy level with a telescope having a few degrees of vertical movement. This increases their capabilities, but they are not considered a transit because the telescope cannot be rotated in a full vertical circle.

Theodolite

A **theodolite** is a transit with higher precision. A good quality transit will measure angles to the nearest minute. A good quality theodolite is capable of measuring angles to the nearest second.

Total Station

Excluding GPS, a **total station** is the instrument of choice for professional surveyors. Total stations are a combination of an electronic transit and an EDM. They have optical crosshairs and can still be used visually with a rod like a transit, but they display angle and distance readings electronically. They also use a built-in or detachable EDM. This allows multiple measurements, such as horizontal angle, vertical angle, and distance, to be recorded simultaneously. The early EDM's used a reflector (prism) to return the signal but later models use reflectorless technology. The distance a total station can measure in reflectorless mode is dependent on the reflectivity of the object being used as the target. Reflectorless technology allows the convenience of single person operation. The use of an onboard CPU enhances the ability of total stations. Many features can be designed into the instrument. Some include storage devices that can be used to record data. They can be programmed to measure in a variety of units, such as feet or meters for distance and degrees, gon and mil for angles. Tones can be used to indicate horizontal and vertical position of the telescope or the condition of the EDM and they can even be designed to automatically compensate for refraction. A time-saving feature on most total stations is the capability to export information to a data logger or portable computer. This eliminates the requirement of writing down the readings and the associated errors. The total station and the data recording devices must be synchronized to insure the data is stored in the desired format.

ACCESSORIES

Many different accessories are used to enhance the surveying process. The following sections discuss a few of these.

Pins, Markers, Flags, and Stakes

Surveyors always need to mark locations. Traditionally, different types of pins, markers and stakes have been used.

Pins

Surveying pins are usually constructed from heavy gauge wire and painted in white and red alternating stripes. They are used when measuring distance with a chain. A set consists of eleven pins.

Markers

Many types of markers are used. Brass, plastic, steel, or aluminum caps are used to identify benchmark and other important points. Some are designed like nails so they can be driven into the ground, rock, or concrete. Others are designed to fit on rebar. Steel ones can also be used that are magnetic so they are easier to find with a metal detector.

Flags

Flags are used first to locate important points before markers or stakes are installed. It is common practice to use different flag colors to represent different components of the survey. Flag colors are also used to identify different utilities.

Stakes

Stakes are usually wood or plastic. They are useful for providing more visibility for a location and provide a writing surface for instructions, identification, elevation, etc.

Field Book

A **field book** is a notebook especially designed for recording surveying notes. The traditional field book, when opened, has a left half and a right half of a page, not two pages as a normal book. Data in field books must follow the four C's: complete, clear, concise, and correct.

To be complete, the field book must contain sufficient information so that a person that was not part of the original survey could find the site and conduct the same survey. Site location has traditionally been by street address, PLSS description, owner's name, or a combination of these methods. Today it is common practice to use GPS to locate the site of the survey.

Completeness also includes insuring the party collected all of the necessary data for the uses of the survey. It is important to insure the information required for the note checks is collected. It is also important to orientate the survey to north, or some other reference, so plots and maps can be drawn correctly.

Being clear refers to readability. The first step in readability is insuring the page is well organized. Traditionally, each segment of information has a specific location on the page, Figure 2-20. The writing must be neat and easy to read, and the data must be well organized and easy to interpret. The reader should be able to find the desired information quickly, and easily understand the purpose for each number. Legibility is a necessity. If the notes are not readable, all of the resources expended to collect them were wasted. Legible notes also reduce the chance for errors when reading the notes.

Conciseness is an important characteristic of survey information because unnecessary information on the page can cause confusion and make it harder for the reader to find the desired information. Abbreviations and symbols/icons are very common. Figure 2-21 is an example of the symbols commonly used to identify the survey party jobs.

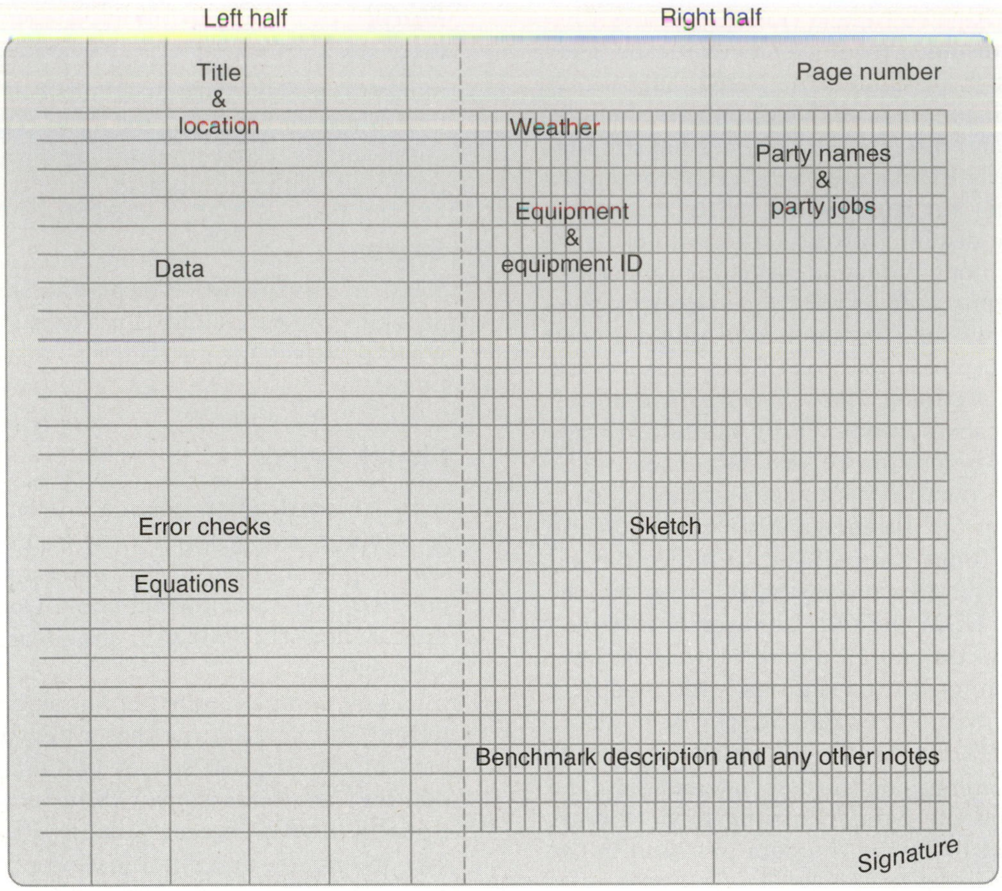

Figure 2-20 Organization of information in a field book.

Correctness cannot be compromised. All of the time and resources expended to collect the information is wasted if the data is not correct. Even worse, incorrect data will lead to incorrect design decisions, which can be very costly. Surveying mistakes can eliminate the profit for the job, cause companies to go bankrupt, and even cost professional surveyors their licenses.

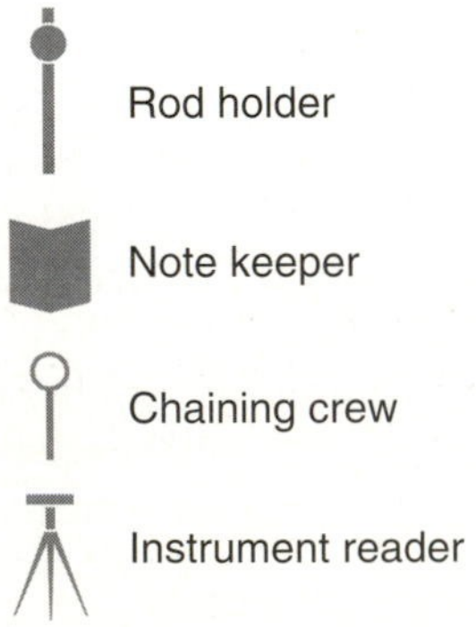

Figure 2-21 Survey party job icons.

Standard Field Book Format

The front cover or first page in a field book is used for the owner's identification. The purpose of identification is to provide a means of returning the book in case it is lost or misplaced. A set of survey data can represent many hours of work and thousands of dollars in expenses, all of which would have to be repeated if the book is lost. The next page(s) contains the index. The index should be located in the front of the book and it should include:

Survey Title	Date of Survey	Page Number(s)

If a book is used for several surveys, a page or series of pages are used for each survey. The information that should be included for each survey is:

- Title
- Location
- Data
- Equations

- Error checks
- Page number
- Weather information
- Party names
- Party jobs
- Equipment list
- Equipment identification
- Sketch of survey
- Benchmark description and location
- Explanatory notes
- Note keeper's signature

Left Half of Page

The title used on the information page must be the same as the title used in the index and it must be descriptive for the survey. An example would be, "Profile of drainage ditch." The location description in conjunction with the sketch must provide enough information to allow the reader to find the site. The data is the information that is recorded during the survey. It is important for the data to be well organized and labeled so that it is easy to read and understand. A table is an excellent way of organizing and presenting data. Any information in the data section that is not original should be labeled as such by using footnotes for equations, etc. For example, when distances are measured by pacing, the recorded numbers are in paces, but distances in paces are not usable. It is common practice to convert the paces to feet by using the individual's pace factor. These distances in feet are not original and should be identified with a footnote and the equation that was used for the conversion. The equation should be included at the bottom of the data table. Mathematical error checks should be included with the data. An example is the note check used with differential and profile surveys.

Right Half of Page

The page number coincides with the number used in the index to aid the reader in finding an individual set of data. Weather information is important for surveys. It may be needed to adjust for errors and may provide information useful for interpreting the data. Temperature data is required to calculate the temperature adjustment for a chain. Temperature and barometric pressure data may be needed to set a total station or EDM. Party names and party jobs are included so if the reader has procedure questions or other questions about the survey they can contact the individual responsible for that part of the data. It is important to include the equipment list and equipment identification. This information may be useful for assigning equipment costs for the job. It also important in case problems occurs with the data. The instrument can be checked or calibrated to determine if it needs adjustment. The sketch is an important part of the notes. Sketches are not expected to be draftsman quality, but they must be clear, concise, and easy to read. They are used to provide more information about the location, the location of and identification used for the stations, the location of the benchmark, and any other information that is pertinent to the survey. The station identifications used in the sketch must be the same as the station identifications used in the data section. Along with the sketch must be a description of the benchmark. This description must be precise. A benchmark description of "utility access cover at the corner of first and west streets" is not complete. A description that reads "the benchmark used was an "X" chiseled in the rim of the telephone utility access located 10 feet south and 20 feet west of the corner of first and West Street" is complete. The sketch must include a north arrow. The last thing that should be included in a set of notes is the note keeper's signature. The note keeper should not sign the notes unless they are sure the notes are complete and correct. Along with the note keeper's signature must be the word "Copied" if the date is not the original date written down as it was collected. Standards of practice for different disciplines may require some differences in the location and methods of recording information.

Technology has changed the way professional surveyors collect data. Electronic instruments have the capability of downloading data with a data logger or portable computer. Another innovation is the ability to take digital pictures and attach them to the data. Computer programs organize the information into what is call an electronic field book. Even when data are collected electronically, it is important to collect weather data and party names, sketch the area, and record other important information.

Rod

A surveying **rod** is a measuring device that is used to determine the distance from the line-of-sight of the instrument, instrument height, to the ground. They were traditionally constructed of wood, but modern ones can be metallic or constructed from other non-metallic materials. A standard surveying rod is essentially a 13-foot wooden scale. Rods are available with different scales. The "Philadelphia style" rods can be read to the nearest 1/100 (0.01) of a foot directly and to the nearest 1/1000 (0.001) of a foot when the target is used, Figure 2-22.

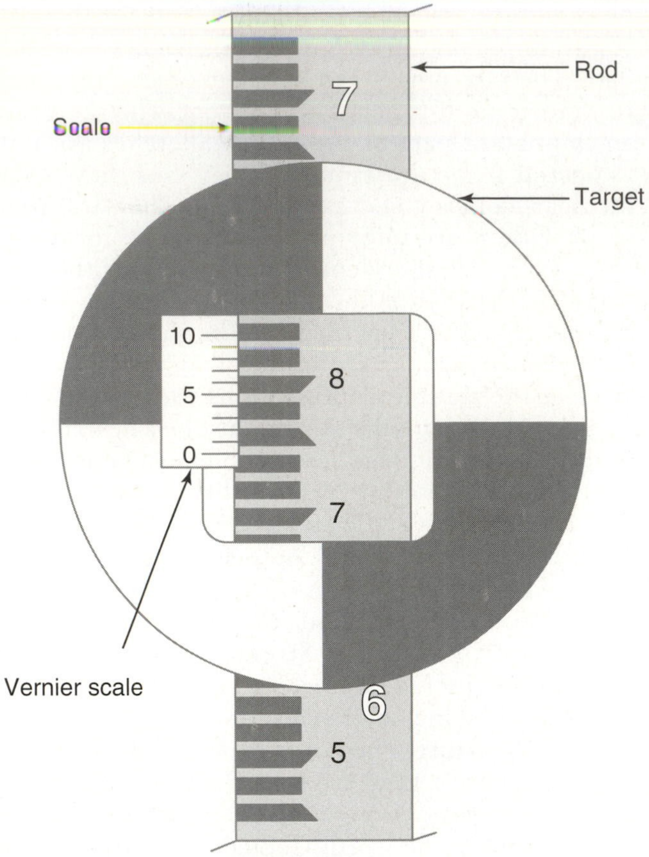

Figure 2-22 Philadelphia style rod and target.

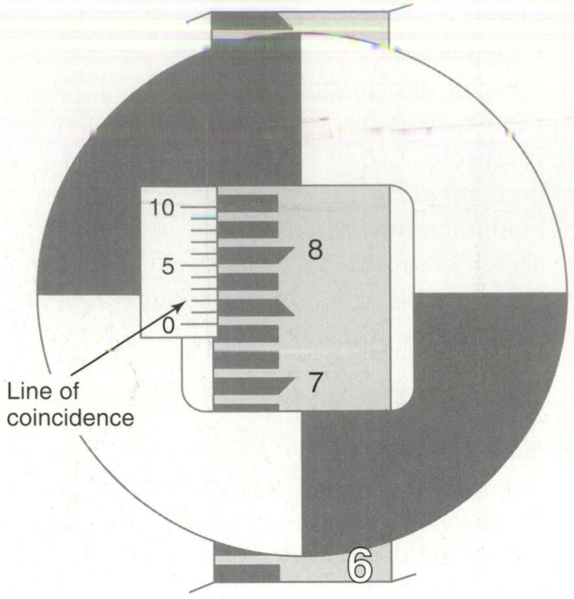

Figure 2-23 Line of coincidence on target Vernier scale.

When a target is used to read a Philadelphia style rod, the rod is read at the zero (0) line of the Vernier scale. If the desired precision is hundredths (0.01) of a foot the distance is read from the rod. When the desired level of precision is thousands (0.001) of a foot, the Vernier scale is used. The Vernier scale is read by finding the line of coincidence, Figure 2-23.

The reading for the rod in Figure 2-23 is 6.742 feet. The whole feet is 6 because the rod starts with 1 foot at the bottom and a small (red) number 6 is visible at the bottom edge of the target. The 7-foot distance is above the target. The reading is 0.7 because the zero (0) line of the Vernier is above 0.7 feet but below 0.8 feet. The hundredths are 0.04 because the zero (0) line on the Vernier scale is above the top edge of the second black line above 0.7 feet, but below the bottom of the third black line. The line of coincidence on the Vernier scale can be difficult to determine on a target because the scale is small. It is more difficult to determine in drawings because printers try to align lines that are slightly offset. In this example the line of coincidence could be either 2 or 3. Quite often when reading scales the reader must just pick the number they think is closest. This example selected the 2, which is 0.002. Therefore, the reading is $6 + 0.7 + 0.04 + 0.002 = 6.742$.

The Philadelphia style rod can be used in four ways; direct reading, indirect reading, extended rod, and high rod.

Direct Reading a Rod

Direct reading occurs when the distance between the instrument and the rod is short enough that the numbers on the rod can be read by the person looking through the instrument. When direct reading the note keeper stays with the instrument and records the rod readings as they are called out by the person on the instrument.

Indirect Rod Reading

As the distance between the instrument and the rod increases, a distance will be reached at which the person at the instrument will not be able to read the numbers on the rod. In this situation the person holding the rod must use a target, a pencil, or some other object to indicate the coincidence of the line-of-sight and the rod. The person on the instrument has the rod person slide the object up or down until it is aligned with the desired crosshair of the instrument. The rod person reads the rod and calls out the results to the note keeper. Indirect rod reading requires the note keeper stay with the rod. Indirect reading is also used when the desired level of precision requires the use of the target. Rod targets are described in more detail in a following section.

Extending the Rod

Extended rod is used when the change in elevation between the instrument position and the rod position is great enough to cause the line-of-sight to pass over the top of the rod. An extended rod must be read directly from the instrument. If the distance is too great to read the rod directly, the high rod technique must be used. This is explained in the next section.

A common error in using the extended rod is failure to extend it all of the way and insure that it locks in place. An extended rod is very difficult to hold vertical in windy conditions. In windy conditions, it is prudent to shorten the distance by using a turning point until the rod can be used retracted.

Using a High Rod

The high rod technique is used when the line-of-sight extends above the rod and the distance between the instrument and the rod makes it impossible to read the rod directly. The high rod technique requires the use of a target. In the high rod technique the target is attached at the 7.0-foot mark and locked in place. The instrument person signals the rod person to extend the rod until the target is aligned with the desired crosshair of the instrument. The rod person locks the rod in place and then lowers it for reading. The high rod technique uses the Vernier scale on the backside of the rod. On a Philadelphia style rod the scale runs up the front and continues down the back. A Vernier scale is mounted on the backside of the rod. When the rod is retracted, the Vernier scale on the backside of the rod will read 7.0 feet. As the rod is extended, distances greater than 7.0 feet are indicated on the backside. The zero point of this scale is used to read the rod to 1/100 of a foot and the Vernier scale can be used to read the rod to 1/1000 of a foot.

Reading a Rod

A surveyor's rod is read differently than a carpenter's tape. A carpenter's tape is read by recording the measurement from the closest line on the tape. When reading a Philadelphia style surveyor's rod, the number of transitions from black to white must be noted. The edge of each black line is one unit, Figure 2-24.

When reading a Philadelphia style rod it is useful to remember the following characteristics:

- Only the top and bottom edges of the black lines are read
- The top edge of a black line is always an even number, and is hundredths of a foot
- The bottom edge of each black line is always an odd number, and is hundredths of a foot

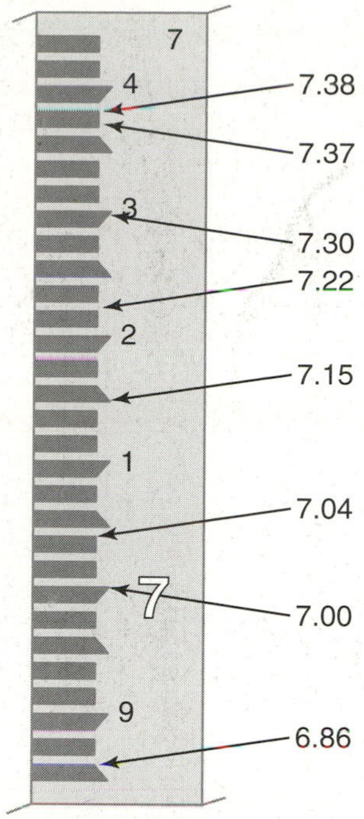

Figure 2-24 Reading a Philadelphia rod.

- A black line with a point on the top edge is a tenth of a foot
- A black line with a point on the bottom edge is 5 hundredths
- Red numbers are whole feet
- Black numbers are tenths of a foot

Common errors in using rods include incorrect rod reading, transposing numbers, and failing to hold the rod vertical. Practice is the best method of controlling reading errors. A rod can be out of vertical in two directions: parallel with the line-of-sight, and perpendicular to the line-of-sight. The person on the instrument can determine if a rod is out of vertical perpendicularly to the line-of-sight by comparing the rod to the vertical crosshair in the instrument. The person on the instrument cannot determine if the rod is not vertical parallel with the line-of-sight. The rod holder must control this potential error. The rod holder can control this error by using a rod level or by rocking the rod. A **rod level** is a bull's-eye or tube type of level mounted on a short frame that fits against one corner of the rod, Figure 2-26. It can be permanently mounted, but the rod holder usually holds it in place. When **rocking the rod,** the rod holder slowly leans the rod towards the instrument and then slowly brings it past center and back towards them.

The practice of rocking the rod is based on characteristics of right triangles. As the illustration in Figure 2-25 shows, the shortest distance, the correct rod reading, occurs when the rod is vertical.

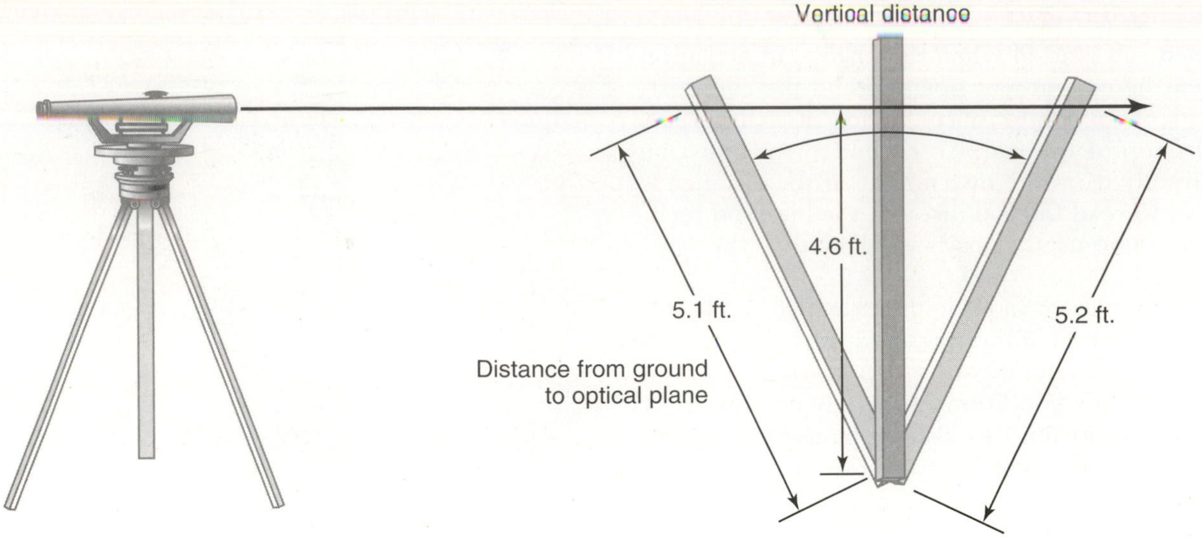

Figure 2-25 Rocking the rod.

The person at the instrument watches the rod readings as the rod is rocked. As the rod is moved off vertical, the rod reading increases and as the rod moves towards vertical, the readings decrease. The correct reading is the minimum reading. It will take some practice for a survey team to determine the best speed to rock the rod and the optimum distance to move the top of the rod.

Target

A **target** is a round or oval device with a rectangular hole in the center and a means to clamp it to the rod. The face of the target is divided into alternating red and white quadrants, and a Vernier scale will be mounted on one edge. It may also include a fine adjustment lever. The target is used when the distance between the instrument and the rod makes it impossible for the instrument person to read the number on the rod or when precision to 1/1000 of a foot is required. Figure 2-22 illustrates a target attached to a rod.

Rod Level

Figure 2-25 illustrates the importance of reading the rod when it is vertical. An option to rocking the rod is using a rod level. A rod level uses a bull's-eye type of level mounted onto an angle bracket, Figure 2-26.

The rod level is either held against or mounted to the rod, and when the bubble is in the center, the rod is vertical.

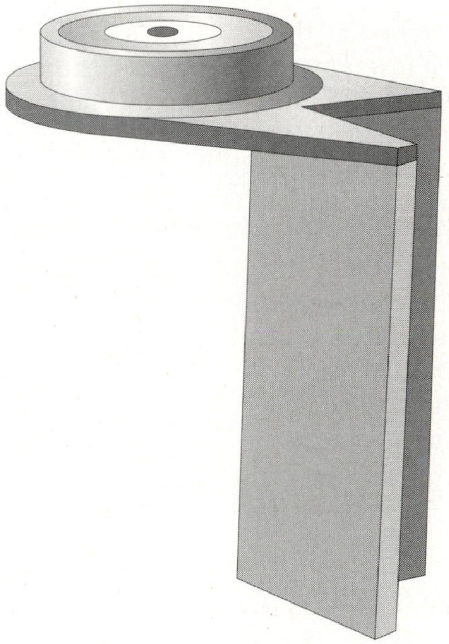

Figure 2-26 Rod level.

Range Poles

Range poles are five- to six-foot or longer tubes or rods with a solid, sharp point on one end. They are painted in alternate red and white wide stripes, Figure 2-27. Range poles are used to provide a visual reference point. They are useful for staying on line when chaining and for marking stations when turning angles with a dumpy or automatic level.

Figure 2-27 Range pole.

Plumb Bob

Plumb bobs are slender, heavy cone-shaped devices with a string attached at the center, Figure 2-28. They are used to establish a vertical line. Vertical lines are useful when transferring a point vertically from the earth to the chain when chaining horizontally and for setting an instrument over the vertex of an angle. Good quality plumb bobs are constructed of brass, have replaceable points, and should last several lifetimes. The string used to suspend the plumb bob should have braided fibers, not twisted. In addition, a sliding lock on the free end of the string eliminates the need for tying knots in the string. When a sliding lock is not available, the taught line knot can be used. Any knots that are used must be removed as soon as possible and should be designed so they do not become too tight to be removed.

Two-Way Radios

Good communication is a necessity for a survey crew. When equipment limited measuring distances and instrument-sight distances to less than 400 feet, verbal communication and hand signals were effective. Modern equipment has extended distances beyond what can be spanned by verbal communication and hand signals. Two-way radios can save many steps and reduce the time required to complete the survey. In addition, they reduce the chance of errors caused by individuals shouting at each other and they allow the transfer of information that cannot be communicated with hand signals. There are still some advantages for two-way radios, but their use has been superseded by cell phones.

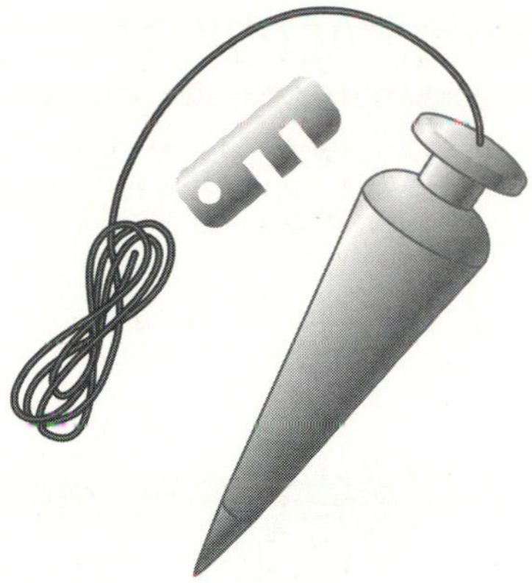

Figure 2-28 Plumb bob and sliding lock.

Summary

In this chapter, the common instruments and equipment used for surveying have been discussed. A good understanding of the names of instruments and equipment, their parts, and how they are used is a requirement of surveying. Attempting to use the wrong piece of equipment for the job, or using a piece of equipment incorrectly, will lead to errors in the data and damaged equipment. The following chapters will explain how this equipment is used to complete common surveys.

Student Activities

1. Identify the parts for the dumpy level.

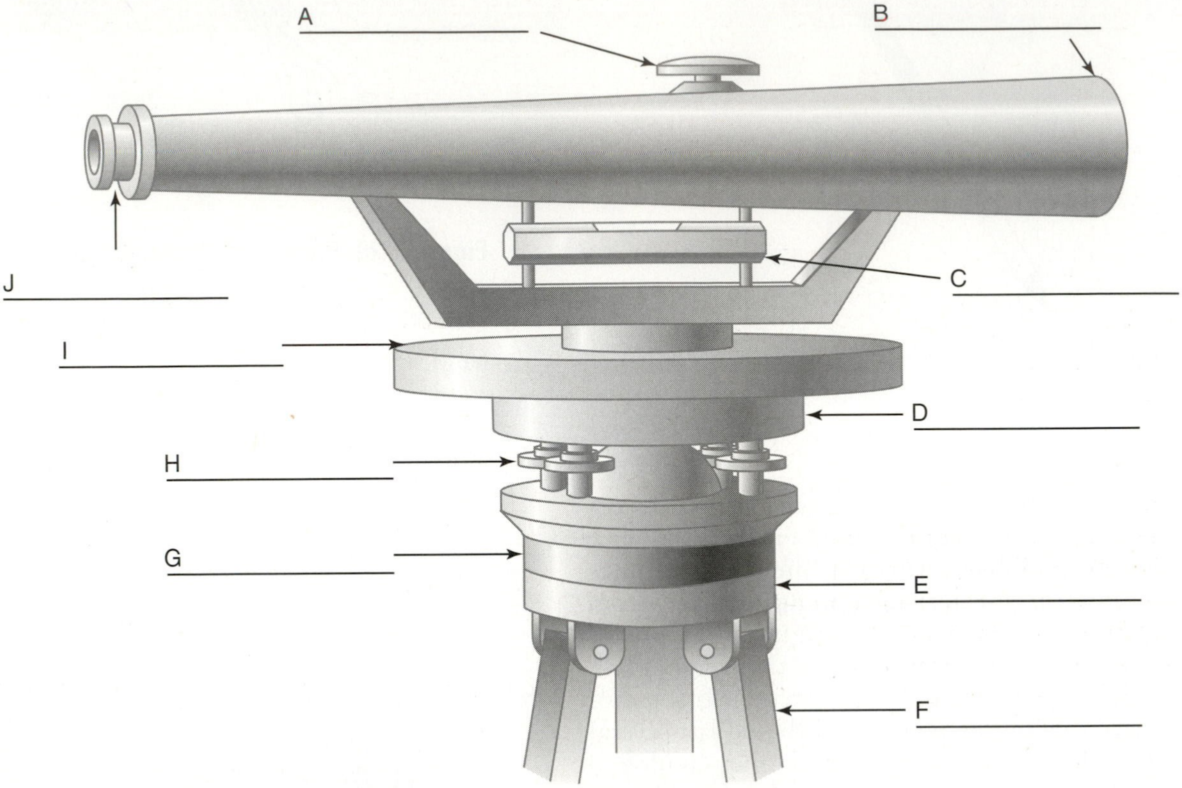

2. Identify the parts of a transit.

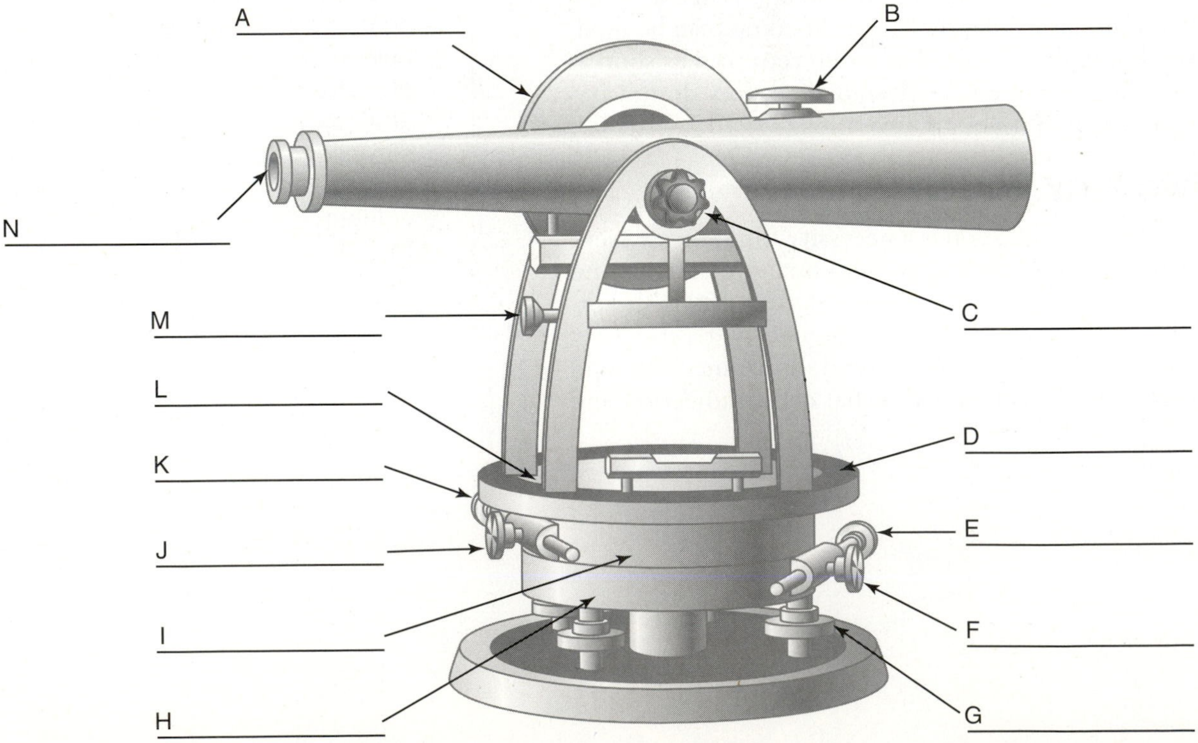

3. Determine the reading for the chain in the illustration:

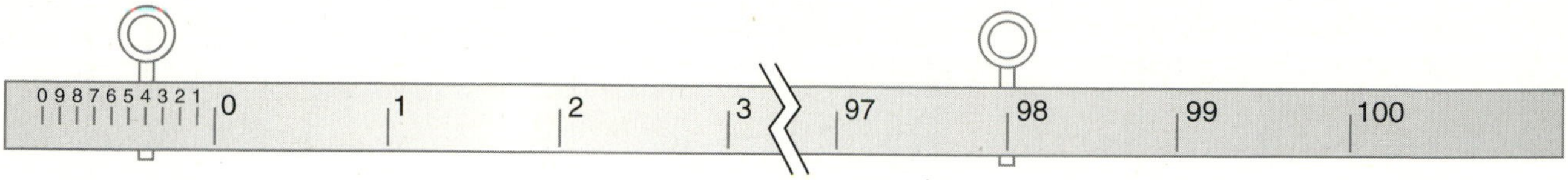

4. Determine the reading for the chain in the illustration:

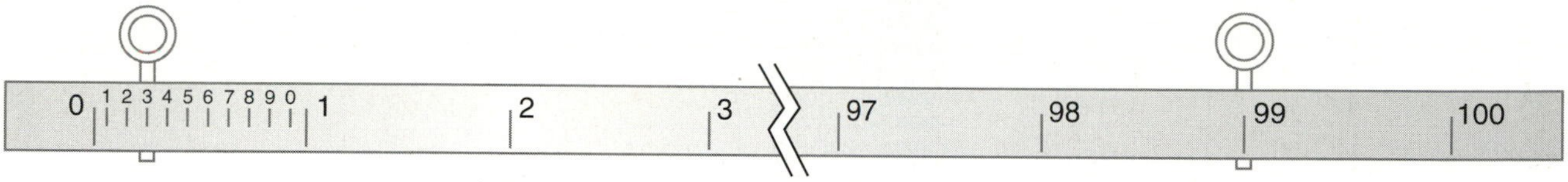

5. Record the rod reading to 0.01 and 0.001 feet for the rod and target in the illustration.

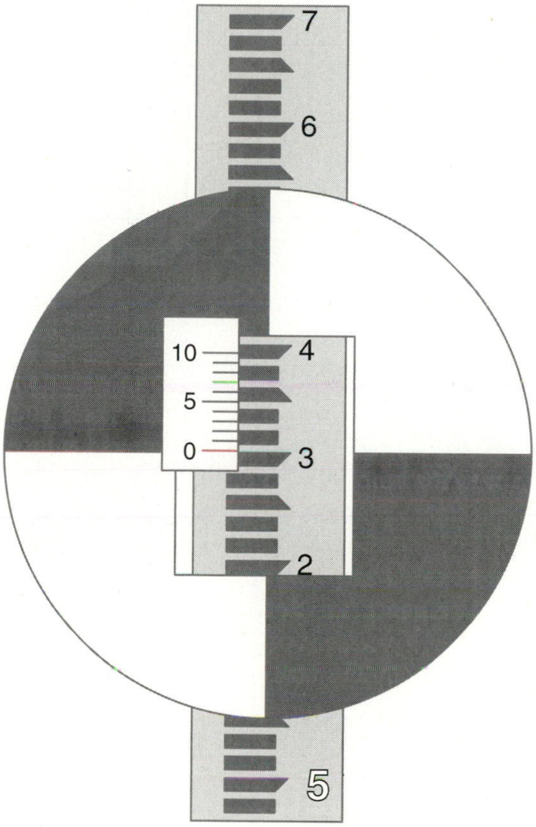

6. Record the readings for the rod in the illustrations.

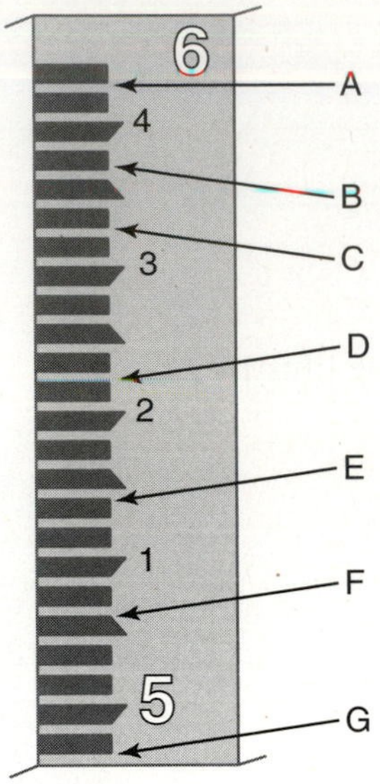

CHAPTER 3

Public Land Survey System

Objectives

After reading this chapter, the reader should be able to do the following:

- Explain the history of the rectangular system of land identification.
- Describe the metes and bounds system of land identification.
- Describe the rectangular system of land identification.
- Define the terms used in the rectangular system of land identification.
- Write the description for parcels of land identified on a map or drawing.

Terms To Know

metes and bounds
Public Land Survey System (PLSS)
initial point
latitude
longitude
baseline
principle meridian
standard parallel
guide meridian
convergence
quadrangle
township
tier
range
section
lots

INTRODUCTION

The need for a system of land identification arose with the development of civilization. Hunter-gatherer societies had no need for such a system because they were nomadic, and there was no advantage to claiming any particular tract of land. As people settled into towns and cities, land identification became more important for the collection of taxes and inheritances. In many small and stable communities, land identification was regulated by tradition. In these societies ownership and the boundary descriptions of the land were oral traditions that were passed from father to son. It was a common occurrence that after several generations the boundaries became easily identifiable because of fences, tree rows, stone walls, roads, etc. that were built. This characteristic of early cities lead to the development of a system of land identification called metes and bounds. The metes and bounds system was adequate for land identification in Europe, but it was a failure in North America.

During the initial settlement of North America by Europeans, it was a common practice for the King of England to give large tracts of land to a group or individuals as a land grant. These land grant boundaries were very vague and often overlapped because of inadequate or nonexistent maps. As individuals attempted to break up and settle these large grants, many problems arose because there were no traditional boundaries or man-made markers to use. By the time of the Revolutionary War the courts were choked with land disputes. To remedy this situation the Continental Congress formed a committee and charged them to develop a system that provided a unique description for every parcel of land. The result was the *Manual of Instructions for Survey of the Public Lands of the United States.* The process is commonly called the Public Land Survey System (PLSS) or rectangular system of land description. This system is explained in the following sections.

METES AND BOUNDS

The term "metes" is an English word that has been used for many years to describe a straight boundary line between two reference points. Geographical features, such as trees and large rocks, were often used as points of reference. The term "bounds" was used to describe an existing geographical boundary such as a watercourse, stonewall, road, or building. Metes and bounds were used for most of the original 13 colonies. An example of a metes and bounds description would read as:

> Commencing at the Hoover Bridge on Elk creek, go northward until reaching the waterfall and then westward to the granite outcrop, then southward to the Hoover road and back to the point of beginning.

This system produces an acceptable identification of the land if the markers remain undisturbed. Unfortunately trees are cut down or die of natural causes, streams change course and human activities change the features of the land surface. Initially there were problems with the system because land grant descriptions overlapped, and individuals gerrymandered boundaries to select the best land and leave out poor areas. The situation became worse as the number of Europeans living in the New World increased and as people spread west of the Appalachian Mountains.

The adoption of the PLSS eliminated the problems associated with the metes and bounds system. The guidelines were published in the "Manual of Instructions for the Survey of the Public Lands" by the Bureau of Land Management. Manuals were first published in the 1850s and continued through the 1900s. The PLSS was first applied in the Ohio River valley. The PLSS was modified further and applied to other land as it was opened for settlement. The last major revision occurred in time to be applied to the Louisiana Purchase. The system has remained unchanged since then. The PLSS established a multiple level grid system based on an initial point, which was located by latitude and longitude. The smallest unit of the grid was a *section*, which is one square mile.

PUBLIC LAND SURVEY SYSTEM (PLSS)

The keys to understanding the PLSS is remembering that it is a rectangular grid applied to a round surface and that one should have a thorough understanding of the terms that are used. The terms used in the PLSS are:

- Initial point
- Baseline
- Principle meridian
- Guide meridian
- Standard parallel
- Convergence
- Quadrangle
- Township
- Tier
- Range
- Section

Initial Point

As the term implies, the **initial point** is the starting point. The PLSS has been applied to the majority of the contiguous states and Alaska, but these areas were not surveyed at the same time, consequently there are many initial points. The survey party would start by selecting a latitude and longitude to use as the initial point. Resurveys have shown that the locations of several initial points were not established very accurately, but the surveyors did the best they could with the equipment they had available. The surveyors identified the initial point using sun and star sights. This was a lengthy, time-consuming process because a large number of shots were required to insure the desired latitude and longitude were located. A good understanding of these two terms will aid in understanding the rectangular system.

Latitude and Longitude

The terms latitude and longitude developed as mapmakers realized they needed a coordinate system that would allow them to locate points on the surface of the earth. A coordinate system requires a reference point or line as a starting point and then it uses one of three methods for describing the location of additional points. These three methods are:

- Three dimensions
- An angle and a distance
- Two angles

The system of latitude and longitude uses two angles.

Latitude

Latitude is the angle of a position north or south of the equator. Visualize a line extending from the equator to the center of the earth. This is the reference line for the angle and the center of the earth is the vertex. A line drawn from the center of the earth to the unknown position on the earth's surface forms the second line of the angle. The latitude is the angle formed by these two lines, Figure 3-1. Common practice has been to express latitude in units of degrees, minutes, and seconds. The use of calculators and computers has increased the use of expressing angles in decimal degrees.

On maps and globes latitude lines are drawn parallel to the equator, Figure 3-2. This also means that all points along a latitude line are equal distance from the poles. In the PLSS these lines are called standard

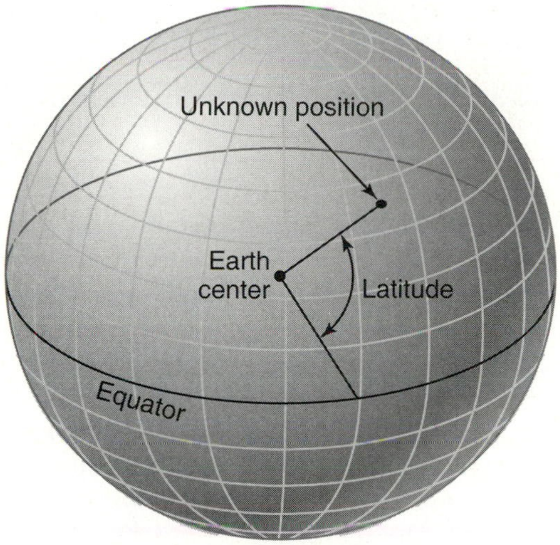

Figure 3-1 Determining latitude.

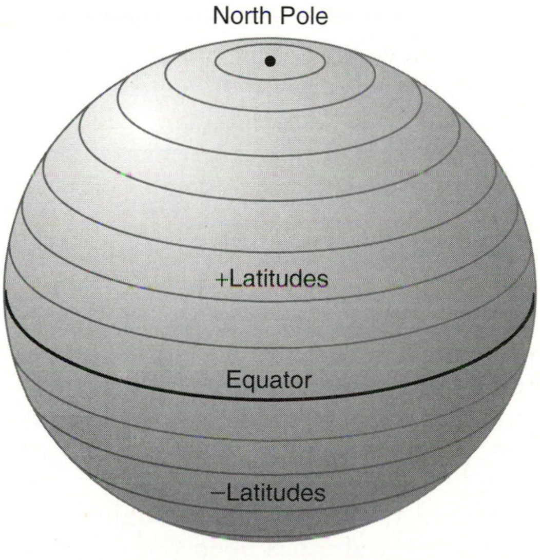

Figure 3-2 Latitude lines.

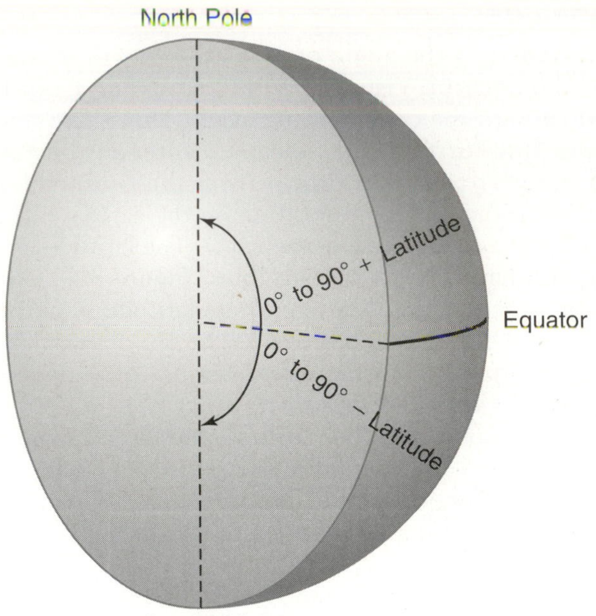

Figure 3-3 Minimum and maximum latitudes.

parallels. The PLSS establishes standard parallels at 24-mile intervals north and south of the baseline. These parallels are unique for each initial point.

Latitudes north of the equation are call North or plus angles (+) and latitudes south of the equator are called South or minus (−) angles.

Zero latitude is at the equator and maximum latitude is at the north and south pole, therefore latitude angles can range from zero to 90 degrees north (+) and zero to 90 degrees south (−), Figure 3-3.

Longitude

Longitude is the angle of a position east or west. Longitude angles are measured along the equator. Mapmakers draw lines perpendicular to the equator at standard intervals along the equator and extend them to both poles. These lines are called meridians. The PLSS uses one principle meridian that is continuous across the survey and guide meridians that have an offset at every standard parallel, every 24 miles. The longitude of a position is determined by where its meridian line crosses the equator. The use of the equator to measure angle of longitude means all points along the longitude or meridian line will have the same degrees of longitude. An important point to remember is that because the meridian lines converge at the poles the distance per degree of longitude is not consistent.

Historically, zero meridian angle, and therefore zero longitude, has been the observatory at Greenwich, England. At this site is a brass disc with a line scribed on it. The longitude angles are numbered eastward and westward from this point, Figure 3-4. The range of longitudes is from zero to 180 degrees west and from zero to 180 degrees east. It is important to record longitudes so that the correct hemisphere is identified. West and east are commonly used. An alternative is to identify east longitudes as plus longitudes (+) and west longitudes as minus (−) longitudes.

The result is a grid system that can be used to locate any point on the earth. The Northern Hemisphere is shown in Figure 3-5.

Baseline

For each survey the **baseline** was established through the initial point and extended east and west until it reached the boundaries of the area being surveyed. Because the baseline follows a latitude, it is parallel to the equator and required the surveyors to lay out the line with the appropriate curvature. As shown in Figure 3-6, all the points on a straight line would not be equal distance from a pole.

In Figure 3-6 lines A through E are all the same length, which keeps the equator an equal distance from the North Pole. The baseline for each rectangular survey is a continuous line parallel to the equator and perpendicular to the principle meridian. The baseline is the reference for north and south for each survey, Figure 3-7.

Principle Meridian

The **principle meridian** is a continuous north-south line that passes through the initial point and extends to the north and south boundaries of the area surveyed. The principle meridian is located by

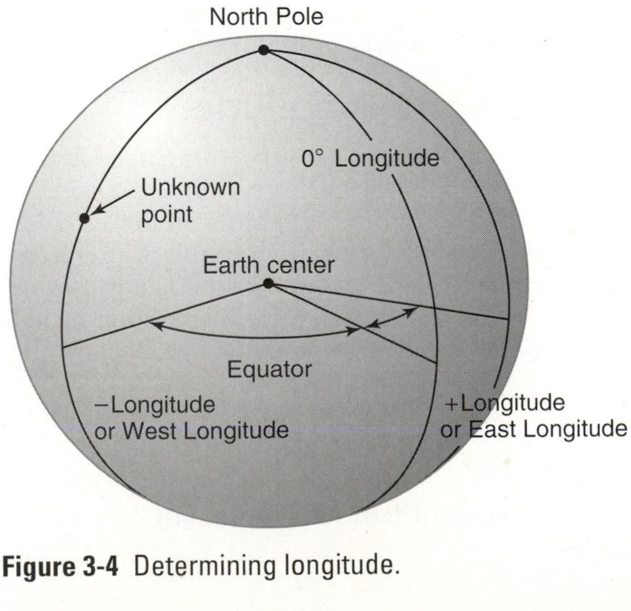

Figure 3-4 Determining longitude.

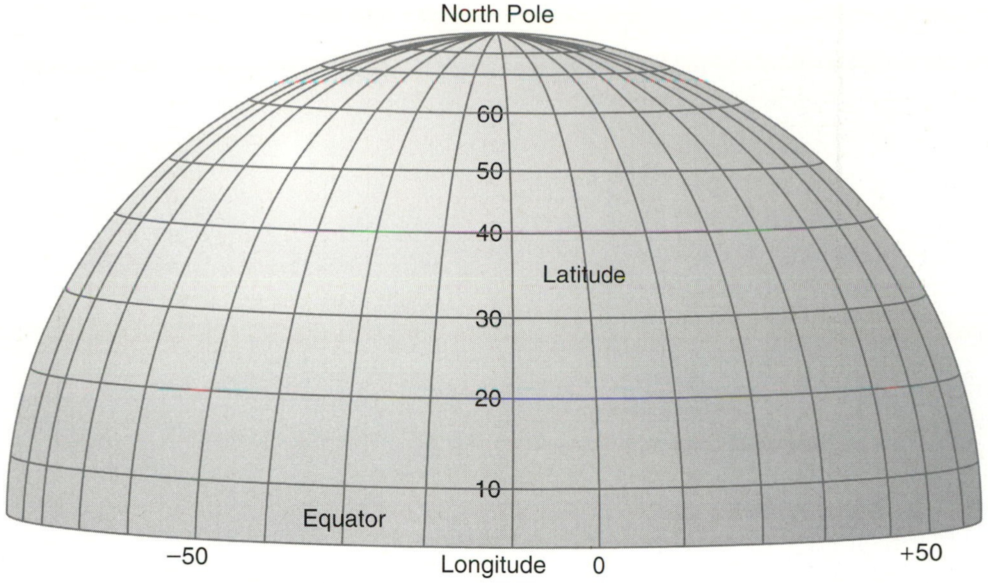

Figure 3-5 Latitude and longitude grid.

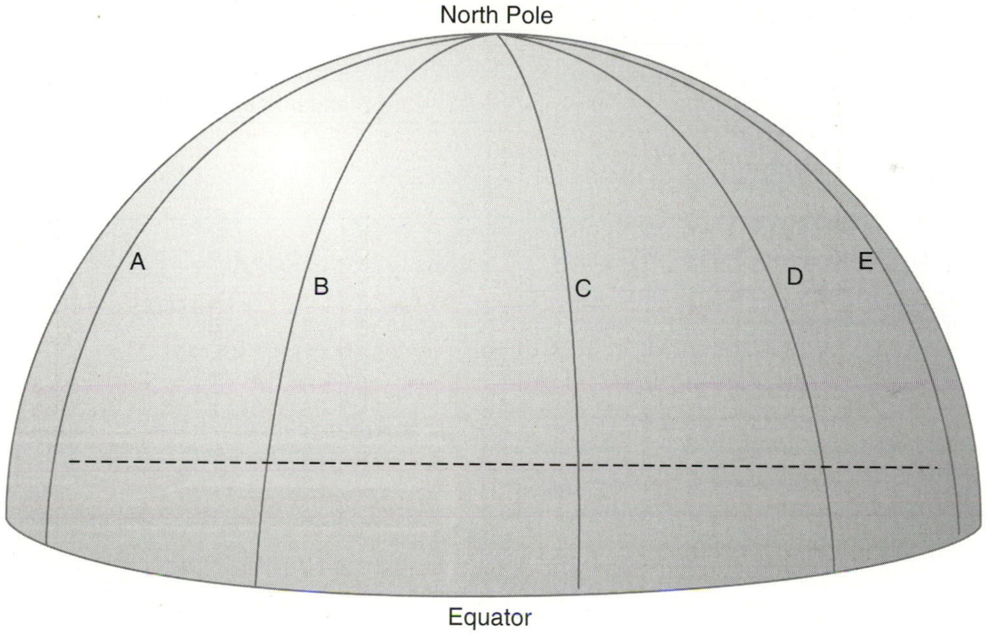

Figure 3-6 Straight line.

degrees of longitude. It is the reference line for identifying east and west for the survey.

After the baseline and principle meridian were laid out, Figure 3-8, the standard parallels and guide meridians were measured and marked.

Standard Parallel

A **standard parallel** is a continuous line that is parallel to the baseline and extends to the east and west boundary of the survey. Standard parallels were established every 24 miles perpendicular to the principle meridian north of the baseline and every 24 miles perpendicular to the principle meridian south of the baseline, Figure 3-9.

Guide Meridians

Every 24 miles along the baseline and at each standard parallel the surveyors established a line perpendicular to the standard parallel and extended it to the next standard parallel. These lines are called **guide**

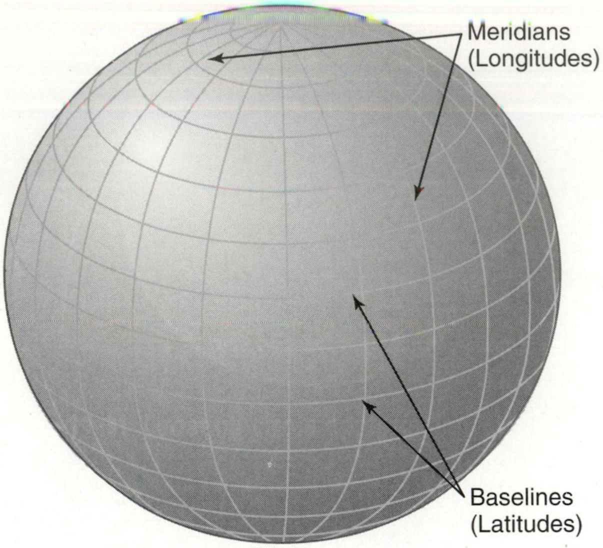

Figure 3-7 Meridians and baselines.

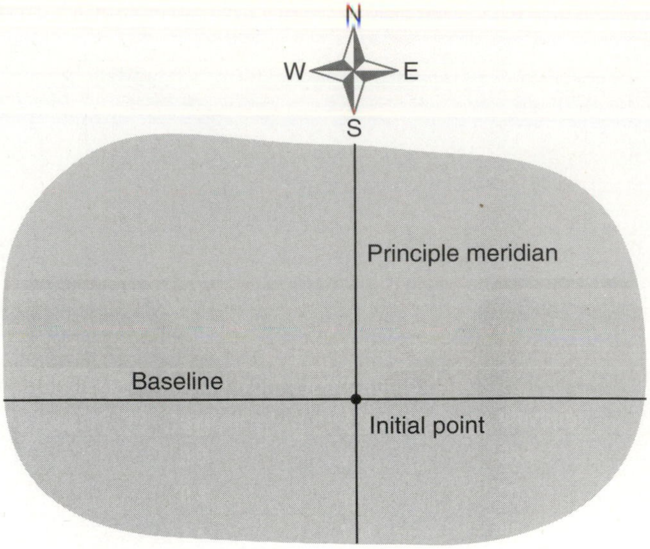

Figure 3-8 Baseline and principle meridian.

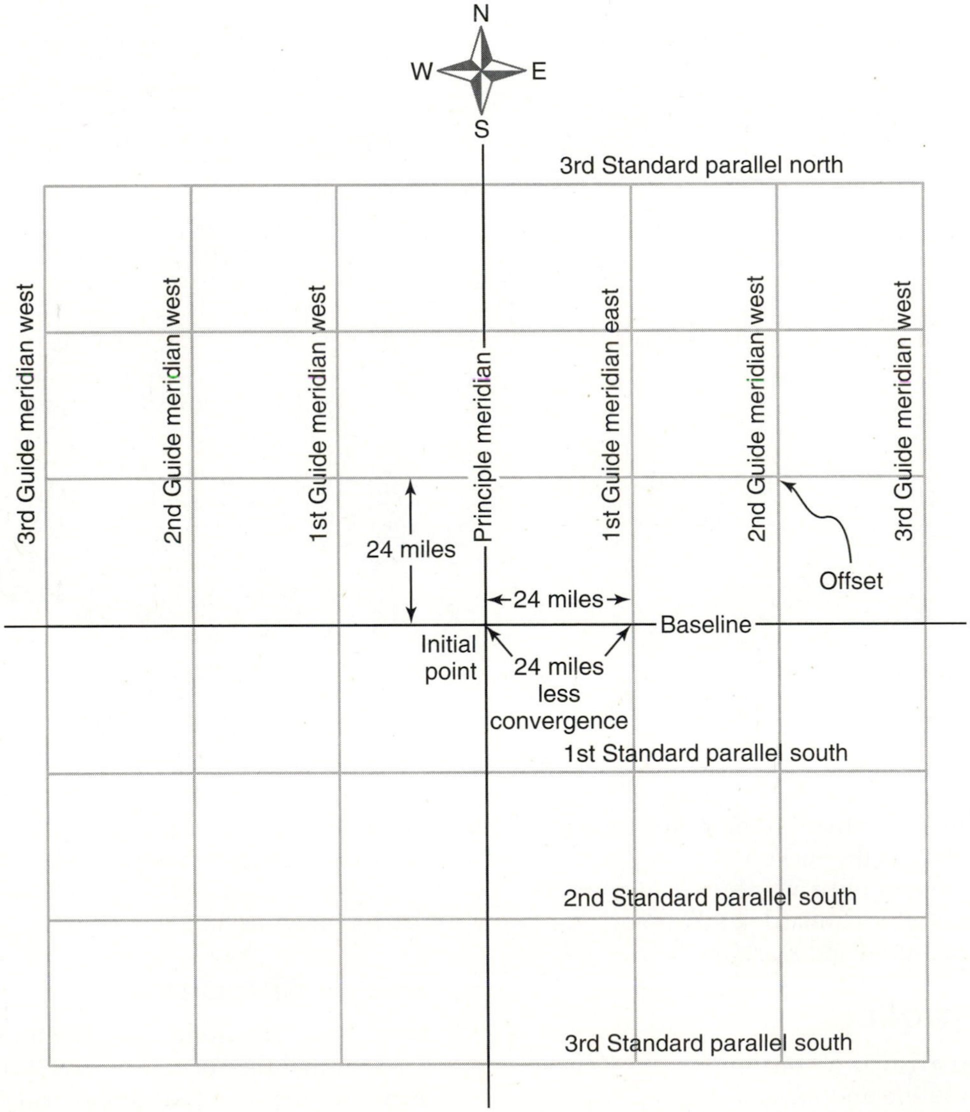

Figure 3-9 PLSS grid.

meridians, Figure 3-9. To maintain a standard size grid the guide meridians are remeasured at each standard parallel. The next section contains additional information on convergence.

Remeasuring the 24 mile interval along each curved standard parallel results in an offset at the next parallel, Figure 3-9. In the Northern Hemisphere, the amount of offset increases as you travel north of the baseline and decreases as you travel south of the baseline. Figure 3-9 is an illustration of the major grid used by the rectangular system.

Any adjustments for convergence were always taken in the last half-mile of the measurement. Each 24-mile square area formed by a standard parallel and guide meridian was divided into 16 townships.

Convergence

Convergence is a phenomenon of all longitude lines. At the equator each degree of longitude is equal to about 70 miles, but they all converge at the poles, Figure 3-10. Because of convergence, the number of miles per degree decreases as you move away from the equator. This is the reason that latitude and longitude is not the best method to use when trying to determine the distance between two points or areas.

The distance between longitude lines decreases as you move closer to either pole, Figure 3-10.

Quadrangle

A **quadrangle** is a rectangular shape that measures nominally 24 miles on each side. The north and south boundaries will be the baseline or a standard parallel, and the east and west sides will be the principle meridian or a guide meridian, Figure 3-9. Each quadrangle is divided into 16 townships. The term quadrangle is not used as part of the PLSS description.

Townships

A **township** is an area that measures six miles by six miles. Townships are identified by tier and range from the initial point. **Tiers** are rows of townships that extend north and south of the baseline. **Ranges** are columns of townships extending east and west of the principle meridian.

In the example in Figure 3-11, the blocked out township is identified as tier six south, range seven west. PLSS descriptions are commonly abbreviated. The previous description would be written as T6S, R7W. Each township is divided into 36 parts called sections.

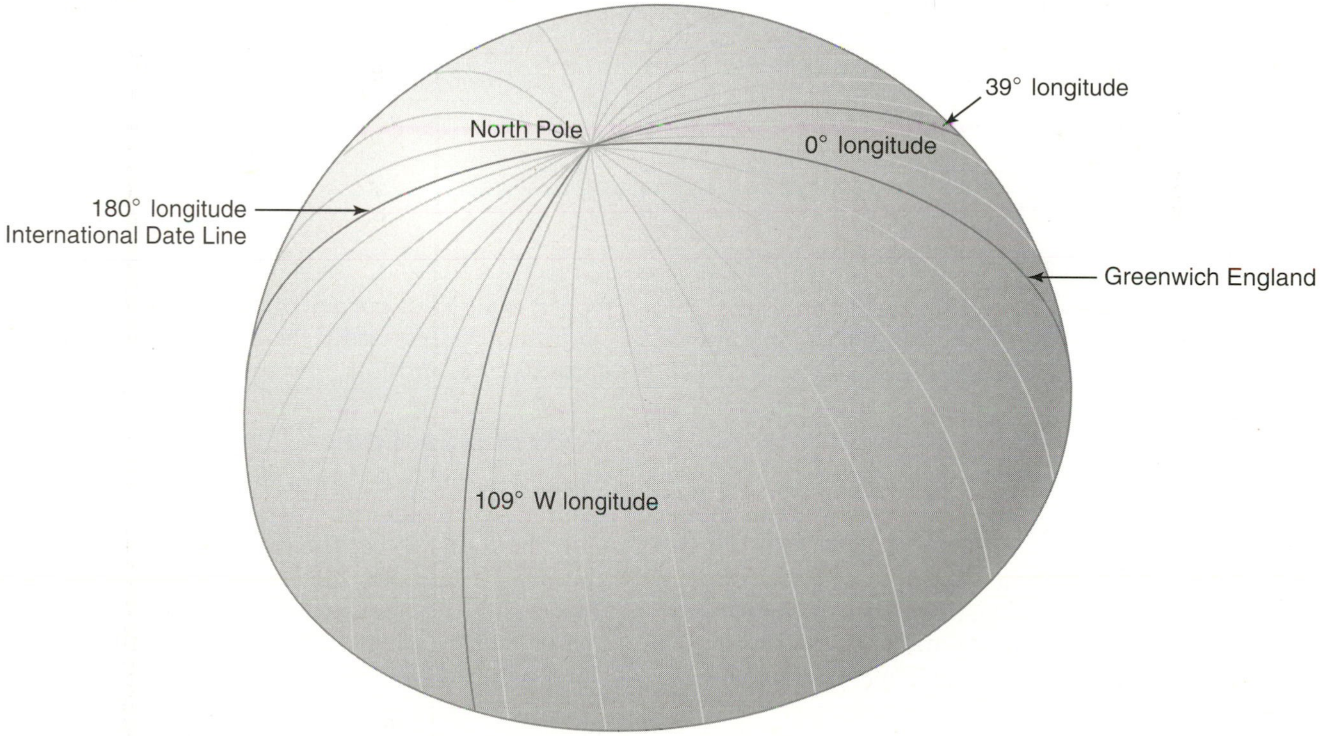

Figure 3-10 Longitude lines.

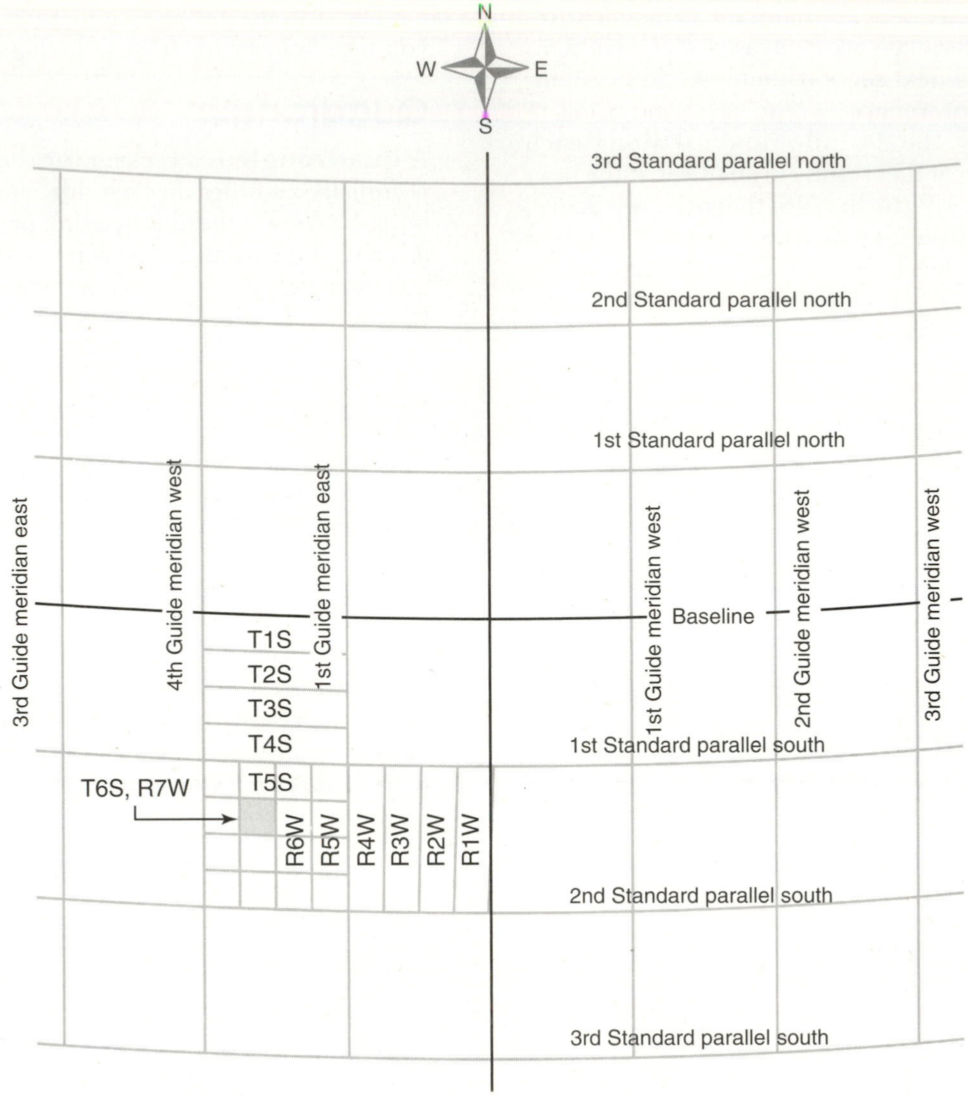

Figure 3-11 Township identification.

Note: The common usage of the PLSS description changes the word tier to township. The common interpretation of T6S, R7W would be township six south, range seven west.

Section

A **section** is an area of land with the nominal dimensions of one mile by one mile. The term nominal dimension is used because to place a rectangular grid on a round surface, adjustments must be made for convergence. Because of convergence, one or more dimensions for some sections will be less than one mile. Nominally each section contains 640 acres. Sections are identified by numbers. Figure 3-12 illustrates the section numbers for two of the townships. The numbering system starts in the upper right hand corner of the township and travels continuously left and right. Sections can be divided into ¼s, ½s, and lots.

Parts of a Section

Multiple combinations of the fractions of ¼ and ½ can be used to identify parcels of land less than one section, Figure 3-13. The number of combinations of the ¼ and ½ fractions that can be used is unlimited. The difficulty is that as the size of the area decreases, the boundary lines of land parcels are less likely to align with the fractional boundary lines. Figure 3-13 illustrates several land parcels in section 9 that are less than one section.

The fractional description can also be used to determine the approximate number of acres in the area.

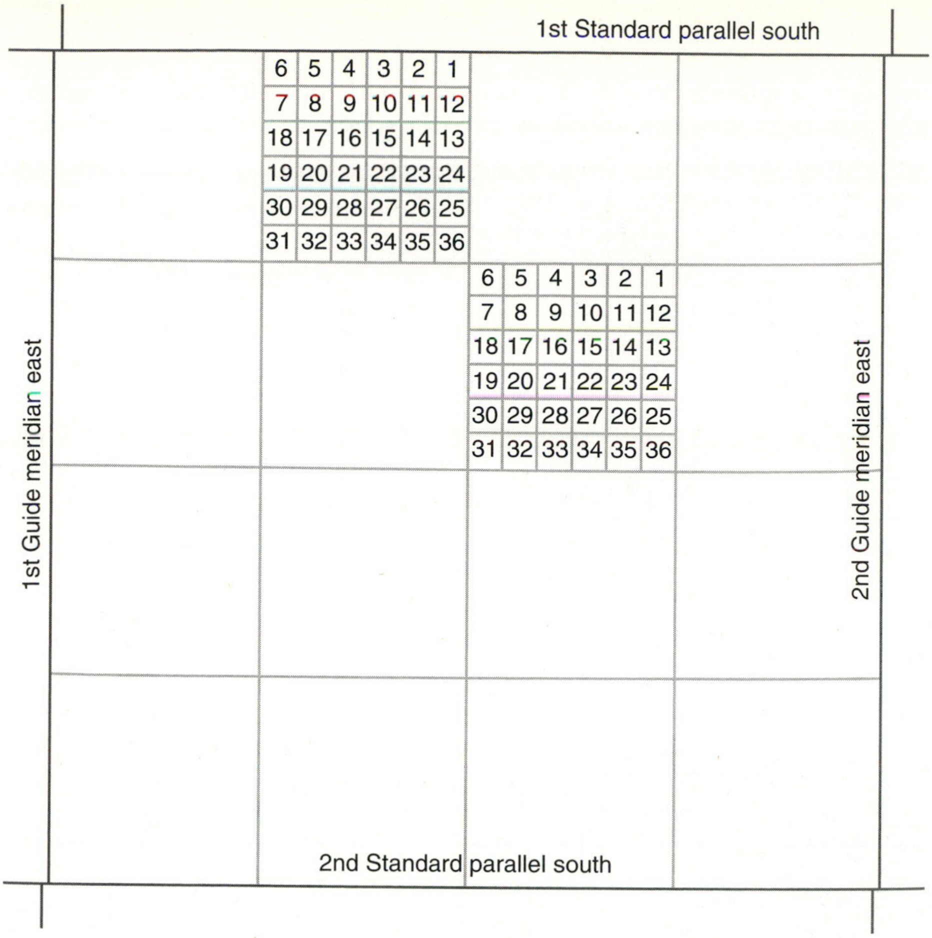

Figure 3-12 Section numbering.

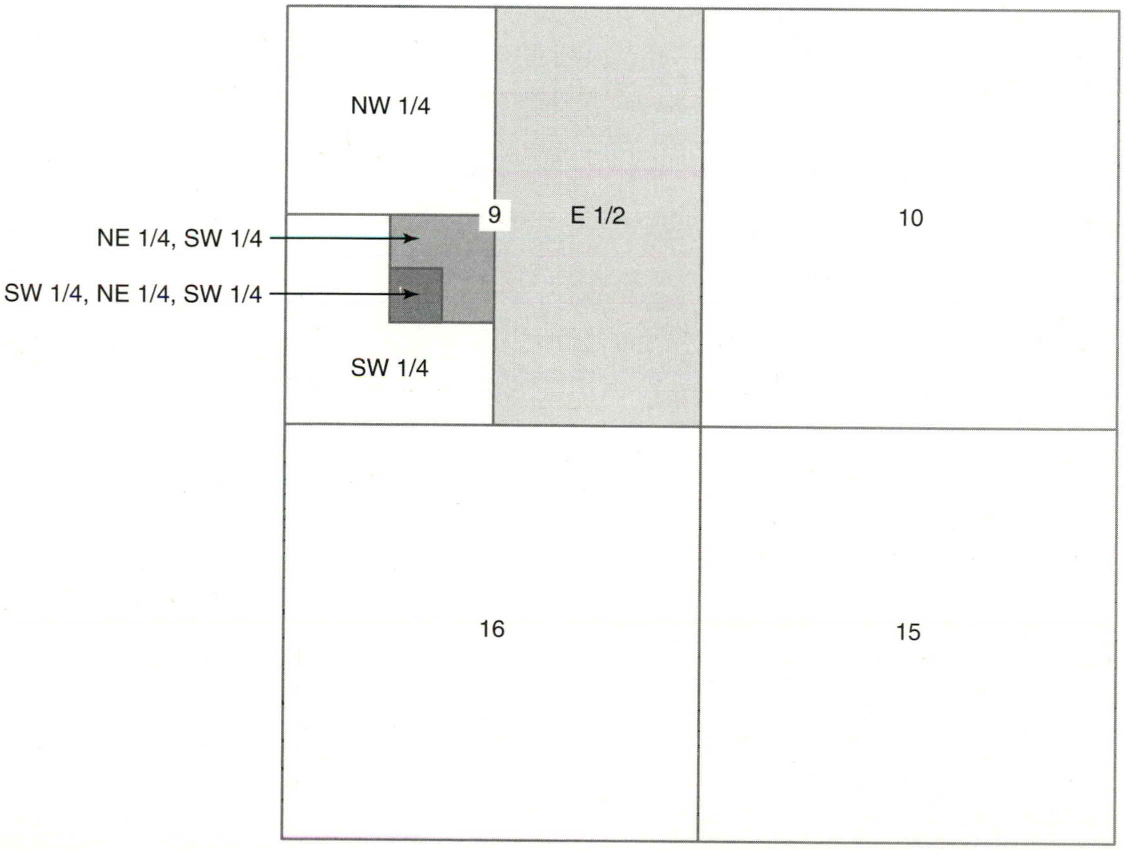

Figure 3-13 Fractions of a section.

This is accomplished by dividing the number of acres in a section, 640, by the product of the fractional denominators in the descriptions. A parcel of land with the description of W$\frac{1}{2}$, SW$\frac{1}{4}$, NW$\frac{1}{4}$ would have an area of 20 acres.

$$\text{Parcel Area} = \frac{640 \text{ acres}}{2 \times 4 \times 4} = \frac{640 \text{ acres}}{32} = 20 \text{ acres}$$

Lots

The rectangular system also provides a means of identifying small, usually irregularly shaped areas caused by meandering river, road, etc. These areas are called **lots** and their boundaries are drawn so that they do not cross over section or $\frac{1}{4}$-section boundaries. Lots are numbered starting with the most eastern lot in the most northern tier as number one and then continuing west through the tier and then east through the next tier south until all are identified. Lot numbering systems may vary from state to state.

Summary

Where the PLSS is used, each parcel of land has a unique description. A unique description prevents confused ownership, incorrect taxes and a way to transfer ownership. Descriptions using the PLSS are written as continuous statements starting with the smallest area of land and ending with the territory. The description of a parcel of land is not unique until the description is tied to the initial point. The information presented in this chapter will be reviewed by studying the following example, Figure 3-14.

The descriptions are:

A: NW 1/4, S8, T6S, R11W, IM, OT

B: NE 1/4, NE 1/4, S8, T6S, R11W, IM, OT

C: SE 1/4, SW 1/4, N2 1/4 and SW 1/4, SE 1/4, NE 1/4, S8, T6S, R11W, IM, OT

D: NW 1/4, SE 1/4, SE 1/4, S8, T6S, R11W, IM, OT

E: S 1/2, SW 1/4, S8, T6S, R11W, IM, OT

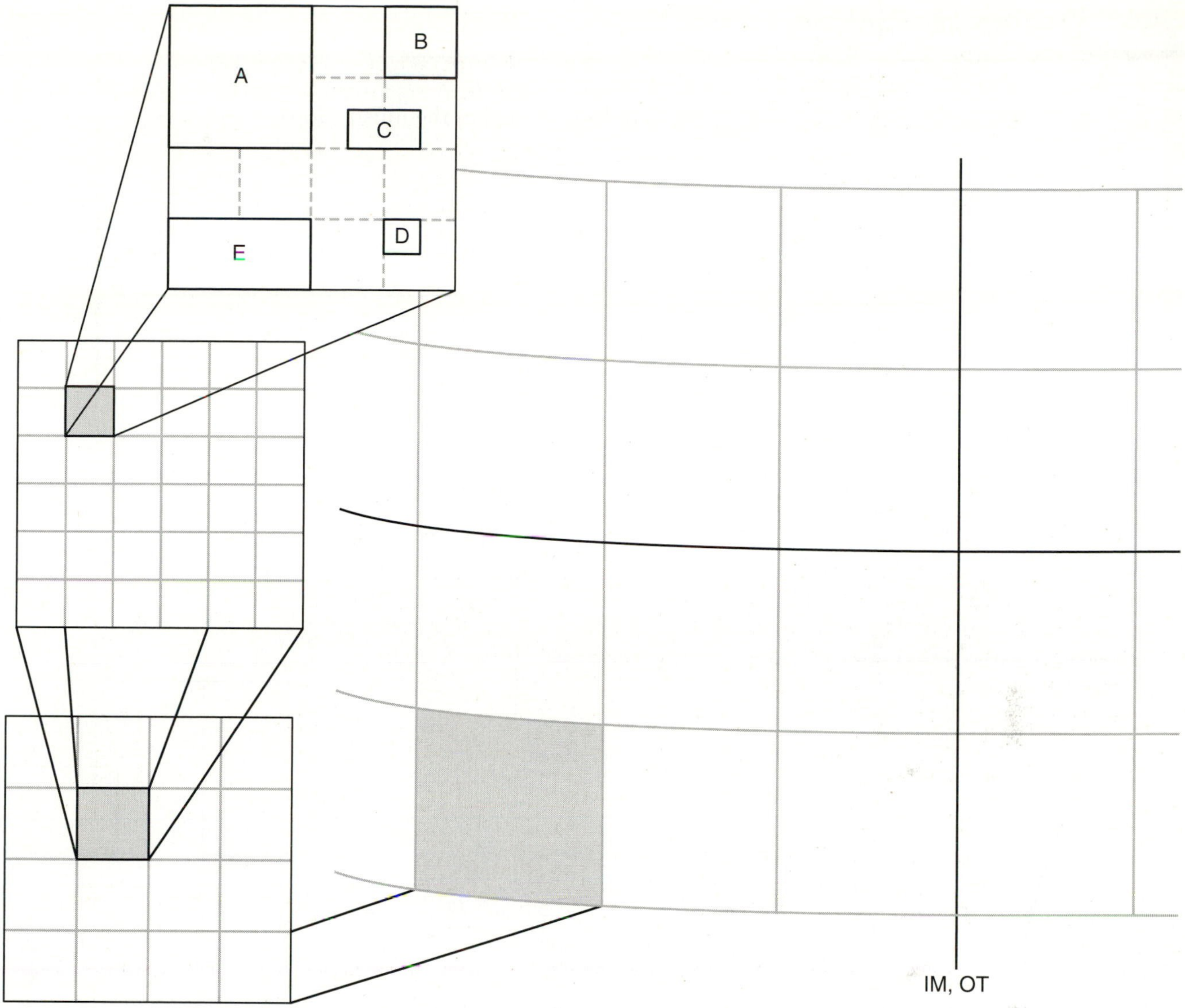

Figure 3-14 PLSS example.

Student Activities

1. Write the descriptions for the parcels of land labeled in the illustration.

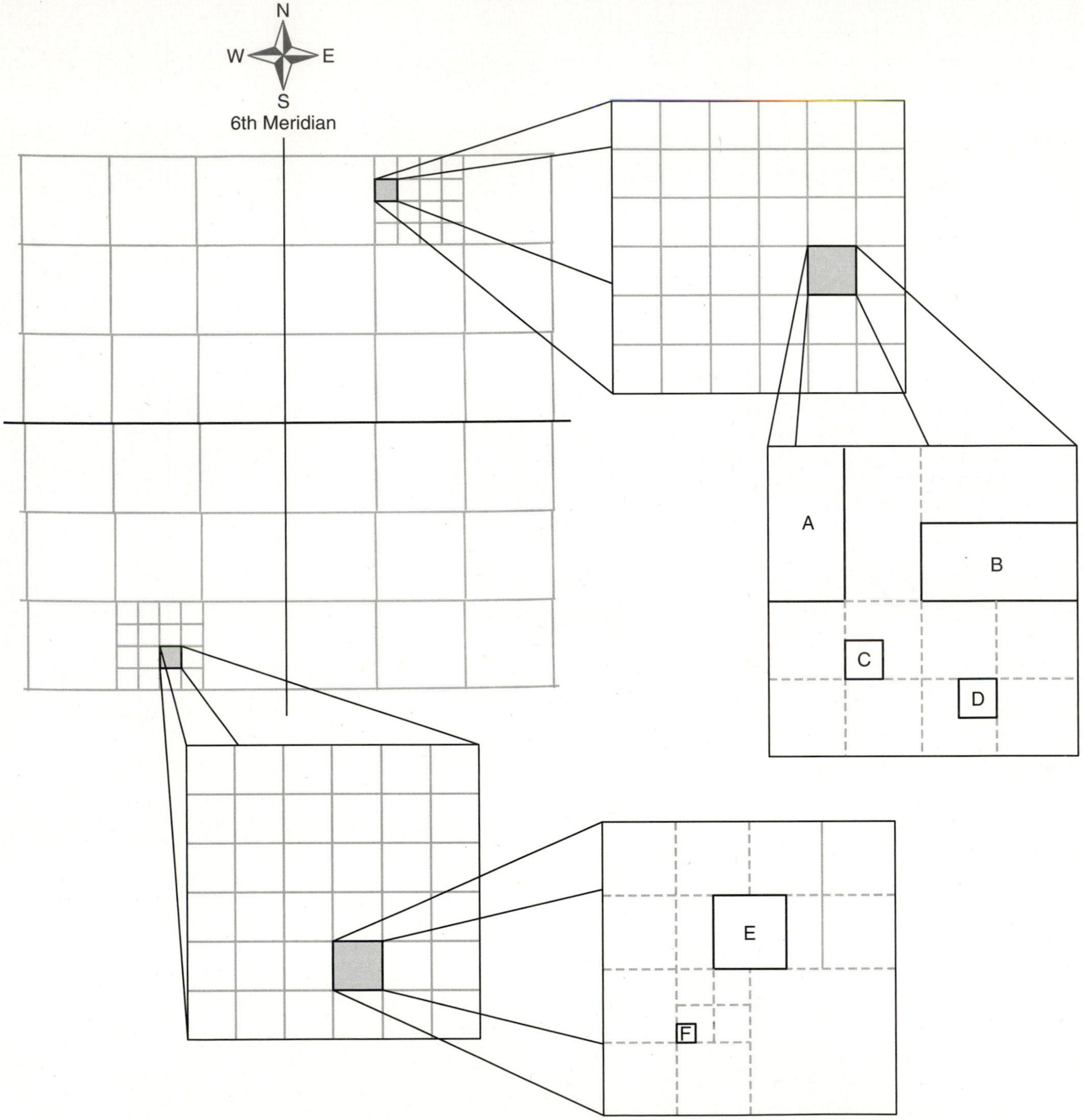

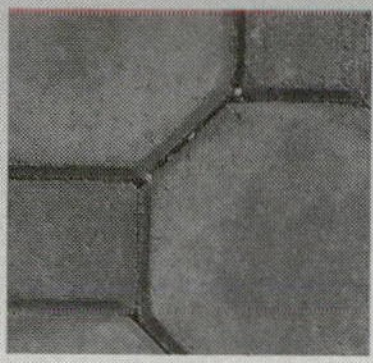

CHAPTER 4

Distance Measuring

Objectives

After reading this chapter, the reader should be able to:

- Select the best method for measuring distance.
- Determine their pace factor.
- Measure distance by chaining.
- Explain the term breaking chain.
- Use an odometer wheel to measure distance.
- Measure distance by stadia.
- Determine correction factors for systematic errors.
- Calibrate a measuring tool.

Terms To Know

pacing	**distance by stadia**	**calibrating**
breaking chain	**half-stadia**	**chaining**

INTRODUCTION

Measuring distance is one of the basic methods used in surveying. For some land measurement jobs, being able to measure distance may be all that is needed. Other types of surveying, such as profile, will require measuring distance as part of the survey. Failing to measure distances correctly will ruin the value of the data collected. The following sections will discuss the methods commonly used for measuring distance.

Distance measuring is based on two principles of geometry: 1) It takes two points to form a line, and 2) The shortest distance between two points is a straight line. Most errors that occur happen because one, or both, of these principles were violated.

In surveying, the term "distance" has two uses. The common use is to measure the displacement (distance) between two or more points. Distance can also be used to define the dimensions of an object. To measure distance accurately, the surveying team must follow the recommended method, practice the best techniques, and use equipment appropriate for the use of the data.

SELECTING THE BEST METHOD

Selecting the best distance measuring method is one of the arts of surveying. When selecting the best method the first consideration must be the use of the data. Measuring a distance to determine the number of boards needed for a fence requires a different level of precision than measuring the distances to completing a legal boundary survey. It is not a good expenditure of resources to use a measuring instrument with a precision of 0.001 feet to determine the number of lengths of pipe needed to install a water line. The opposite is also true. It is not good practice to use a measuring device with a precision of one foot when laying out a foundation for a building or establishing the boundary lines for a parcel of property. This would result in a large error and an unsatisfactory result. It is usually the responsibility of the person in charge of the survey crew to select the method(s). A list of factors that should be considered include:

- Use of the data
- Environment
- Equipment available
- Expertise of individuals
- Personal preference
- Topography
- Client specifications
- Regulations
- Standard practice

The use of the data is always the number one consideration, but the influence of the other factors on the decision making process will vary from job to job. Different surveyors may select different methods and still end up with good data. What is important is that someone must make the decisions before the equipment is removed from storage, and definitely before any data is recorded.

HORIZONTAL DISTANCE OR SLOPE DISTANCE

Another decision that should be made before any distance measurements are recorded is whether the job requires horizontal distance or slope distance, and if horizontal distance is required will it be collected directly or calculated. When horizontal distance is desired and slope distance is recorded, one additional bit of information must be collected. This decision is influenced by the uses of the data and the amount of resources required to collect the information. A job may require horizontal data, but the variability of the topography may make it extremely difficult to record horizontal distance. In this situation, slope distance and the required additional data, would be collected and horizontal distance would be calculated. Refer to the following section for additional information on converting slope distance to horizontal distance.

High precision surveying, such as property surveys and locating benchmarks, usually requires horizontal measurements. The use of survey quality EDM's has almost eliminated this decision because they either measure horizontal distance or calculate it from slope distance and vertical angle.

Slope Distance

A distance measured by following the contours of the earth's surface is a slope distance or surface distance. Converting slope distance to horizontal distance requires collection of one additional piece of information.

- The difference in elevation
- The percent slope
- The vertical angle

With any one of these pieces of additional information, the horizontal distance can be calculated. The mathematical process used to determine horizontal distance depends on the additional information obtained.

Difference in Elevation

Using the difference in elevation sets up a situation in which the Pythagorean Theorem can be used, Figure 4-1. The difference in elevation is the rise, which is equivalent to the vertical height of a right triangle. The hypotenuse is the slope distance and the length of the horizontal side of the right triangle is the horizontal distance. An unknown side of a right triangle can be calculated using the Pythagorean Theorem if the lengths of the other two sides are known.

Study Figure 4-1 and Figure 4-2. These figures illustrate a situation in which the slope distance is 26.20 feet, the elevation of one station is 4.90 feet, and the elevation of the second station is 2.40 feet.

Two of the sides of the triangle are known, the unknown is the horizontal distance.

$$a^2 = b^2 + c^2$$

$$b^2 = a^2 - c^2$$

$$b = \sqrt{a^2 - c^2}$$

$$= \sqrt{26.20^2 - 2.50^2}$$

$$= 26.08 \text{ ft.}$$

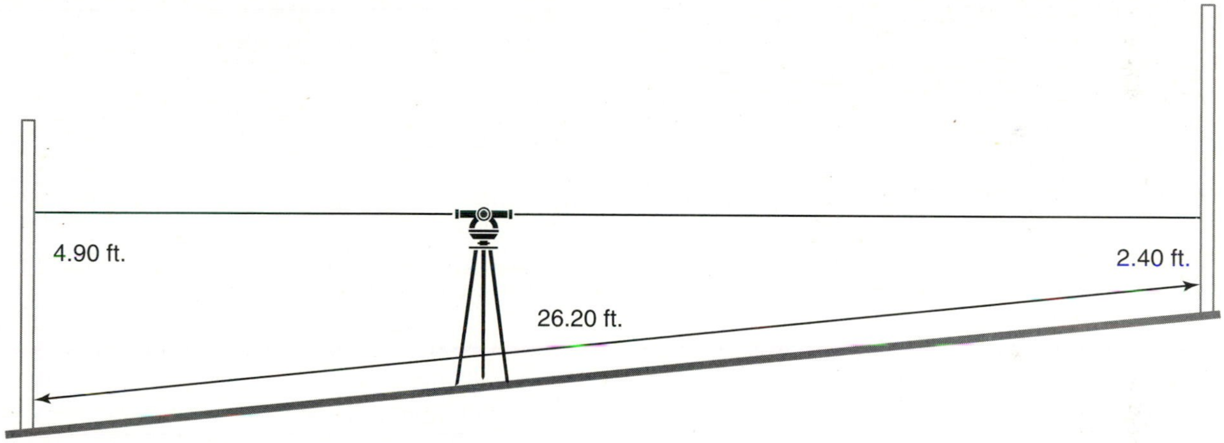

Figure 4-1 Converting slope distance to horizontal distance using the difference in elevation.

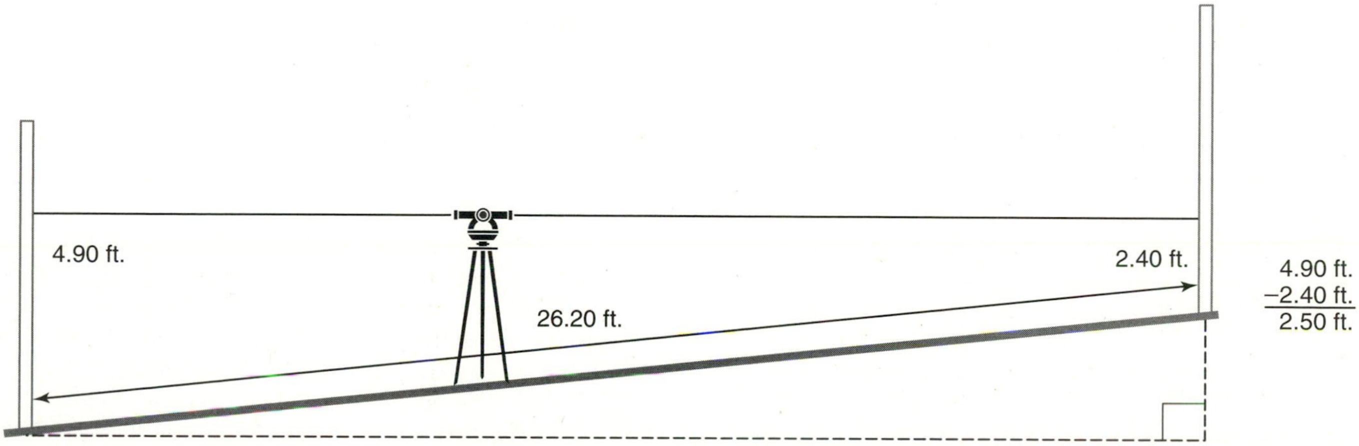

Figure 4-2 The principles of a right triangle applied to determining horizontal distance.

When the slope distance is 26.20 feet and the difference in elevation is 2.50 feet the horizontal distance is 26.08 feet.

Percent Slope

The percent slope can also be used to convert slope distance to horizontal distance. The first step is to use the percent slope to determine the amount of rise between the two stations. Assume the slope is 8.0 percent and the slope distance is 28.5 feet. The amount of rise for the length is:

$$\text{Slope} = \frac{\text{Rise}}{\text{Run}} \times 100$$

$$\text{Rise} = \frac{\text{Slope} \times \text{Run}}{100}$$

$$= \frac{8.0 \times 28.5 \text{ ft.}}{100} = 2.28 \text{ ft.}$$

The second step is to use the rise (2.28 ft) and the Pythagorean Theorem to calculate the horizontal distance.

$$b = \sqrt{a^2 - c^2}$$

$$= \sqrt{26.2^2 - 2.28^2}$$

$$= 26.10 \text{ or } 26.1 \text{ ft.}$$

When the slope distance is 28.5 feet with an 8% slope the horizontal distance will be 26.1 feet.

Vertical Angle

The third method that can be used to convert slope distance to horizontal distance is to use a vertical angle. This method will require an instrument that can measure vertical angles, such as a transit or builder's level. In addition, because angles are used, this method requires the use of a trigonometric function. For example, if the vertical angle for a slope distance of 50.3 feet is 6.5 degrees, what is the horizontal distance? The calculations are:

$$\text{Cosin } \theta = \frac{\text{Adjacent}}{\text{Hypotenuse}}$$

$$\text{Adjacent} = \text{Cosin } \theta \times \text{Hypotenuse}$$

$$= 0.9936 \times 50.3 = 49.97\ldots \text{ ft. or } 50.0 \text{ ft.}$$

When the slope distance is 50.3 feet and the vertical angle between the two stations is 6.5°, the horizontal distance will be 50.0 feet. More information on trigonometric functions is included in Chapter 7.

The surveyor has several options when slope distance is measured and horizontal distance is calculated. All of these options require the collection of one additional bit of information. The decision on the method that is used should be decided before starting the distance measuring survey. Notice that for these three examples, the difference between slope distance and horizontal distance is small. For low-precision surveys the difference between slope distance and horizontal distance can usually be ignored for slopes less than 5%.

HORIZONTAL DISTANCE

The horizontal distance between two points is the distance between those points measured on a horizontal plane (see definitions in Chapter 1). This can be accomplished by using an instrument that can be leveled, or using a chain and a level. These methods are included in the next sections.

EQUIPMENT AND METHODS FOR MEASURING DISTANCE

This section will discuss common equipment and methods used for measuring distance and their corresponding level of accuracy and precision. Using inappropriate equipment commonly results in having measurements with the wrong level of precision or accuracy. The distance measuring equipment and methods that will be discussed include:

- Pacing
- Chaining
- Odometer wheel
- Stadia
- Rangefinders
- Electronic distance measuring (EDM)

Pacing

To measure a distance by pacing, the person counts the number of paces (steps) required to travel the distance. Measuring distances by pacing is very useful for estimating distances and as a check for large errors when using other distance measuring equipment. The process is described in the name. The number of paces times the individual's pace factor (PF), length of step, determines the distance.

Experts disagree on the length of pace that should be used. Some recommend that a person practice and use a standard three-foot pace. The major problem with this method is that unless you are at least six feet tall this is a very uncomfortable and tiring pace for long distances. The alternative method is to practice taking a uniform, comfortable pace that is different from a normal walking step because the length of a walking step is quite variable. This variability is discussed in more detail in the section on limitations of pacing. The essential factor is the uniformity of the pace length.

When measuring distance by pacing, an individual must know his or her pace factor (PF). The pace factor is calculated by pacing a known distance, 100 to 300 feet, several times, determining the average number of paces for the length and then dividing the distance by the average number of paces.

$$\text{PF} = \frac{\text{Distance}}{\text{Average number of paces}}$$

For example: If an individual recorded 32, 33 −1/2 and 33 paces walking a 100 foot distance, then:

$$\text{Average Number of Paces} = \frac{32 + 33 \ - 1/2 + 33}{3}$$

$$= \frac{98.5}{3} = 32.833\ldots \text{ paces}$$

$$\text{PF} = \frac{100 \text{ ft.}}{32.833\ldots \text{ paces}}$$

$$= 3.45\ldots \text{ or } 3.0 \frac{\text{ft.}}{\text{pace}}$$

This individuals pace factor is 3.0 feet per pace.

With practice and careful attention to details, an individual on a uniform surface should be able to achieve an accuracy of 1/50 to 1/100. This designation of accuracy means that at 1/50 the error will be plus or minus 1/2 a pace for every 50 paces traveled. The amount of error can be reduced if several measurements are taken and then averaged. Another limitation is precision. The precision of pacing is the length of the individual's pace. For most individuals, this results in a precision of 2.5 to 3.0 feet.

Limitations of Pacing

The accuracy of pacing is dependent on a uniform length of pace. Many factors influence the length of a person's step when walking. A few of these are:

- Topography
- Shoes
- Time of day
- Height of vegetation
- Soil surface

Topography influences the length of a pace because the length of a pace will be shorter walking uphill and longer when walking downhill. When walking in shoes with low heals the length of a step is longer than when walking in shoes with higher heels. In the morning when we are fresh we tend to have a longer stride than in the afternoon when we are tired. When walking in tall vegetation the length of stride will be shorter than when walking in short vegetation. On a hard dry surface we will take longer steps that on a soft or wet one. There are other factors that influence the length of our steps, but these five reinforce the fact that for pacing to achieve a reasonable level of accuracy, the individual must practice a pace that is different from his or her normal walking step on the same surface as the one being measured.

Chaining

Chaining is the process of measuring distances using a surveyor's chain. Historically, chaining was the preferred method for measuring distances accurately, but modern electronic means have replaced chaining for professional surveyors. Chapter 2 contains a discussion of the principles and types of chains that are available. This section will concentrate on the techniques of chaining.

Chaining Slope Distance

To obtain consistent and accurate results, the chaining party must adopt an organized method. Different surveying crews will develop and adopt different methods, but general recommendations for measuring slope distance include:

1. Establish the line of travel with flags, range poles, or stakes.
2. Lay out the chain with the zero end at the starting point and the 100-foot mark in the direction of travel. Some surveyors prefer the head person take the zero end of the chain. Either method will work.
3. The rear person takes one pin from the ring and gives the head person the remaining ten.
4. The head person stretches the tape with the appropriate tension and the rear person has the head person move the chain to the left or right until it is aligned with the direction of travel.
5. The head person should lift the chain off the ground, pull it tight, and carefully lay it on the surface and place a pin at the 100-foot mark.
6. The rear person leaves the first pin in the ground so the team can return to the same spot when the team closes the loop back to the starting point. The chain is pulled forward until the zero mark is aligned with the pin placed at 100 feet. The rear person aligns the head

person in the direction of travel. The chain is lifted and pulled tight again.

7. The person at the head places a pin in the ground at the 100-foot mark. The rear person pulls the remaining pins at the zero mark each time before the chain is moved.
8. The process is repeated until the destination is reached.
9. If the head person runs out of pins before the destination is reached, the pins are transferred from the rear person to the head person and a tally mark is recorded in the field book. Each mark represents 10 pins or 1000 feet.
10. When the end of the run is reached the partial chain and partial foot is recorded.
11. The chain is switched, the pins are transferred back to the rear person, and the party chains back to the beginning.

Chaining True Horizontal Distance

The steps for chaining true horizontal distance are the same as slope distance. The difference is how the chain is held. To measure horizontal distance the chain must be held horizontal and a plumb bob is used to transfer the measurement from the chain to the ground. A hand level is also used to insure the chain is horizontal. In step five of measuring slope distance the head person raises the chain off of the ground to insure it is straight and then lays it back on the ground. In horizontal chaining the head person, if traveling downhill, raises the chain off the ground until it is horizontal and then uses a plumb bob to mark the point on the ground. The rear person raises the chain and uses the plumb bob if the party is chaining uphill. A pin is inserted at the point located by the plumb bob. The chain is moved and the process is repeated. A potential error when chaining horizontal is inconsistent sag in the chain. Accurate results require a consistent chain tension. A tension of 15 pounds is appropriate for most chains. A small spring scale is available for this purpose.

From a practical standpoint, this method is limited to slopes of 5% or less. With a 100-foot chain and 5% slope, the down slope end of the chain will be five feet above the ground, Figure 4-3.

In this situation the down slope person must be able to hold the end of the chain five feet above the ground, apply the appropriate tension to remove the sag from the chain and mark a point on the ground with the plumb bob. Slopes of 3% or less are the limit for most people. A procedure called breaking chain is used when the amount of slope makes using a full chain length impractical. Breaking chain is explained in a later section.

Measuring Less than a Full Chain

Measurements that are less than a full chain and less than a whole foot require additional procedures. The measurement is started by the rear person holding the zero foot mark on the last pin. The chain is pulled tight and the head person notes the next foot mark greater than and less than the pin position. When an add chain is used, the chain is moved to the smaller foot mark. When a cut chain is used the chain is shifted to the larger foot mark. The person at the rear of the chain reads the partial foot. If an add chain is used the partial foot is added to the foot mark at the head of the chain. If a cut chain is used the partial foot is subtracted from the foot mark at the head of the chain.

Breaking Chain

The **Breaking chain** method reduces the length of the chain to a common unit less than 100 feet, which reduces the change in elevation at the down slope end of the chain. If a 50-foot length is used, then on a 5% slope the elevation of the down slope end of the chain will be only 2.5 feet, Figure 4-4.

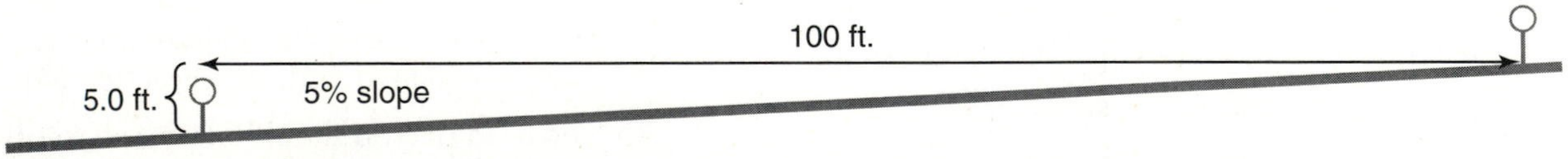

Figure 4-3 Change in elevation with 5 percent slope.

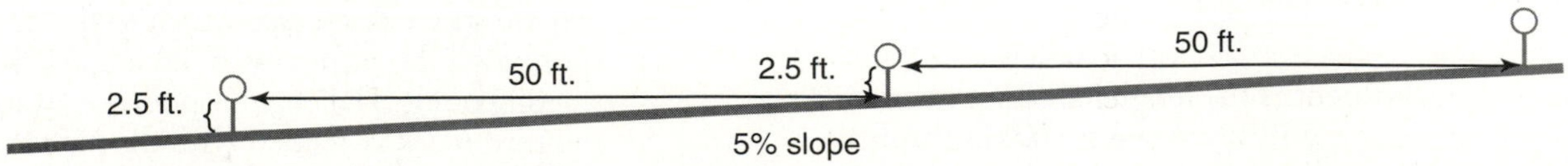

Figure 4-4 Breaking chain.

If the breaking chain method is used, it is very important that the length of the chain used is recorded in the field book.

Chaining Errors

Chaining provides many opportunities for error. These must be carefully managed to produce accurate results. Sources of error include:

- Not staying on line,
- Using the incorrect tension,
- Failure to use temperature correction,
- Errors in reading the chain,
- Measuring slope distance and assuming it is horizontal distance,
- Incorrect pin count.

Limitations of Chaining

One of the limitations of chaining is the requirement that the team must be able to walk the distance. This will require that a path be cut through brush, weeds, trees, etc. It also poses a problem for measuring across water and congested roadways. Another limitation is the number of people required. In open terrain, at least three people should be used. In terrain that requires the team to cut a path through vegetation such as trees and brush, the number of people in the party will increase dramatically. It was not unusual for early territorial survey parties in forested regions to have a survey party of 50 or more individuals when chaining.

Odometer Wheel

Odometer wheels are a very useful instrument for low-precision measurements of distance. The variations in construction of odometer wheels were included in Chapter 2. This section will discuss the use of odometer wheels when measuring distance.

Mechanical odometers record distance with a series of interlocking wheels. The internal mechanisms of the wheels are designed so that the second and succeeding wheels advance one unit (number) with every complete revolution of the first wheel to the right. With experience, an operator can usually determine the unit of measure by studying the numbers on the wheels as the odometer is rotated.

Figure 4-5 illustrates a three wheel mechanical odometer. When the wheel on the right completes one revolution, the middle wheel advances one number. When the middle wheel completes one revolution, the wheel on the left advances one number. The measurement from a mechanical odometer with three wheels records will be in 100s, 10s, and whole units. The precision of the odometer in Figure 4-5 is not known, but if it is one foot, the distance measured is 185 feet. If the unit of measure is not indicated on the odometer it can be determined by measuring a known distance.

The odometer illustrated in Figure 4-6 is different. A wider partition between the right hand wheel and the next wheel to the left or a wheel that is a different color indicates that the right hand wheel is somehow different. In this design, the right hand wheel records tenths of the unit of measure. This can be determined by noticing the numbers on the right hand wheel range from 0 to 9. Nine numbers = ten spaces.

If the precision of this odometer wheel is 1/10 of a foot, the correct reading for this odometer is 185.0 feet.

In the odometer wheel in Figure 4-7, the last wheel has eleven numbers, twelve graduations. This

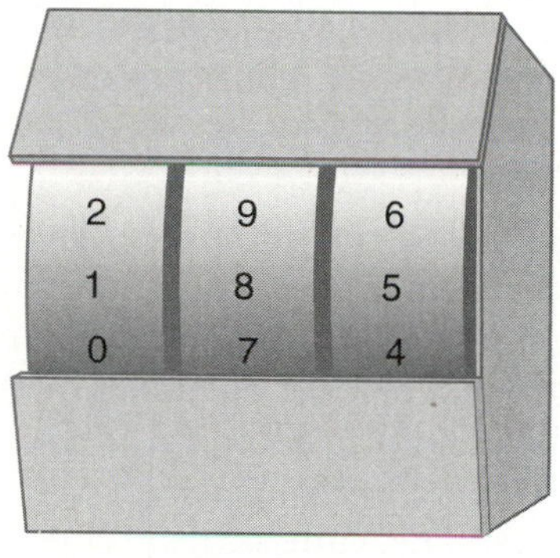

Figure 4-5 Three wheel odometer.

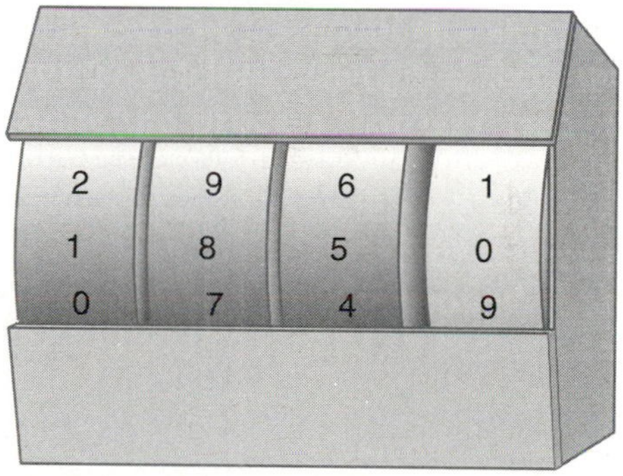

Figure 4-6 Four wheel odometer.

Figure 4-7 Foot and inch odometer.

would indicate the odometer wheel measures in feet and inches. In this example, the reading is 185 feet 0 inches.

Another variation in construction is the linkage between the wheel and the odometer. Some models use a set of gears to connect the odometer to the wheel. This design has a readout that constantly changes as the wheel rotates. To use this type of odometer wheel, the wheel is lifted off the ground, the odometer is reset to zero, and then the wheel is lowered to the ground and rolled along the line of travel. As the odometer is being reset, it is important to hold the handle in the position that it will be used when measuring a distance. Rotating the handle to a comfortable position after the odometer is reset has the same effect as turning the wheel. This will cause an error in the measurement.

An alternative design uses a trip lever on the odometer. When a trip lever is used, the operator must be more diligent when starting a measurement. The trip lever is activated whenever a pin mounted in the rim passes by the odometer. An error will occur if the trip lever is not located properly before the measurement is started. If the owner's manual states that the pin should be ahead of the trip lever when starting, and it is behind it, then the measurement will have an error equal to the distance between the pins. The same is also true if the owner's manual states the pin should be behind the trip lever when starting and it is ahead of it.

To measure using this type of odometer wheel the operator must rotate the wheel until the pin and trip lever are in the correct position. Then the odometer is reset to zero and the wheel lowered to the ground. Rotating the handle after the trip lever is set in the correct position will cause an error in the measurement.

Another concern when using an odometer wheel is the line of travel. The course must be well marked and the operator must be diligent to walk in a straight line between stations. Any deviation will cause an error in the data. In rough terrain or for routes with poor visibility the use of range poles will help the operator stay on course.

Advantages and Disadvantages of Odometer Wheels

The number one advantage of odometer wheels is their ease of use. The only requirements are to reset the odometer before starting and walking in a straight line.

One disadvantage of odometer wheels is the same as chaining: the operator must be able to walk the route. They are also limited to measuring surface distance. In the example illustrated in Figure 4-8 the difference is 1.3 feet for a distance of 250 feet. It is obvious that an odometer wheel is not practical for measuring across water. In Chapter 2 the information on odometer wheels explained the relationship between the roughness of the terrain, height of vegetation and recommended wheel size. Errors of 1:100 have been determined for odometer wheels calibrated for hard surfaces when they are used on grass. An odometer wheel with a small diameter wheel used in rough terrain or tall vegetation will have a greater error.

With proper care and attention to details, accuracy of 1/200 of the distance can be obtained. This means that if a distance of 200 feet is measured, the actual distance will be somewhere between 199.5 and 200.4 feet.

Distances can be measured using the principles of the odometer wheel when a commercial odometer wheel is not available. Calculating the circumference of a wheel (cir = πr^2), such as a bicycle wheel, and counting the number of revolutions required to travel the unknown distance will determine the length.

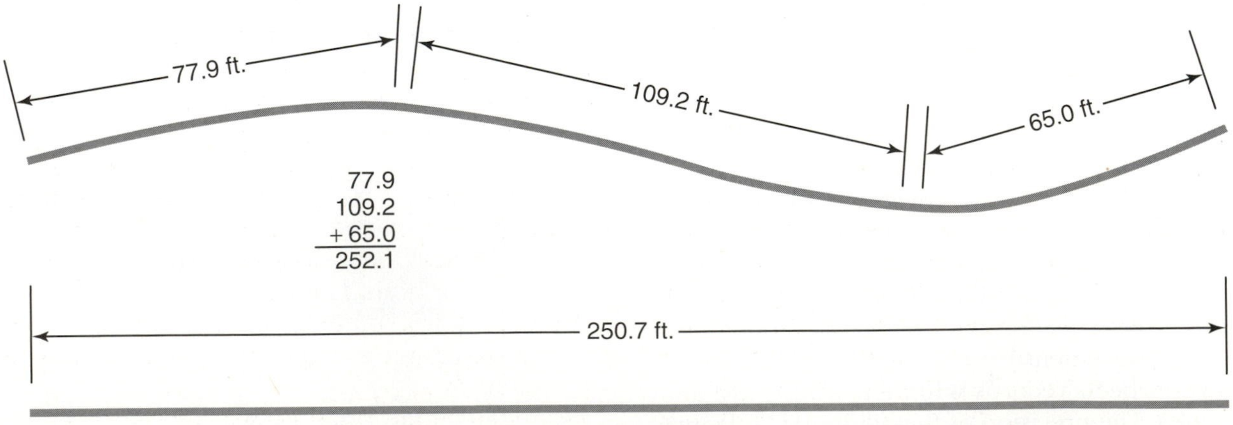

Figure 4-8 Difference between slope distance and horizontal distance.

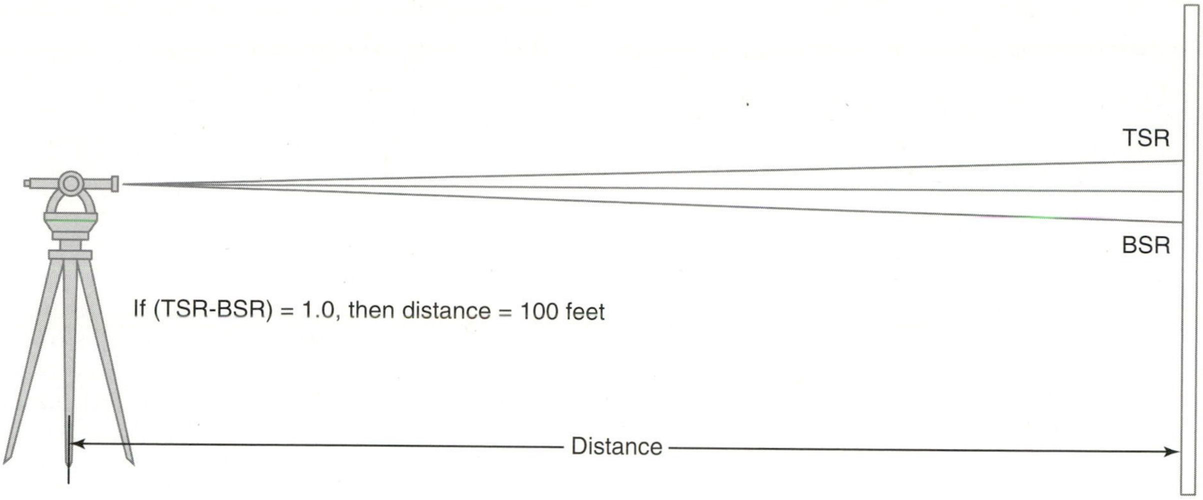

Figure 4-9 Stadia principles.

Distance by Stadia

In the discussion of instruments in Chapter 2 stadia crosshairs were discussed. To measure **distance by stadia** the manufacturer of the instrument installs an additional horizontal crosshair equal distance above and below the elevation crosshair. The crosshair above the elevation line is called the top stadia crosshair and the crosshair located below the elevation crosshair is called the bottom stadia crosshair. Sighting across these crosshairs creates two lines of sight that diverge at a known rate. The difference between rod reading sighting across the top stadia crosshair (TSR) and the rod reading sighting across the bottom stadia crosshair (BSR) multiplied by the stadia factor of the instrument equals the distance. Expressed as an equation:

$$\text{Distance} = (\text{TSR} - \text{BSR}) \times \text{SF}$$

TSR = Top Statia Reading
BSR = Bottom Statia Reading
SF = Stadis Factor

The distance is determined by multiplying the instrument factor times the difference between the top and bottom stadia readings, Figure 4-9.

Instrument Factor

Most instruments have an instrument factor of 100. An instrument with an instrument factor of 100 is designed so that the line-of-sight across either one of the stadia crosshairs diverges from the line-of-sight across the center crosshair at a rate of 1/2 foot per 100 feet. This can also be expressed as one foot difference between the top stadia reading and the bottom stadia reading per 100 feet of distance, Figure 4-10.

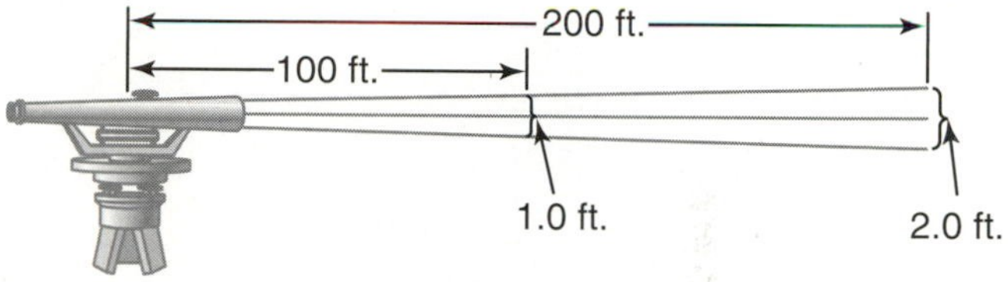

Figure 4-10 Instrument factor.

Measuring Distance by Stadia

To measure distance by stadia the instrument is set up and leveled. The rod is read sighting across the top stadia crosshair. The rod reading is recorded in the appropriate cell in the data table. Next the rod is read sighting across the bottom stadia crosshair. This number is also recorded in the appropriate cell in the data table. Subtracting these two numbers gives the stadia interval. The stadia interval multiplied by the instrument factor, usually 100, produces the horizontal distance.

An instrument with stadia capabilities has a central vertical crosshair and a central horizontal crosshair for reading elevations. The stadia crosshairs may look identical to the center one, but in some instruments, the stadia hairs may be different thickness or just a short line centered on the vertical crosshair. These changes help to reduce the chance of one of the common errors when using an instrument—reading the wrong horizontal crosshair. Figure 4-11 illustrates what the instrument person could see when looking through the eyepiece of an instrument with stadia crosshairs.

Determine the distance for the stadia readings illustrated in Figure 4-12.

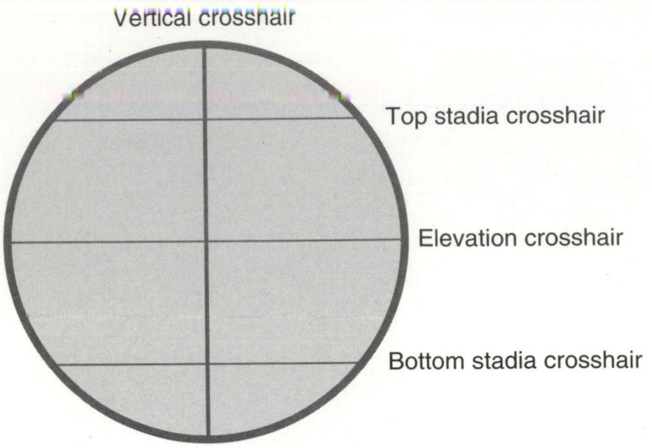

Figure 4-11 Instrument crosshairs.

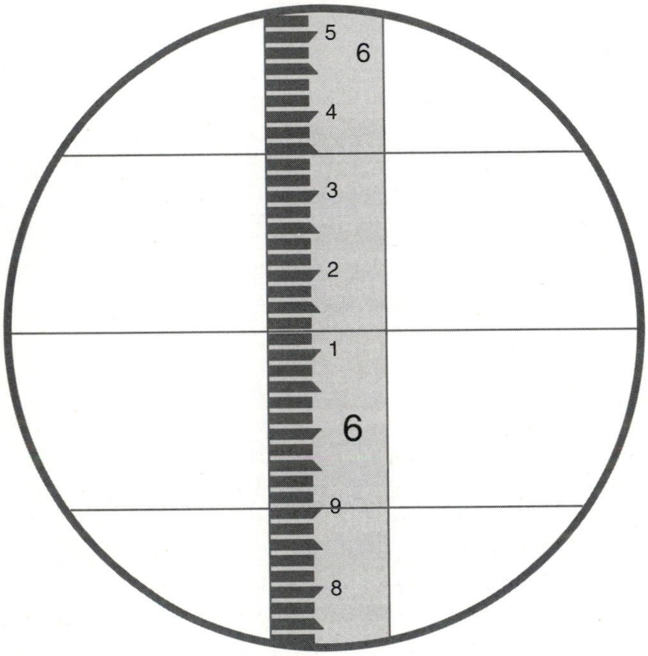

Figure 4-12 Stadia example.

In this example the top stadia reading is 6.35 and the bottom stadia reading is 5.90. The distance is:

$$
\begin{aligned}
\text{Dist.} &= (\text{TSR} - \text{BSR}) \times \text{IF} \\
&= (6.35 - 5.90) \times 100 \\
&= 0.45 \times 100 \\
&= 45 \text{ ft.}
\end{aligned}
$$

The rod is 45 feet away from the instrument.

Note: The precision of the answer is whole feet because the rod readings are multiplied by 100. For smaller precision the target must be used. The target has a Vernier scale that provides a means for reading the rod to the nearest 1/1000 of a foot, 0.001. This increases the precision of the measured distance to the nearest tenth of a foot. For example, if the TSR = 6.342 and the BSR = 4.681 then the distance is:

$$
\begin{aligned}
\text{Dist.} &= (\text{TSR} - \text{BSR}) \times \text{IF} \\
&= (6.342 - 4.681) \times 100 \\
&= 166.1 \text{ ft.}
\end{aligned}
$$

Half-Stadia

Occasionally when using stadia to measure distances either the TSR or the BSR reading is obscured by overhanging obstacles or items on the ground. When only one of the stadia crosshairs and the elevation crosshair can be read the half-stadia method is used. The half-stadia method substitutes the elevation rod reading for the obscured stadia reading. When only one stadia reading and the elevation reading are used the difference must be multiplied by two. One stadia reading and the elevation reading are half of the stadia interval, Figure 4-13.

The stadia method of measuring distance should not have an error greater than 1/500. To achieve this result, the recommended procedures for using the instrument must be followed, the rod must be read correctly, and the rod must be read to 0.001 feet by using the target.

Advantages of Stadia

Distance by stadia has several advantages over pacing and chaining. The stadia method uses an instrument and rod. If the instrument is set level, which it always is, then the measurements are horizontal distances. Secondly, all that is required is a line-of-sight. It is not necessary to be able to walk the distance as in pacing, chaining, and the odometer wheel. Distances can be measured over the top of topographic features such as roads, bodies of water, ditches, etc.

Disadvantages of Stadia

The main disadvantage of the stadia method for measuring distance is that it requires an instrument and rod. In addition, the precision is limited to whole feet if the rod is used, and 0.01 of a foot if the target and Vernier are used.

EQUIPMENT ACCURACY

Accurate equipment is a necessity. Use, abuse, and circumstances can reduce the accuracy of measuring instruments. Normal wear causes movements to loosen, which can cause alignment errors. Abuse damages parts, which reduces the instrument's accuracy. Some equipment, such as EDM's, must be adjusted every time they are used for environmental conditions.

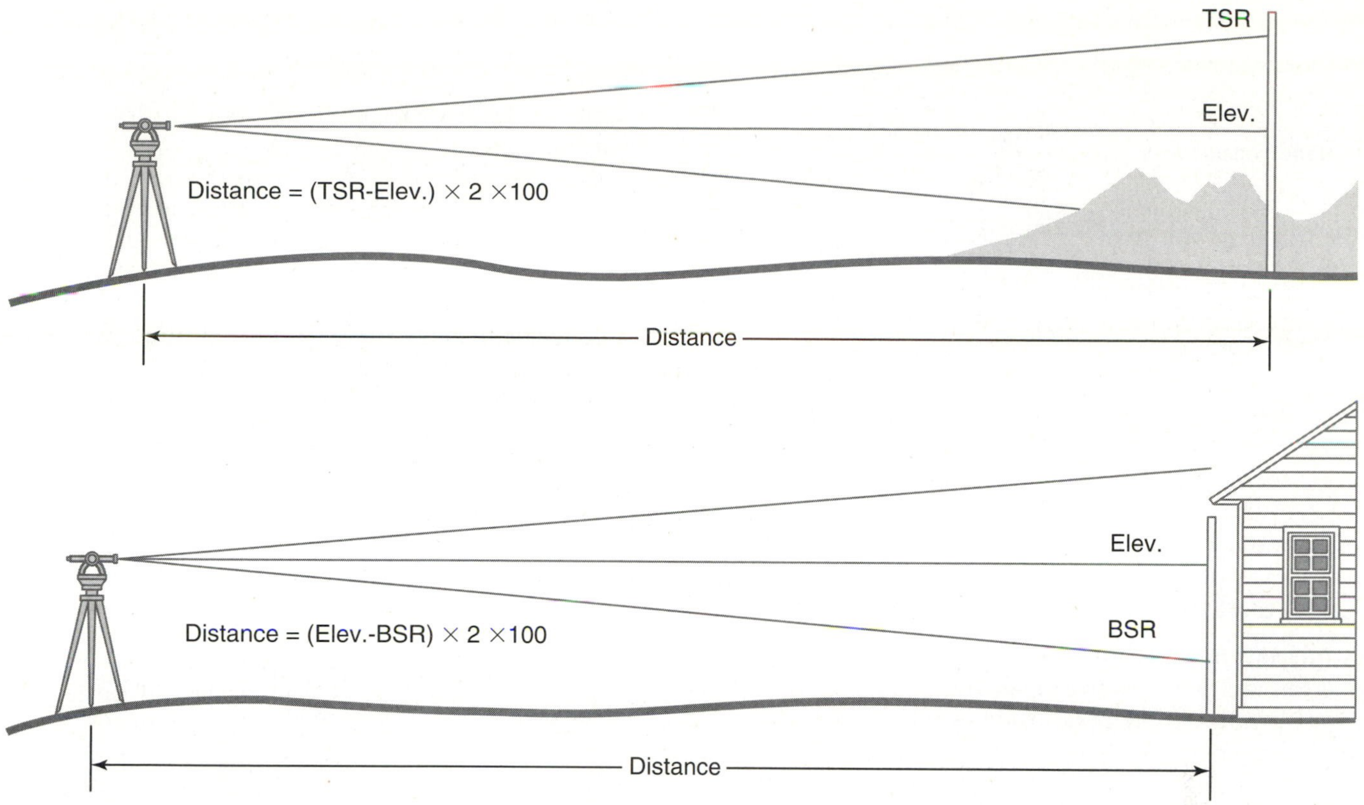

Figure 4-13 Half-stadia.

Errors in accuracy can be random or consistent. Random errors are not predictable and are managed by following recommended methods and procedures. Consistent errors can be managed by calculating a correction for the instrument. To insure measuring equipment is producing accurate results it must be checked on a regular basis. Professional equipment must be sent off to companies that specialize in this service. Other equipment can be checked and if the amount of error is small and is consistent, a correction can be calculated and used to adjust the measurements. An instrument with a large error or an inconsistent error should be removed from service and repaired. The process of evaluating the accuracy of an instrument is called calibrating the instrument.

Calibrating an Instrument

Distance measuring instruments are calibrated by making multiple measurements of a known, or sometimes called standard, distance and computing the difference between the known distance and the measurements. Before determining the correction, the data should be analyzed to determine if the error is acceptable. A correction can be determined for any quality of instrument performance, but the corrected results could still be inaccurate if the measurements from the instrument were very variable. The variability of consecutive measurements for a distance measuring instrument should be no more than one or two of the smallest units. For example, assume a distance measuring instrument produced readings of 99.4, 101.5, 100.1, and 104.5 feet for a standard distance of 100 feet. For an instrument with a precision of a foot, this instrument has excessive variability and should be sent in for repairs.

Assume an instrument with a precision of 0.1 feet produced readings of 100.0, 100.1, 100.2, 100.1 and 100.2 feet for a standard distance of 100 feet. The variability of readings for this instrument would be acceptable and a correction factor could be determined. A distance measuring instrument that produces variable measurements should be carefully evaluated to determine if the variability exceeds the required precision of the survey.

CORRECTION FACTOR

Correction factors for distance measuring instruments have units of feet of error per foot of distance. This value is determined by determining the error and dividing the error by the standard distance. The error is determined by averaging several measurements of the standard distance and subtracting the average from the standard distance. Using the acceptable instrument readings from the previous section:

$$\text{Average measurement} = \frac{100.0 + 100.1 + 100.2 + 100.1 + 100.2}{5}$$

$$= \frac{500.6\text{ ft.}}{5} = 100.12\text{ ft.}$$

$$\text{Error} = 100\text{ ft.} - 100.12 = -0.12\text{ ft.}$$

$$\text{Correction factor} = \frac{-0.12\text{ ft.}}{100\text{ ft.}} = -0.0012\frac{\text{ft. error}}{\text{ft. distance}}$$

Once the correction factor is determined it can be used to correct the readings recorded by the instrument.

Note: In this example the error is negative. It is important to include this sign throughout the calculations.

Correcting a Measurement

A measurement is corrected by multiplying the correction factor times the measured distance and adding this value to the measured distance. Expressed as an equation:

$$\text{Corrected distance} = \text{Measured distance} + (\text{Measured distance} \times \text{Correction factor})$$

Assume the instrument in the previous example recorded a distance of 556.7 feet. The actual distance would be:

$$\text{Corrected Distance} = 556.7\text{ ft.} + \left(556.7\text{ ft.} \times -0.0012\frac{\text{ft. of error}}{\text{ft. of distance}}\right)$$

$$= 556.7 + (-0.668)$$

$$= 556.03 \text{ or } 556.0\text{ ft.}$$

Note: Because the error was negative, the correction was subtracted from the measured distance. Intuitively this makes sense because the average of the recorded measurements was longer than the standard distance.

All distance measuring tools such as EDM's, optical rangefinders, etc. can be calibrated using the same method. The calibration process requires comparing the performance of a measuring tool against a standard. If the variability is acceptable, a correction factor can be determined and it can be used to correct the readings recorded by the instrument.

SUMMARY

A person has many options for measuring distance. The task is to select the best method for the use of the data and the resources that are available. With the information presented in this chapter and Chapter 1 a person should be able to select a method for measuring distance and produce results that have an acceptable level of error. It is important to have a good understanding of distance measuring because differential, profile, topographic, and many other surveying methods use these principles.

Student Activity

1. Determine the correct distance for a measured distance of 345.67 feet for an odometer wheel that measured 100.98, 100.99, 100.97, and 100.99 feet over a standard distance of 100.00 feet.
2. Determine the pace factor for an individual that recorded 39, 40, and 39.5 paces in a 100.0 foot distance.
3. Use the stadia method to determine the distance in the illustration. Assume instrument factor is 100.

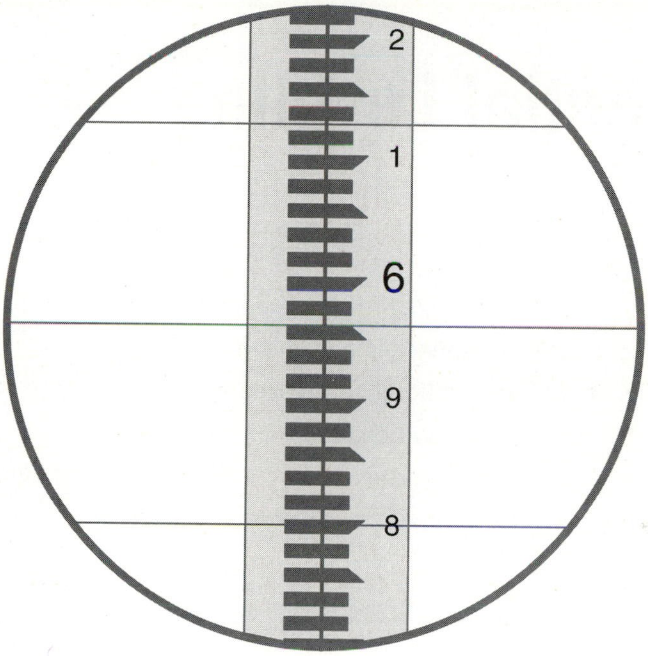

CHAPTER 5

Differential Leveling

Objectives

After reading this chapter, the reader should be able to:

- Explain the principle of leveling.
- Complete a differential leveling survey.
- Establish a new benchmark.
- Complete the three checks for error used with differential leveling.
- Explain the term "allowable error of closure."

Terms To Know

allowable error of closure
leveling
reference line
reference plane
control point
balancing the sights
differential leveling
instrument height
height of instrument
closing the loop
note check

INTRODUCTION

In Chapter 1, the term level was defined as *"being perpendicular to a vertical line."* An object is also said to be level when it is parallel to the horizon. The concept of level is not unique to surveying. When a picture is hung on the wall, care is taken to insure that it hangs straight (level). Kitchen appliances are "leveled" when they are installed. As a carpenter builds forms for placing concrete, they are leveled, or checked to see if the top of the forms are the same elevation. In Chapter 1 the term level was also used to describe a category of tools, Figure 5-1.

The application of the term level is dependent on a tube of liquid with an air bubble called a spirit level. The air bubble is centered in the tube when the tube is horizontal, or level. Spirit levels are mounted on surveying instruments such as a dumpy level parallel to the line-of-sight of the instrument so that when the spirit level is "level" the line of site of the instrument is horizontal. The following sections will explain the principles of leveling and how these principles are used in differential leveling.

LEVELING

Leveling is the process of determining if an object is perpendicular to a vertical line or if two or more objects are at the same elevation. A person can attach two pictures to the wall and be satisfied that they are at the same height, but there is a good chance that another person will disagree with their placement. It is very difficult to tell if two or more objects are level without the use of a tool or instrument. One alternative when mounting pictures to the wall is to measure down from the ceiling and set both pictures the same distance. This will insure the pictures are parallel to the ceiling, but they will only be level if the ceiling is level. This illustrates the importance of a level reference line, which will be discussed in more detail in the following sections.

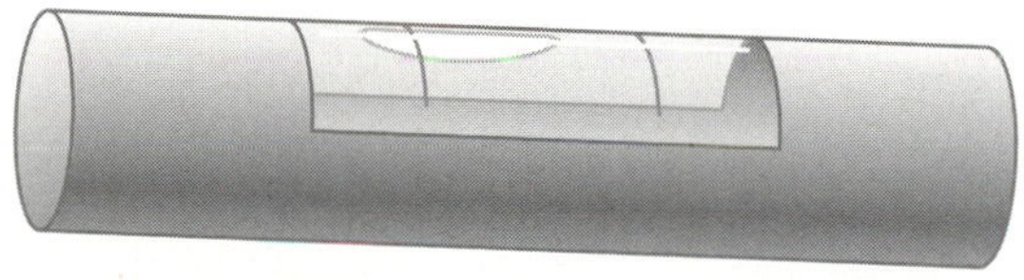

Figure 5-1 Spirit level.

The principles of leveling can be illustrated with a simple type of level—a garden hose. Clear, vented, graduated tubes are attached to each end of a garden hose. When the hose and tubes are filled with water and the tubes are held in a vertical position, the water will be at the same elevation at both ends of the hose. We would say that a straight line across the top of the water at each end of the hose is level, Figure 5-2. This is a very primitive tool for leveling and should be used with care, but the distance over which the hose level can be used is only

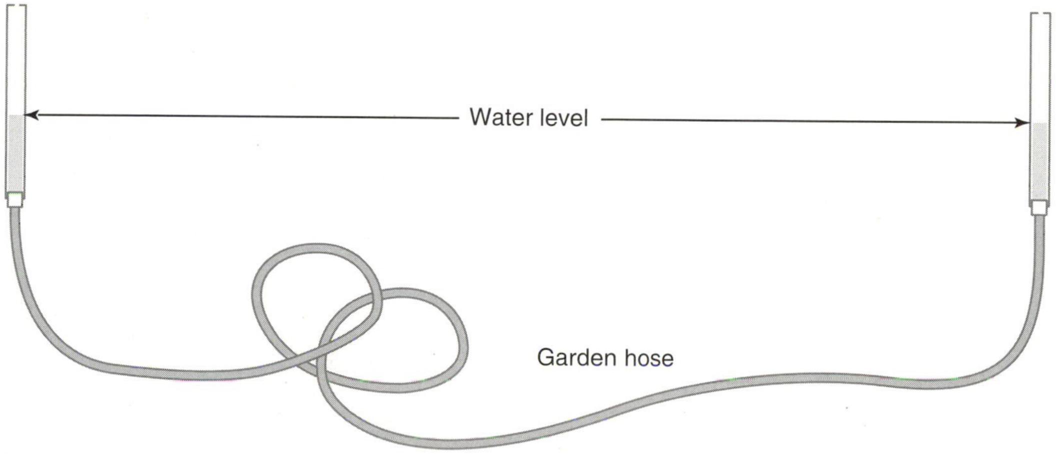

Figure 5-2 Garden hose level.

limited by the length of hose the operators are willing to manipulate.

If the tubes are held near different objects, the top of the water will indicate if the objects are level, or at the same elevation, Figure 5-3.

A commercial version of the hose level has graduations on the tubes at both ends of the hose. This style of hose level can be used to determine the difference in elevation between two points. Figure 5-3 is an illustration of a hose level being used to determine the difference in the height at two ends of a wall. In this case, the difference in height is 2.9 feet. The hose level is not practical for surveying large areas, but the level of precision is sufficient for general contract work such as leveling a building pad, or establishing the grade for a sewer line.

TIP

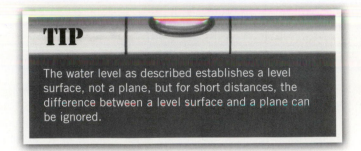

The water level as described establishes a level surface, not a plane, but for short distances, the difference between a level surface and a plane can be ignored.

LEVELING WITH AN INSTRUMENT

Try to visualize an instrument set up so the centerline of the instrument is at the same elevation as the top of the water at one end of the hose in Figure 5-4. When the horizontal crosshair of the instrument telescope

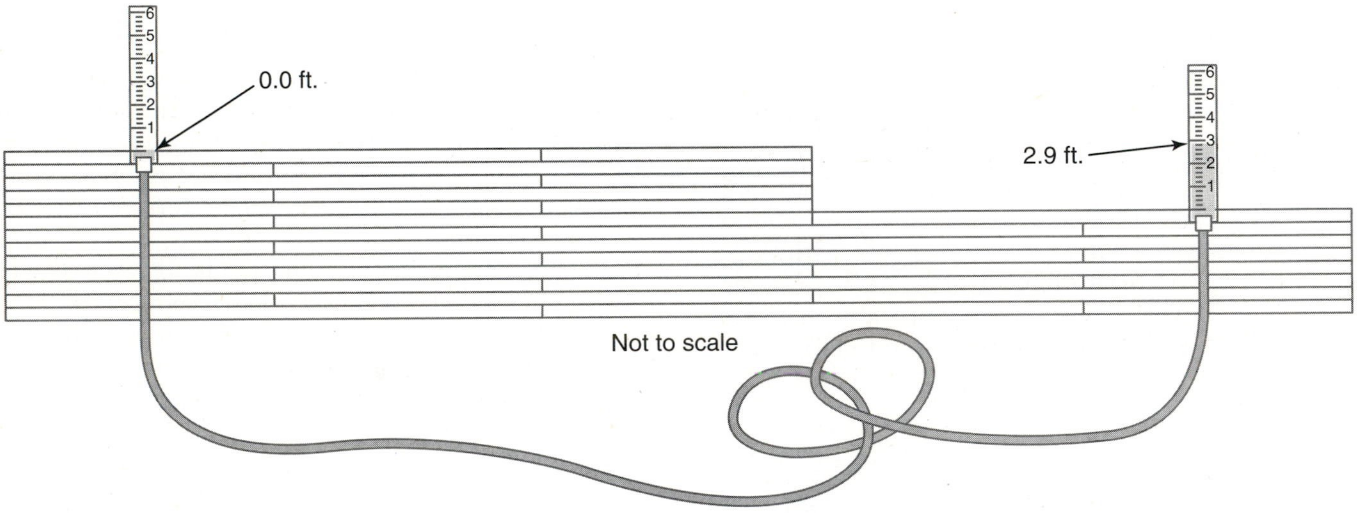

Figure 5-3 Using a hose level.

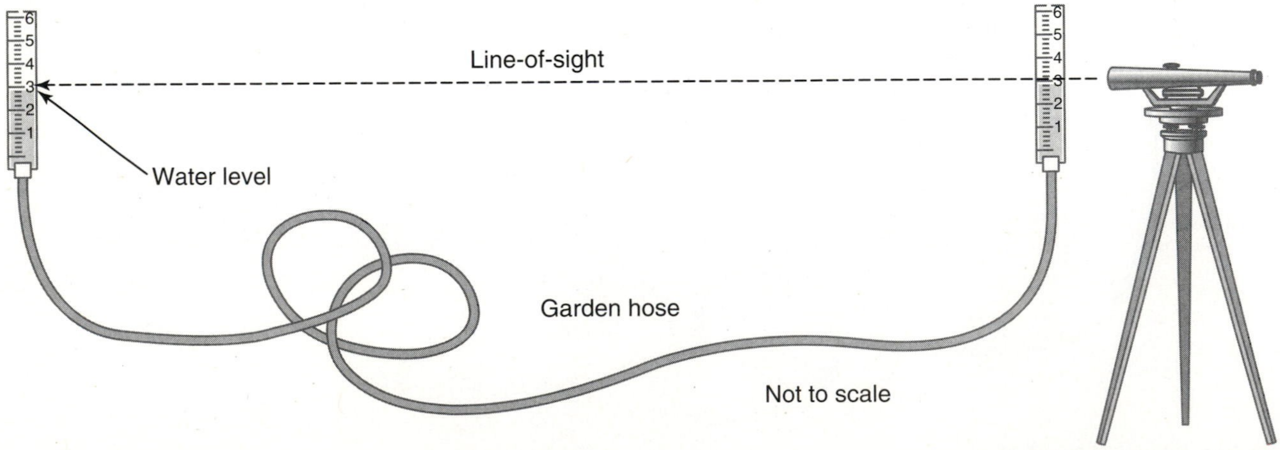

Figure 5-4 Comparison of hose level and instrument line-of-sight.

is at the same elevation as the water in the first tube, then the line-of-sight through the instrument will strike the second tube of a short hose at the height of the water. The line-of-sight through the telescope of the instrument establishes a level reference line.

The line-of-sight of an instrument is horizontal, if it is set up correctly, therefore it can be used to compare the relative elevation of two or more objects. The rod is placed on one object, or station, and the center crosshair is read on the rod. The rod is then placed on the second object, or station, and the center crosshair is read for the second time. Subtracting the rod readings results in the difference in elevation between the two stations. The rod simply measures the distance from the line-of-sight, the **reference line**, to the top of the object the rod is resting on. If the same reference line is used, then different rod readings can be compared. If the first rod reading is greater than the second, then the first object is at a lower elevation than the second object.

The opposite is also true. If the first rod reading is less than the second rod reading, the first object is higher than the second object. If the difference between the rod readings is zero, then the two objects are at the same elevation.

This is the principle of leveling used for differential and other types of surveys. Assuming the instrument is set up correctly, the line-of-sight through the telescope establishes a horizontal reference line that can be used to compare the elevations of two or more objects/stations.

Reference Line

Figure 5-5 is an illustration of three points above and below a reference plane. When point A is 5.0 feet above the horizontal plane and point C is 7.5 feet below the plane, the vertical distance between point A and point C is:

$$5.0\text{ ft.} + 7.5\text{ ft.} = 12.5\text{ ft.}$$

In addition, because point A is 5.0 feet above the plane and point B is 3.5 feet above the same plane, then the difference in height, elevation, between point A and point B must be:

$$5.0\text{ ft.} - 3.5\text{ ft.} = 1.5\text{ ft.}$$

The key to understanding this concept is to remember that points A, B, and C are on different planes. Study Figure 5-6. This illustration shows that the 5 and 7.5 foot distances shown in Figure 5-5 are the distances

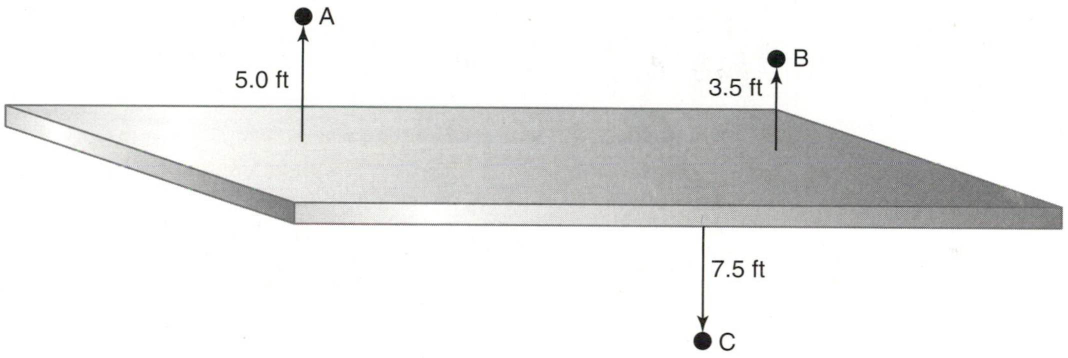

Figure 5-5 Relationship of points to a plane.

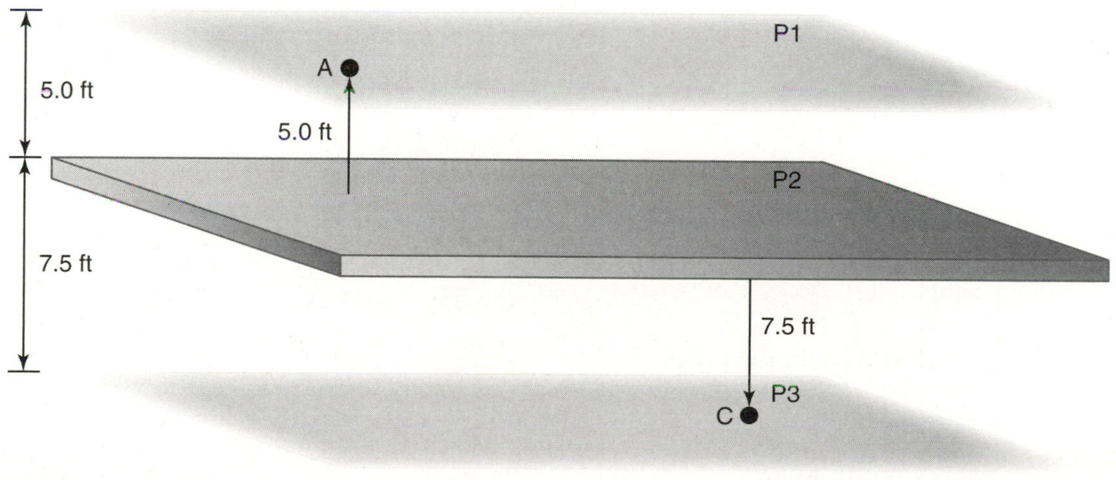

Figure 5-6 Multiple planes.

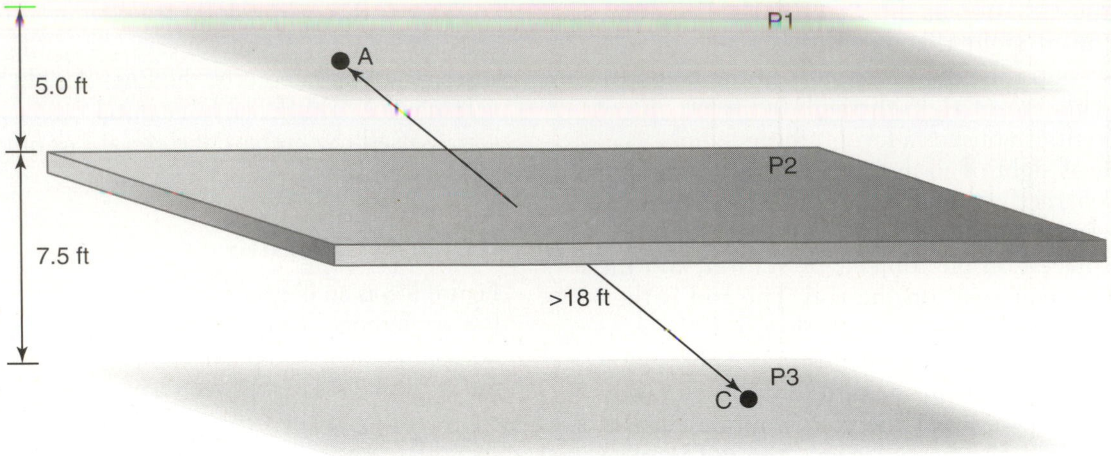

Figure 5-7 Distance from point to point.

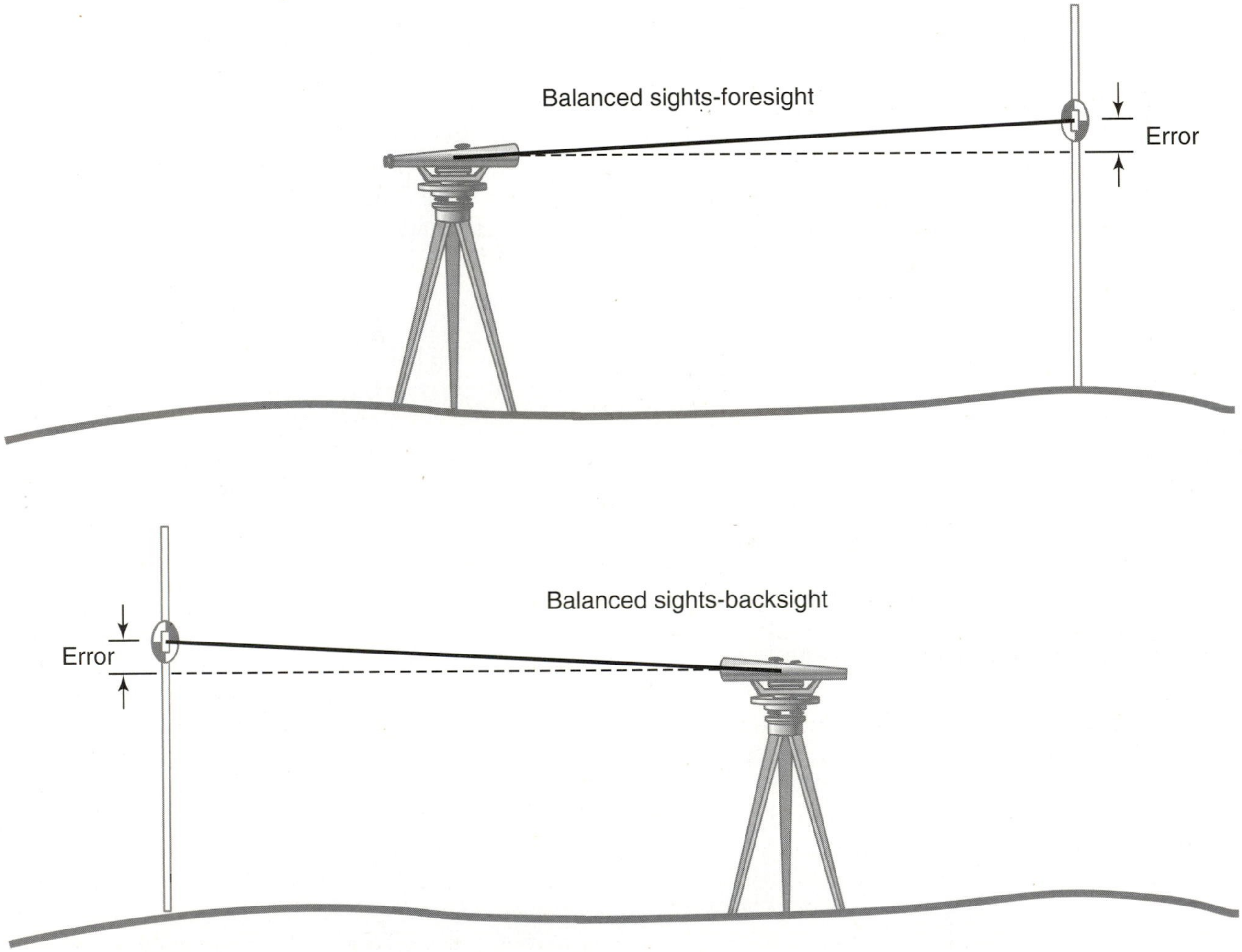

Figure 5-8 Balanced sights.

between the three planes that contain the points, not the distance between the points. The distance between Points A and C is illustrated in Figure 5-7.

The distance, difference in elevation, between plane P1 and plane P3 is 12.5 feet, but because points A and C are not on the same vertical line the distance between them is greater than 12.5 feet, Figure 5-7.

In surveying, the instrument is used to establish a reference line or reference plane. The surveying rod is used to measure the vertical difference between

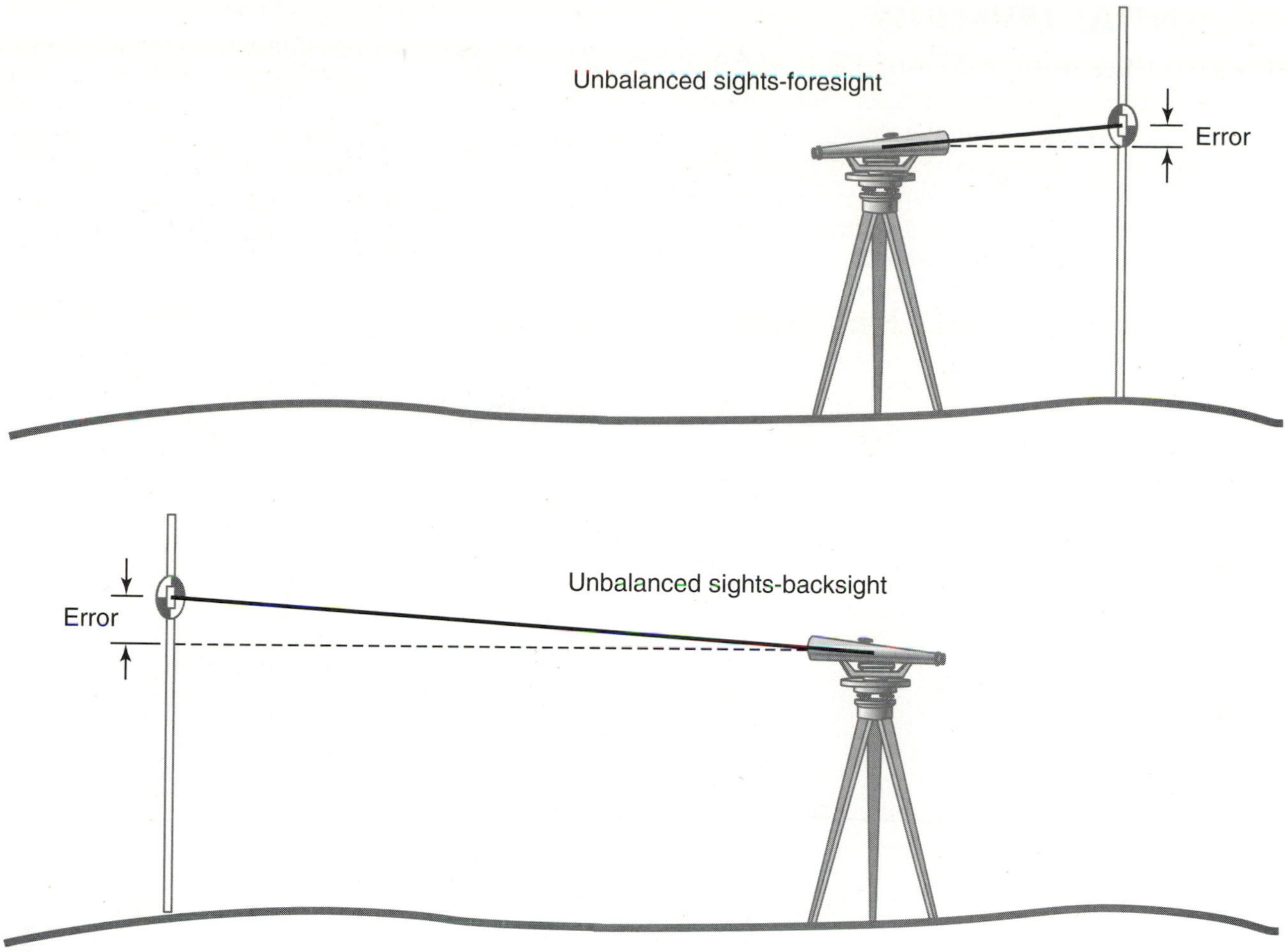

Figure 5-9 Unbalanced sights.

the reference line or plane and the points being surveyed. As long as the reference line or plane is horizontal, all of the points measured from the reference line or plane can be compared.

This is important to remember when surveying, every time the instrument is set up and leveled, the plane established by the line-of-sight through the telescope will be at a different elevation. When rod readings from multiple instrument positions need to be combined, they must have at least one common point. These common points are often called control points.

Balancing the Sights

Balancing the sights is a technique used to manage the error that can occur if the instrument is not level.

Figure 5-8 is an illustration of the effect of an instrument that is not level. In the illustration the dashed line represents the horizontal line-of-sight for a level instrument and the solid line represents the actual line-of-sight because the instrument is not level. Note that because the instrument is set up halfway between the two stations, Figure 5-8, the error between the rod reading and the correct rod reading is the same in the backsight as in the foresight. When keeping notes, the error would be added in one column and subtracted in another. The error cancels out and the elevation difference between the two stations would be correct. Now study Figure 5-9.

In this illustration, the instrument is farther from the rod for the backsight than for the foresight. Consequently, the error between the rod reading and correct reading is much larger for the backsight than for the foresight. In this situation, the error in the backsight would be greater than the error in the foresight. Not all of the error is canceled. The value for the elevation of station two would not be correct. Establishing the instrument position halfway between the two stations will, under most conditions, reduce the effect of an error caused by the instrument not being level.

DIFFERENTIAL LEVELING

Differential leveling is used to determine the difference in elevation between two or more points. One common use is to establish the elevation of a new benchmark referenced to an existing benchmark. It is also used to compare the elevation of several points or objects. Figure 5-10 illustrates using differential leveling to check the forms for a footing.

If the three rod readings in Figure 5-10 are the same, then the top of thc form is level. There is no difference in the elevation for the three corners of the footing.

Establishing Benchmarks

Another use of differential leveling is establishing the elevation of a new benchmark. When the existing benchmark and the location of the new benchmark can be seen from one instrument position, the procedure is very simple. The instrument is set up halfway between the points and leveled. The first rod reading is taken on the existing benchmark (backsight) and the second rod reading is taken on the new benchmark (foresight). The backsight reading is added to the elevation of the existing benchmark to establish the height of the line-of-sight, reference line. This is called the **instrument height** (IH), or **height of instrument** (HI). The foresight is subtracted from the instrument height. The result is the elevation of the new benchmark.

Turning Points

As defined in Chapter 1, a turning point is a temporary benchmark or a reference point. They are used to extend the survey along the route when the backsight and foresight are not accessible from one instrument position, Figure 5-11.

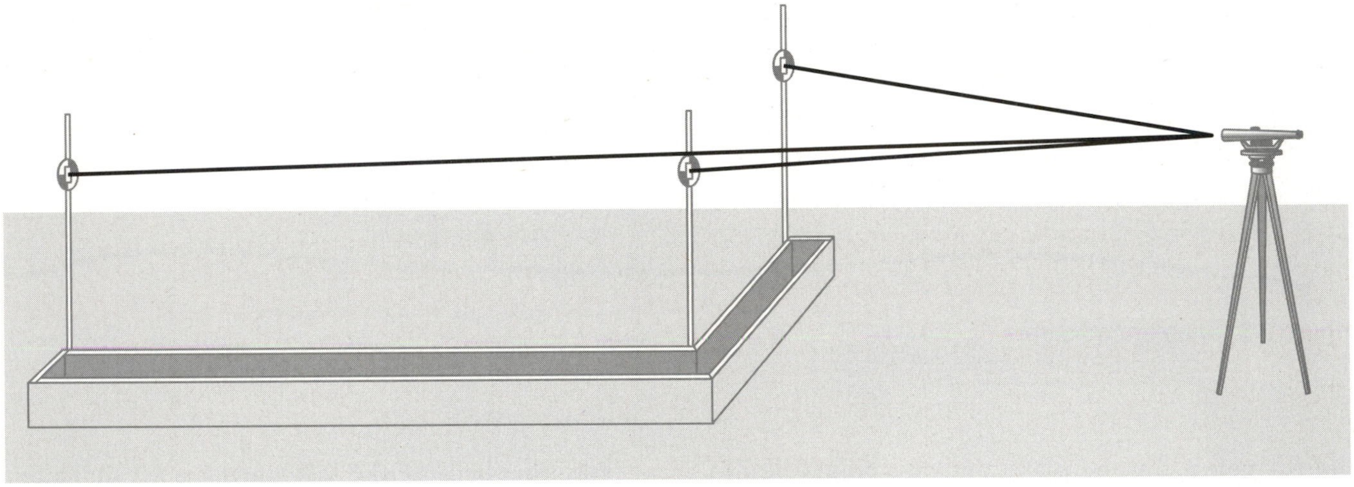

Figure 5-10 Using differential leveling.

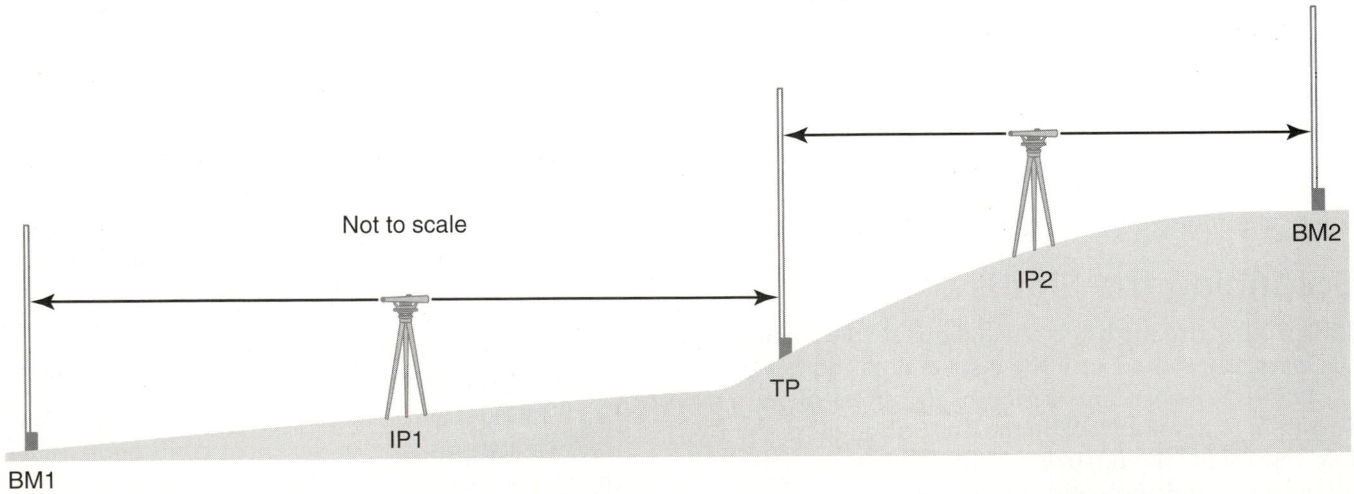

Figure 5-11 Turning point.

Turning points are temporary benchmarks: they must be a stake or some type of structure that has a stable elevation. The three primary factors causing the need for turning points are:

- The line-of-sight intersecting the ground
- Blocked view
- Exceeding the focal length of the instrument

The use of a turning point is illustrated in Figure 5-12. In this example, the line-of-sight will intersect the ground before reaching station B. A turning point is established between stations A and B. This allows the instrument to be moved and reset at a higher elevation.

Note: In this situation a hand level is very useful for determining the set up location of the instrument. Checking the backsights and foresights with the hand level could save an instrument set up.

Turning points are also used when the line-of-sight is blocked by trees, structures, and other objects. The third use of turning points is due to the limitations of the instrument. All instruments have a maximum recommended distance that should be used, called the focal length. Exceeding the focal length will increase the error in the readings. If the distance between the two stations being surveyed is greater than two times the maximum focal distance for the instrument, then a turning point must be used.

The elevation of turning points is calculated as if they were benchmarks. A backsight is recorded for the benchmark and a foresight is recorded for the turning point, Figure 5-13. With this information, the elevation of the turning point can be determined. The instrument is moved to IP2 and the process repeated. When the instrument is at IP2 the backsight on the turning point and the foresight on the next station are used to calculate the elevation of the station.

The example in Figure 5-13 is completed using only one turning point. There is no limit on the number of turning points that can be used to complete a survey. Just remember, each instrument set up requires a few minutes of time and increases the opportunity for error. If several turning points will be needed, the surveyor should reconnoiter the route before collecting data to determine the best locations

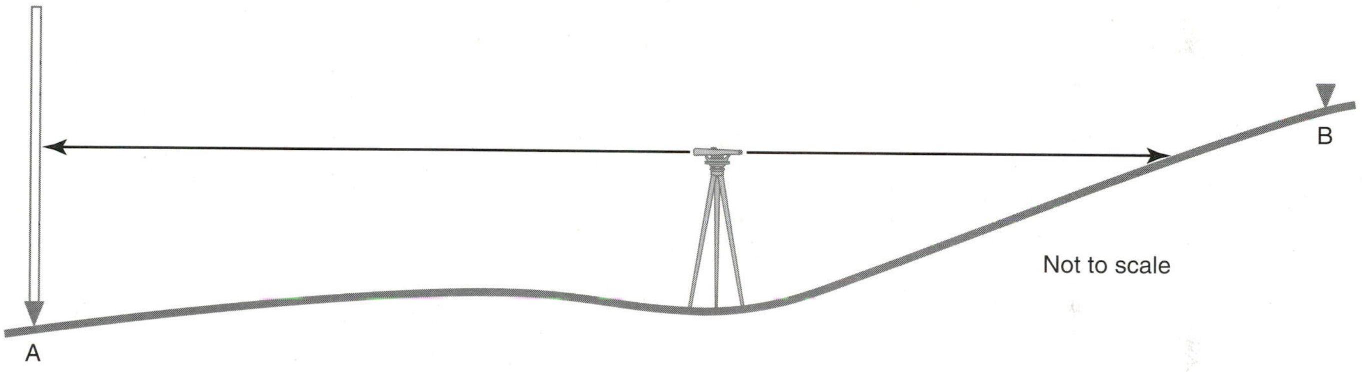

Figure 5-12 Differential survey requiring a turning point.

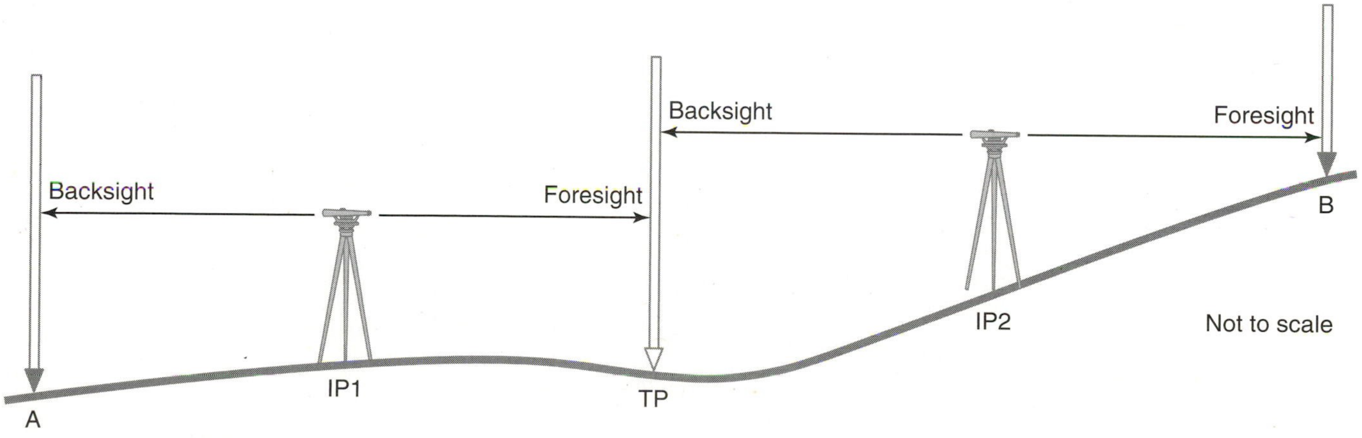

Figure 5-13 Nomenclature when using turning points.

for instrument positions and turning points. An experienced surveyor can accomplish this by walking the route; for the less experienced person, using a hand level to check changes in elevation and pacing distances will reduce the number of turning points that will be required.

The purpose of differential leveling is to compare the elevation of two or more points. The data collected will not allow the user to define or draw the topography between the two points. The primary influence of the topography, when differential leveling, is the difficulty it presents for completing the survey.

It is much easier to complete a differential leveling survey on a flat open field, than if the survey must traverse hills, forests, creeks, etc. The time and resources required to establish a new benchmark will depend on the distance between the existing and new benchmark, and the terrain between the two points.

Differential Example

The purpose of this example is to show how to establish the elevation of station B when the elevation of station A is 100.00 feet. In this example both stations can be seen from one instrument position, therefore the instrument is set up halfway between station A and station B. The instrument is leveled and a backsight of 8.47 feet is recorded at station A, Figure 5-14.

The instrument is rotated to align on station B and a foresight of 6.11 feet is recorded, Figure 5-15. Because station A and station B are both visible from one instrument position, sufficient data has been collected to determine the elevation of station B based on the elevation of station A.

When the elevation of station A is 100.00 feet, the elevation of station B is 102.36 feet.

$$100.00\text{ ft.} + 8.47\text{ ft.} = 108.47\text{ ft.}$$
$$108.47\text{ ft.} - 6.11\text{ ft.} = 102.36\text{ ft.}$$

This example is very simple and therefore the amount of data generated is small and easy to manage. As the complexity of surveys increases the amount of information generated also increases and the task of keeping the data organized becomes more difficult. Tables are a good way of organizing data and one has been developed to reduce the chance of random

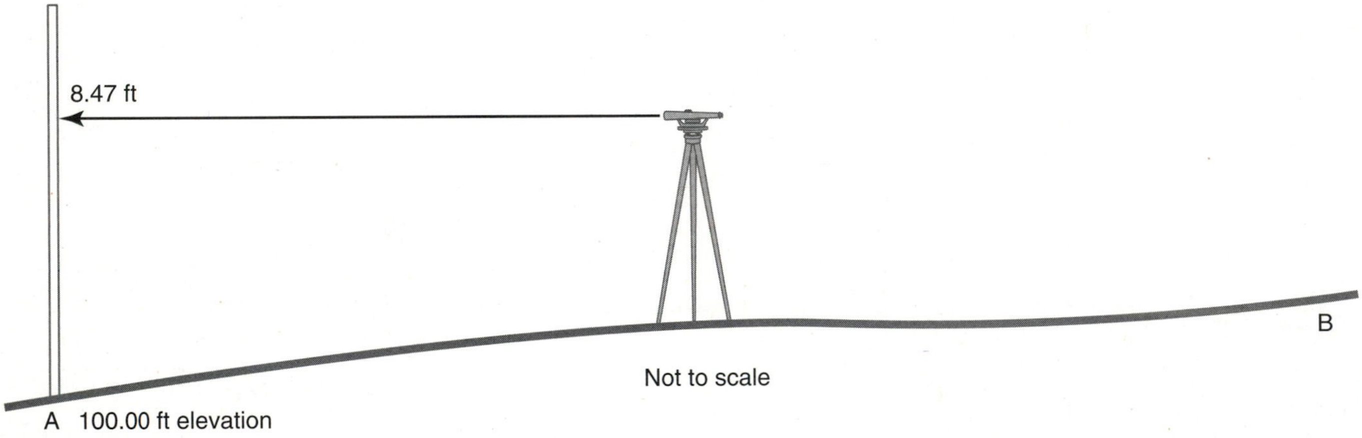

Figure 5-14 Differential example-backsight.

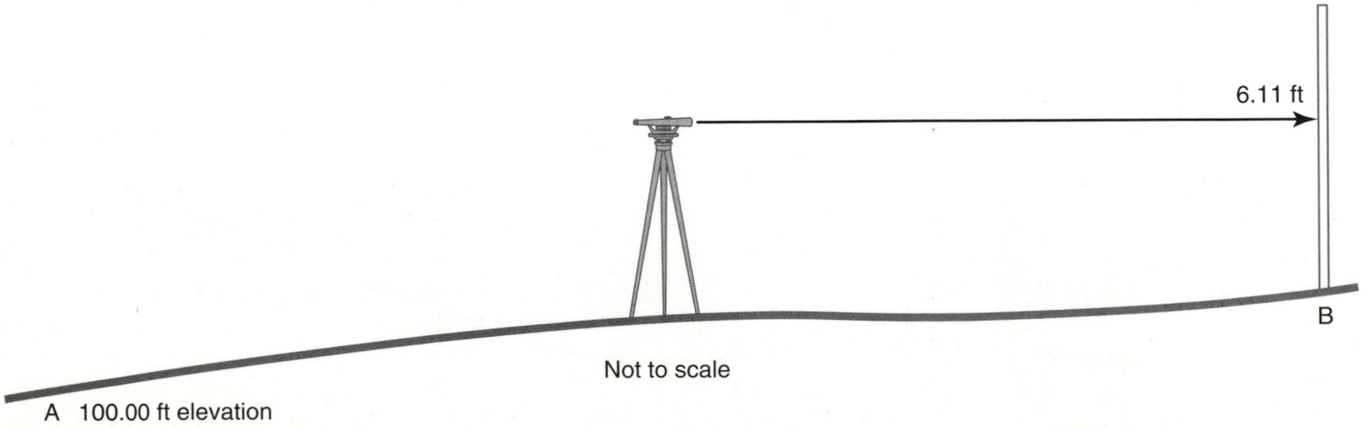

Figure 5-15 Differential example-foresight.

errors in managing differential data, Table 5-1. The use of this table helps organize the data, making it easier to read. It also reduces errors in recording and completing calculations.

Differential Data Table

The table used for differential leveling data uses columns and rows, Table 5-2. Columns are used for the station identification, backsights, foresights, height of instruments and elevations and rows are used for each stations. For the example, the backsight rod reading of 8.47 feet on station A in Figure 5-14 is placed in the row A and column BS.

The rod readings are recorded in the columns labeled BS (backsight) and FS (foresight). The numbers in the columns labeled HI (height of instrument) and ELEV (elevation) are calculations. The calculations are governed by two equations:

$$HI = ELEV + BS$$

$$ELEV = HI - FS$$

In Table 5-1 the elevation of 102.36 feet for station B is correct only if no errors were made during the survey or in the math. It is not safe to assume there were no errors.

The table illustrated in Table 5-1 is one of several formats that have been developed for recording differential survey data. An alternative style is illustrated in Table 5-2. The advantage of this table is that the elevation of the first station "A" is on a separate line from the elevation of the instrument "HI." Table 5-2 illustrates that it is an acceptable practice to leave blank spaces and or blank lines in the data table if it reduces the chance of confusing numbers or makes the table easier to read. The form used in Table 5-1 will be used for the remainder of this text and any references to table format will refer to the style used in Table 5-1.

TIP

It will help keep the math correct in a differential data table if you remember that the addition always occurs with numbers in the same row, and the subtraction is always from a number in a previous row.

Table 5-1 Differential example

STA	BS	HI	FS	ELEV
A	8.47	108.47		100.00
B			6.11	102.36

STA = Station

BS = Backsight

HI = Instrument height

FS = Foresight

ELEV = Elevation

Table 5-2 Alternate format for differential information

STA	BS	HI	FS	ELEV
A				100.00
IP1	8.47	108.47		
B			6.11	102.36

IP = Instrument position

A good way to begin checking data is the eye ball test. Look at the data. Does everything look correct? For example, differential surveying notes should have the same number of backsights as foresights. The FS of first station and the BS of the last station should be blank.

Three Checks for Error

Three checks for error are used to determine the quality of differential survey data. These are closing the loop, the note check, and allowable error of closure. All three checks should be completed even if the data appears to be correct.

Closing the Loop

In differential leveling, closing the loop means surveying back to the starting point. Any difference in the elevation of the starting point initially and upon closing is error, because the elevation of the starting point should not change during the survey. This difference is called closure error or error of closure. A small amount of closure error is normal for most surveys. The survey must be closed before the note check and allowable error of closure can be completed.

When closing the loop it is not necessary to follow the same route as the survey. The closing loop is another differential survey, but this time the starting point is station B and the survey is completed to station A.

The first step is to move the instrument from the position used to record the foresight on station B. The purpose of moving the instrument is to reduce the chance of repeating an error. If the instrument is not moved, the backsight used to start the closing loop will be the same rod reading as the foresight used to determine the elevation of benchmark two. The surveyor may be tempted to use foresight that completed the route as the backsight for the closing loop.

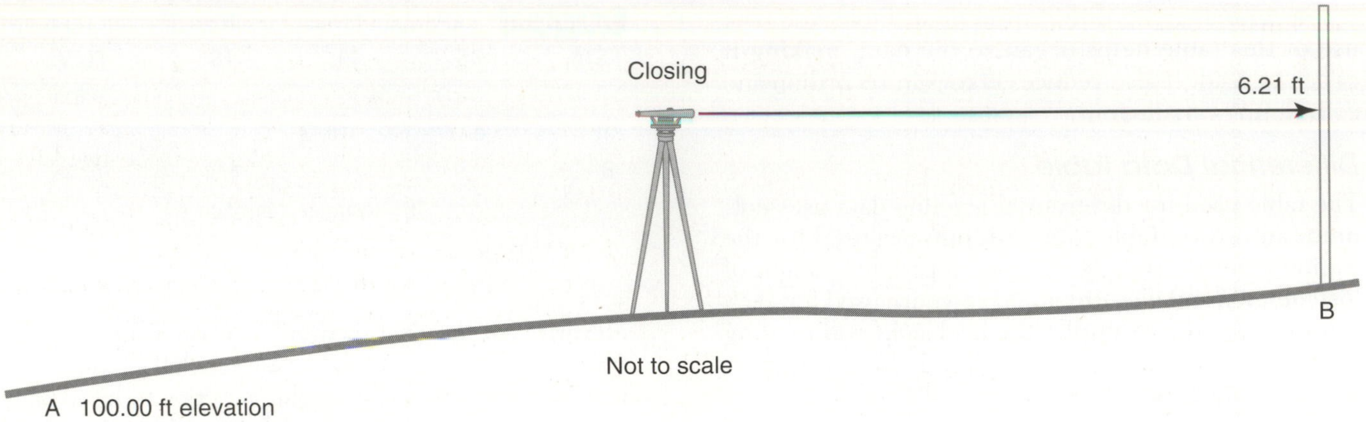

Figure 5-16 Closing backsight.

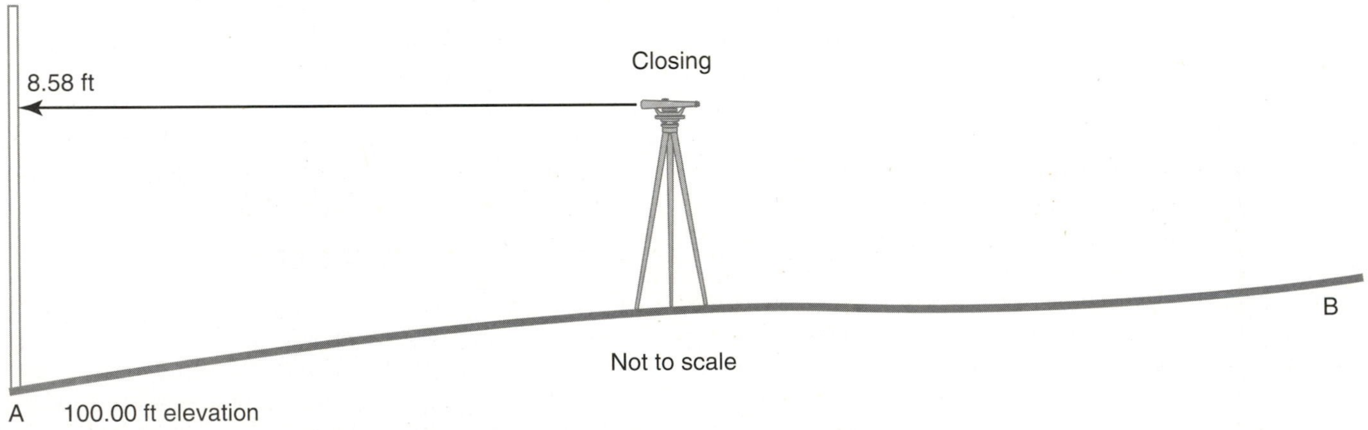

Figure 5-17 Closing foresight.

This would cause any error in the foresight reading to be repeated in the backsight.

Closing the loop for the differential example is illustrated in Figure 5-16. The instrument is leveled and a backsight of 6.21 ft. is recorded on station B.

The closing loop is completed by rotating the instrument and recording the foresight reading of 8.58 ft. for station A, Figure 5-17.

These numbers are recorded in the table and the math completed.

The calculations in Table 5-3 show that the elevation of station A was 100.00 feet at the start of the survey, but 99.99 feet at closing. An error of 0.01 occurred sometime during the survey. The presence of this error means the calculated elevation of station B (102.36) has some uncertainty.

Table 5-3 Differential example

STA	BS	HI	FS	ELEV
A	8.47	108.47		100.00
B	6.21	108.57	6.11	102.36
A			8.58	99.99

Note Check

The second check for error is the note check. The **note check** is used to determine if the closure error is a surveying error or if it was caused by a math error in the data table. If the closure error in the data table is caused by a mistake in the math, the table is recalculated. This may or may not remove the closure error. The note check should be completed every time, even when the data does not have a closure error.

The note check is completed using the following equation:

$$|\Sigma BS - \Sigma FS| = |\Sigma BM_i - BM_c|$$

This equation reads, "absolute value of the sum of the backsights minus the sum of the foresights equals the absolute value of the initial elevation of the benchmark minus the closing elevation of the benchmark." If the equation *is not* true, the data has a math error. When this occurs, the note keeper must recalculate the notes and find the error before the third check for error is completed.

In this example the note check results in:

$$|14.68 - 14.69| = |100.00 - 99.991|$$

$$0.01 = 0.01$$

The note check statement is true. The difference in the starting elevation of Station A and the ending elevation of Station A is not caused by a math error in the notes. The closure error is caused by a survey error. The source of the error could be an incorrect reading of the rod, the rod not being plumb, instrument not level, etc. An error in the survey does not mean the data is not usable. One additional check must be completed. This is called the allowable error of closure.

Allowable Error of Closure

The concept of allowable error of closure developed because engineers and surveyors realized that it is not realistic to expect surveys to be completed without any error. If perfection is not possible, then a level of acceptability must be established. The **allowable error of closure** sets a standard for determining if the survey closure error is acceptable. The closure error should be very small, but some error is normal. The allowable error of closure establishes the amount of closure error the survey can have and still be acceptable.

The allowable error (AE) of closure is determined by the equation:

$$AE = K\sqrt{M}$$

AE = Allowable error

K = Constant (1.0 to 0.001)

M = Distance traveled in miles

The allowable error is equal to the constant K times the square root of M. The variable M is the total distances traveled during the survey, in miles.

Note: Survey distances are usually measured in units of feet. There are 5280 feet per mile. The total distance is the distance of the survey plus the distance of the closing the loop.

The value for K ranges from 1.00 to 0.001, depending on the precision requirements of the survey. The K value is a range because not all surveys need to meet the same standard. A K value of 0.01 to 0.001 is used for precise surveys such as boundary surveys and for projects that have stringent error requirements, but a K value of 1 or 0.1 is acceptable for agricultural, most construction and landscaping surveys.

The method used for measuring distances during the survey must be appropriate for the class of survey. A chain or accurate EDM should be used for a precise survey. Pacing may be acceptable for a lower precision of survey. The decision must be based on the intended use of the data.

If we assume the distance in the previous example between station A and station B is 524.2 feet and the appropriate value for K is 0.1, then the allowable error is:

$$AE = 0.1\sqrt{\frac{524.1 \text{ ft.} \times 2}{5280}}$$

$$= 0.02$$

In this example the actual error, 0.01, is less than the allowable error, 0.02. The correct conclusion is that the survey is acceptable. The survey has errors, but it is acceptable based upon the K value that was used. This is demonstrated by recalculating the allowable error using a K value of 0.01.

$$AE = 0.01\sqrt{\frac{524.1 \text{ ft.} \times 2}{5280 \text{ ft.}}}$$

$$= 0.002$$

When a K value of 0.01 is used the closure error, 0.01, is greater than the allowable error, 0.002. In this situation the survey *would not* be acceptable. The surveyor would be required to repeat the survey. The value for K must be established before the survey is conducted.

Differential Leveling Example with Turning Point

Figure 5-18 illustrates a differential survey that requires one turning point. For this example the distance between stations A and B is 545.7 feet.

In this example, a single setup of the instrument is not workable. The backsight on station A is not a problem, but the foresight to station B intersects with the ground before reaching station B. This situation requires the use of at least one turning point. The first step is to determine the number of turning points that are needed. An evaluation of the site indicates the survey can be completed with one turning point. The next step is to determine the best location. In this example, the turning point was placed at the bottom of the slope.

Next, the instrument is set up halfway between station A and the turning point, and leveled. The first rod reading is a backsight on station A, Figure 5-19.

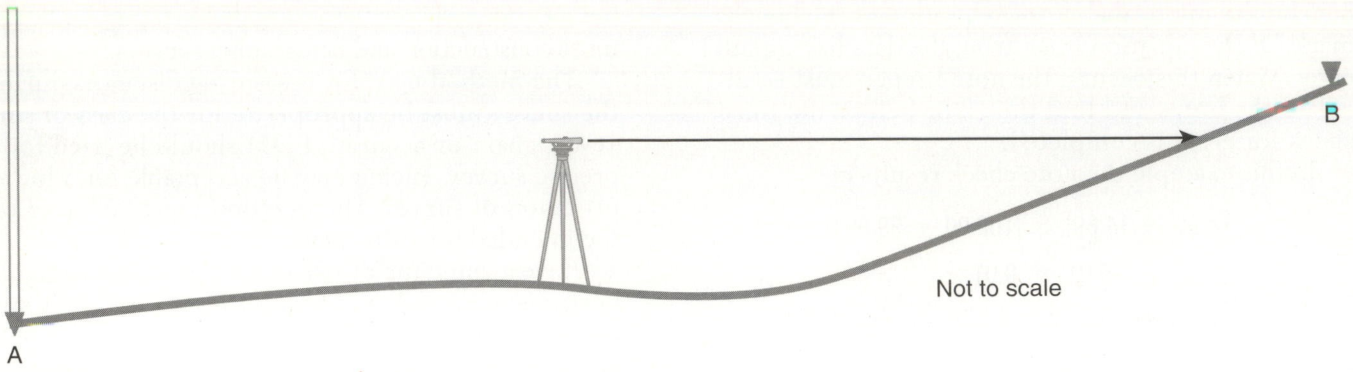

Figure 5-18 Survey requiring at least one turning point.

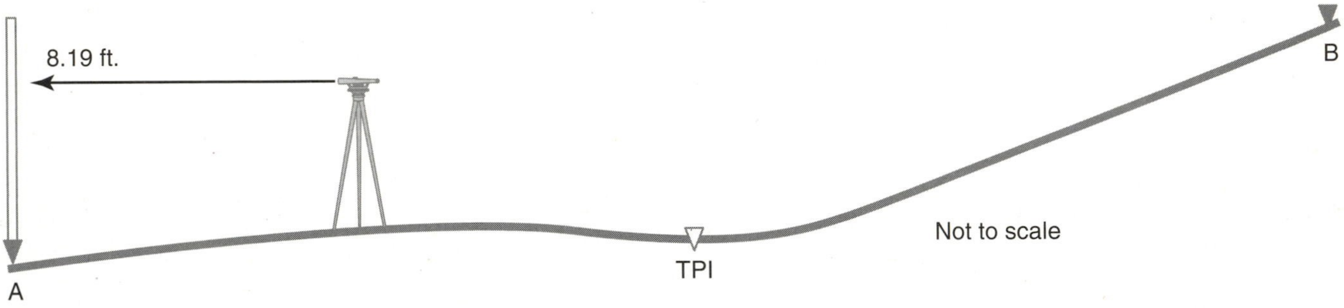

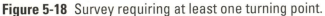

Figure 5-19 Backsight on station A.

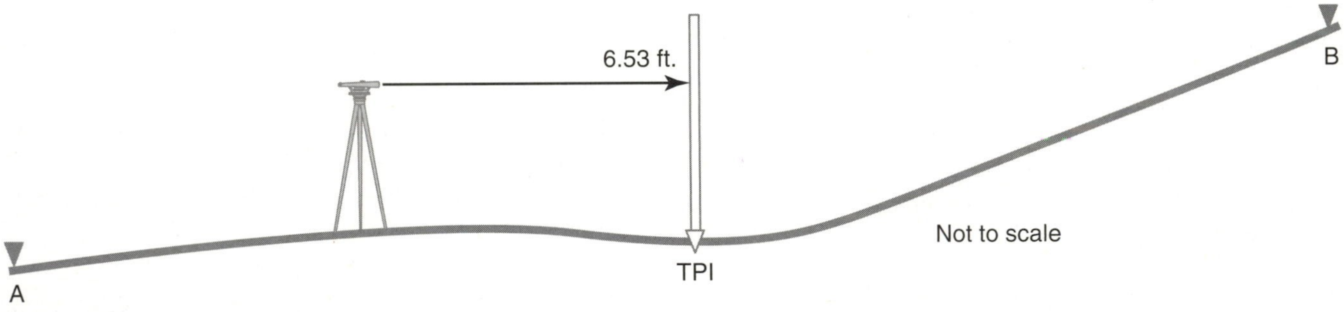

Figure 5-20 Foresight on turning point.

Table 5-4 Differential example with backsight and height of instrument

STA	BS	HI	FS	ELEV
A	8.19	108.19		100.00

This rod reading is recorded in the appropriate cell in the data table and used to determine the height of the instrument, Table 5-4. In this example, an elevation of 100.0 feet is assumed for station A.

The next step is to rotate the instrument and record the foresight on the turning point, Figure 5-20.

The rod reading of 6.53 ft. is recorded in the turning point one (TP1) row of the foresight column and used to determine the elevation of turning point one, 101.66 feet, Table 5-5.

Next, the instrument is moved to a location approximately halfway between turning point one and

Table 5-5 Differential example with foresight and elevation

STA	BS	HI	FS	ELEV
A	8.19	108.19		100.00
TP			6.53	101.66

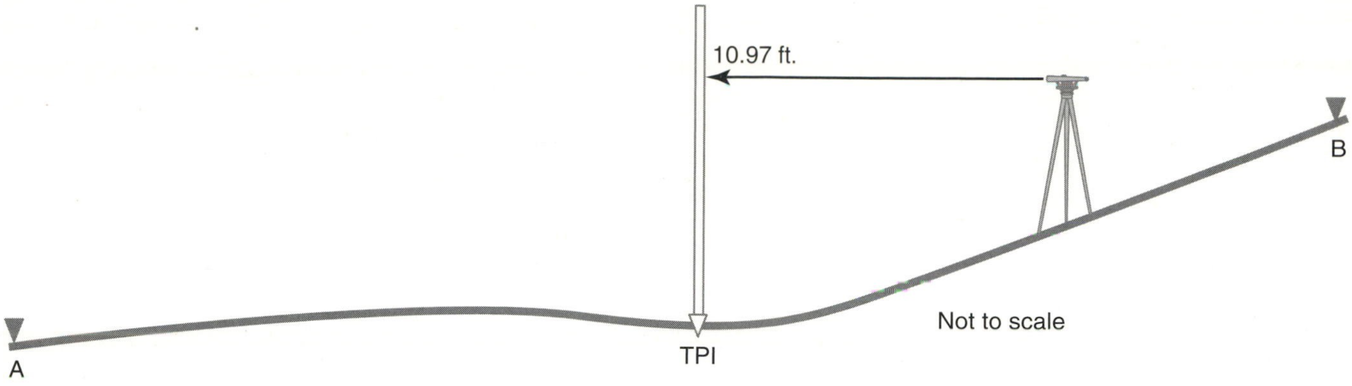

Figure 5-21 Backsight on turning point one.

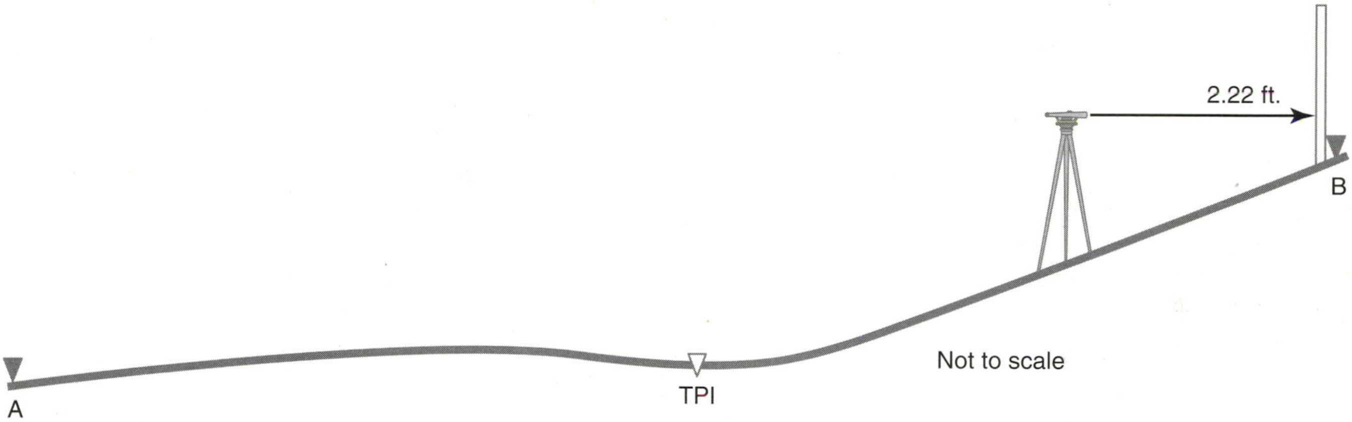

Figure 5-22 Foresight on station B.

Table 5-6 Differential example with backsight and turning point elevation

STA	BS	HI	FS	ELEV
A	8.19	108.19		100.00
TP1	10.97	112.63	6.53	101.66

station B. Each time the instrument is moved, it must be leveled and a backsight recorded to reestablish the height of the instrument, Figure 5-21.

The reading of 10.97 ft. is recorded in the TP1 row of the backsight column and is added to the elevation of TP1, 101.66 ft., to establish the height of the instrument, 112.63 ft., Table 5-6.

The last sep to complete the survey is to rotate the instrument and record the foresight on station B, Figure 5-22.

The reading of 2.22 feet is recorded in Table 5-7 and subtracted from the height of the instrument, 112.63 ft., to determine the elevation of station B, 110.41 ft.

The elevation of station B has been determined, but the survey is not complete until the three checks for error are conducted. The first is closing the loop, Figure 5-23.

In this example of closing the loop, all of the information is included in one figure. In the field, each reading would be recorded in the backsight-foresight sequence as in the first portion of this example. Table 5-8 includes this information with the note check and check for allowable error.

Table 5-7 Differential example with foresight and station B elevation

STA	BS	HI	FS	ELEV
A	8.19	108.19		100.00
TP1	10.97	112.63	6.53	101.66
B			2.22	110.41

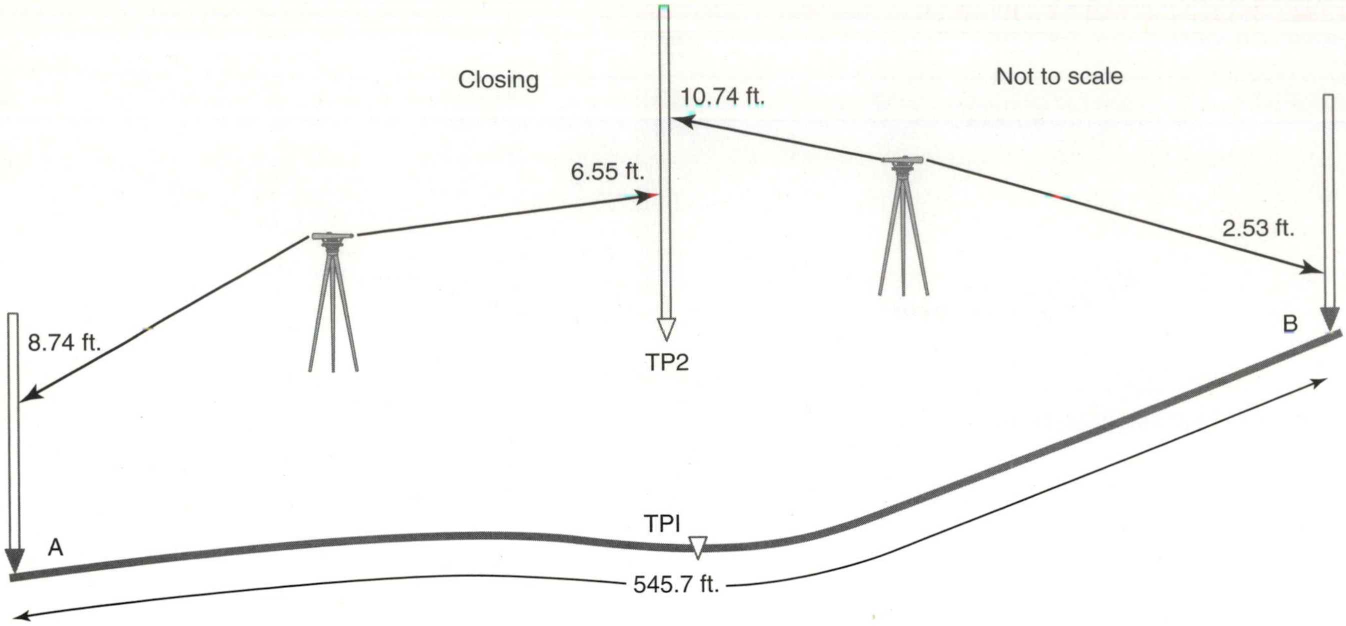

Figure 5-23 Closing loop.

Table 5-8 Completed notes for differential example

STA	BS	HI	FS	ELEV
A	8.19	108.19		100.00
TP1	10.97	112.63	6.53	101.66
B	2.53	112.94	2.22	110.41
TP2	6.55	108.75	10.74	102.20
A			8.74	100.01
Σ	28.24		28.23	

$$|28.24 - 28.23| = |100.00 - 100.01|$$
$$0.01 = 0.01$$
Notes acceptable
$$AE = K\sqrt{M}$$
$$= 0.1\sqrt{1091.4/5280}$$
$$= 0.04$$
$$0.01 < 0.04$$
Error of closure is acceptable

The closing loop shows the survey as an error of closure of 0.01 feet. This note check is true which means the error of closure is a surveying error, not a math error in the data table. The error of closure, 0.01 ft., is less than the allowable 0.02. The survey is acceptable.

Summary

Differential leveling is a comparison of the elevation for two or more stations. This principle forms the basis for several other types of surveys. In the next chapter, we will learn that a profile survey is a differential survey with additional foresights. Differential leveling is also used in topographic surveys and traverses. These will be discussed in the following chapters.

Student Activities

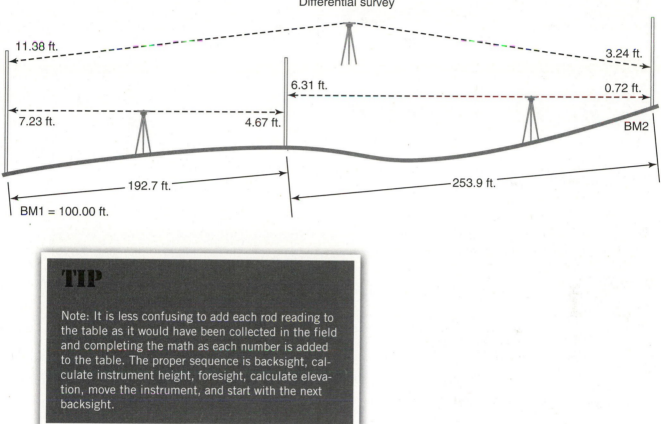

1. Complete the notes for the differential survey in the illustration.

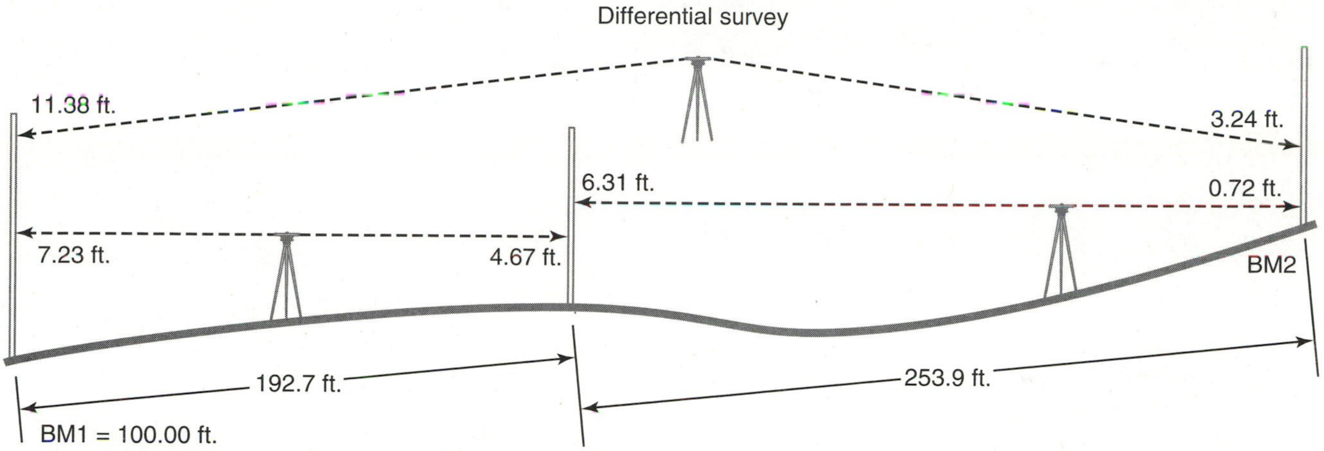

> **TIP**
>
> Note: It is less confusing to add each rod reading to the table as it would have been collected in the field and completing the math as each number is added to the table. The proper sequence is backsight, calculate instrument height, foresight, calculate elevation, move the instrument, and start with the next backsight.

2. Determine the appropriate number for the table cells labeled A through F in the following table.

STA	BS	HI	FS	ELEV
BM1	A	964.75		960.00
TP1	1.12	B	3.20	961.55
TP2	3.62	956.26	C	952.64
BM2	D	961.37	2.97	953.29
TP3	4.69	964.78	1.23	E
BM1			4.77	960.01
Σ	F		22.25	
		0.01	=	0.01

CHAPTER 6

Profile Leveling

Objectives

After studying this chapter, the reader should be able to:

- Explain the difference between true and intermediate foresights.
- Explain the use of three-wire reading, double reading, and plunging the telescope.
- Keep profile notes with and without an intermediate foresight column.
- Plot profile data on a graph.

Terms to know

profile leveling
intermediate foresights
plunging the telescope
spot elevations

INTRODUCTION

A profile is a cross section or side view. **Profile leveling** establishes a side view or cross sectional view of the earth's surface. The primary use of profile surveys is for surveying for routes such as sidewalks, streets, fences, retaining walls, and utilities. The data generated by a profile survey can be used to determine the slope between points, depths of trenches, ground clearance for overhead power lines, and many other applications.

Profile leveling has many similarities to differential leveling. Both profile and differential leveling use backsights, foresights, benchmarks, and turning points. The primary differences are the use of intermediate foresights (IFS) and using distances for station identification.

INTERMEDIATE FORESIGHTS

Intermediate foresights are rod readings on stations with unknown elevation just like the foresights in differential surveying, except they are rod readings at stations which will not be used as turning points or benchmarks. The function of intermediate foresights is to define the topography of the ground and locate man-made and natural features that are important to the survey. The addition of intermediate foresights increases the complexity of the data table in the field book and the opportunity for errors.

Intermediate foresights are not used during the note check. If they are included in the note check, the equation will always be false, indicating a math error in the data table when no math error exists. Therefore, when the profile data is recorded the intermediate foresights must be distinguishable from the true foresights.

Two different methods are commonly used to sort the foresights in a data table. One method includes adding a column to the notes. In this method an intermediate foresight (IFS) column is included in the notes and all of the intermediate foresights are recorded in this column. This column is not included in the note check. In this method, the foresights are separated in the field as they are being recorded. Table 6-1 is an example of the method that adds an additional column.

In the second method used to sort foresights, the intermediate foresights are recorded in the same column with the true foresights. When summing the foresight column during note check, the intermediate foresights are ignored, and only the true foresights are added. One technique that will help do this correctly is placing the true foresights in a box. Table 6-2 is an example of this method. Note that the true foresights are stations that are paired with backsights. The note keeper must remember that the first backsight and the last foresight are also a pair. For the example, in Table 6-2 the backsight of 4.3 feet on station 0.0 that starts the survey and the foresight of 8.0 feet on station 0.0 that closes the survey are a pair. The 8.0 foot foresight on station 0.0 is included in the note check.

When the single column method is used, the numbers in the boxes are the only foresights that are included for the note check. Either method will work. The key is to be able to distinguish the difference between intermediate and true foresights. In the first method the foresights are separated in the field as the data is being collected. When the second method is used, the sorting can occur in the field or during the note check.

Table 6-1 Profile data table with IFS column

STA	BS	HI	FS	IFS	ELEV
0.0	4.3	104.3			100.0
67.9				3.6	100.7
100.7				2.7	101.6
145.8	6.2	108.5	2.0		102.3
178.4				4.6	103.9
235.0				5.8	102.9
345.6	3.6	107.6	4.5		104.0
TP2	2.3	108.0	1.9		105.7
0.0			8.0		100.0
Σ	16.4		16.4		

0.0 = 0.0

Notes Ok

$$0.1\sqrt{\frac{345.6 \times 2}{5280}} = 0.04$$

$0.0 < 0.04$

Survey acceptable

Table 6-2 Profile data table without IFS column

STA	BS	HI	FS	ELEV
0.0	4.3	104.3		100.0
67.9			0.0	100.70
100.7			2.7	101.6
145.8	6.2	108.5	2.0	102.3
178.4			4.6	103.9
235.0			5.6	102.9
345.6	3.6	107.6	4.5	104.0
TP2	2.3	108.0	1.9	105.7
0.0			8.0	100.0
Σ	16.4		16.4	

0.0 = 0.0
Notes Ok

$$0.1\sqrt{\frac{345.6 \times 2}{5280}} = 0.04$$

$0.0 < 0.04$
Survey acceptable

Table 6-3 Data table for three-wire leveling

STA	BS	HI	FS	IFS	AVG	ELEV
0.0	4.5	104.5				100.0
100.2				5.6		
				4.1		
				2.7	4.1	100.4
145.8				6.2		
				4.3		
				2.3	4.3	100.2
200.3	5.0	104.2	5.3			99.2
0.0			4.2			100.0
Σ	9.5		9.5			

9.5 − 9.5 = 100.0 − 100.0
0.0 = 0.0
Notes Ok

$$0.1\sqrt{\frac{200.3 \times 2}{5280}} = 0.03$$

$0.0 < 0.03$
Survey acceptable

The second problem with intermediate foresights is insuring their accuracy. In differential leveling, each station is used as a backsight and a foresight. This acts as a check for accuracy because a rod reading error will result in excessive error of closure. Excessive error of closure means the survey must be repeated, but at least the error is discovered before the data is used for additional calculations or for designing a project. Intermediate foresights are not included in the note check; therefore, any errors in the intermediate rod readings will not be detected. Errors in intermediate foresight rod readings will result in incorrect station elevations. Such errors can be very costly if they are not caught before the data is used.

Error Management when Using Intermediate Foresights

Several techniques have been developed to reduce the chance of rod reading error when using intermediate foresights. The three common techniques are three-wire leveling, multiple reading, and plunging the telescope. It is important to remember that although these methods are recommended for intermediate foresights, they can be used with any rod reading when extra diligence in managing errors is warranted.

Three-Wire Leveling

Three-wire leveling, also called three-wire reading, gets its name because it utilizes the elevation crosshair as well as the two stadia crosshairs. In this method all three horizontal crosshairs are read and averaged. The average is used as the elevation for the station. See Table 6-3.

In Table 6-3, the elevation of 100.4 feet for station 100.2 is determined by subtracting the average of 5.6, 4.1, and 2.7 from the height of the instrument, 104.5.

This method provides a quick check for accuracy because the top and bottom stadia crosshairs are an equal distance from the elevation crosshair. If the average of the three crosshairs does not equal the value of the center crosshair, there is a chance that an error has occurred. It must be remembered that measuring devices are not perfect and the level of precision is plus or minus one-half of the smallest unit. A variation in readings that is within this range is not considered an error. Assume an instrument that measures to $1/100^{th}$ of a foot, 0.01 feet, recorded the following readings:

TSR	7.42
Elevation	5.02
BSR	2.61

$$\frac{7.42 + 5.02 + 2.61}{3} = \frac{15.05}{3} = 5.01666\ldots \text{ or } 5.02$$

The reading is acceptable because the rounded average of the three readings is equal to the elevation of the center crosshair. Consider the next set of readings:

TSR	7.34
Elev	6.01
BSR	4.66

Is this data acceptable?

$$\frac{7.34 + 6.01 + 4.66}{3} = 6.00$$

$$6.01 - 6.00 = 0.01$$

In this example the average does not equal the elevation crosshair reading, but the data are acceptable because the error, 0.01, is equal to one unit of the precision of the data.

In this example, three-wire leveling was not used for the true foresights. If the requirements were for a high-precision survey, then three-wire leveling should also be used for the true foresights and backsights.

Multiple Reading

When the multiple reading is used, the instrument person reads the rod and calls out the reading to the note taker. The instrument operator then rotates the telescope a few degrees to either the left or the right and then back to the rod. The rod is then read and recorded again. For precise surveys reading the rod ten or more times is not unusual. All the readings are recorded in the data table and averaged. For precise surveys the surveyor must also evaluate the consistency of the readings. All of the readings should be within the least count of the rod. For example, if the rod is read directly the least count is 0.01 feet. Rod readings of 5.03 and 5.02 would have an acceptable consistency.

In this example, multiple reading was not used for the true foresights or backsights. If this survey was required to meet the standards for a high-precision survey, multiple reading should also be used for the true foresights and backsights.

Plunging the Telescope

Plunging the telescope requires a transit, or total station. In this method the rod is read with the telescope level and in the standard position. Then the telescope is plunged, vertically rotated 180°, which also requires rotating the instrument horizontally 180°, and the rod is read the second time. Both readings are recorded in the notes. The operator must use the vertical angle scale to carefully set the vertical angle to 0 or 180 degrees, depending on which one the instrument uses for horizontal zero. Failing to set the vertical angle correctly will cause an error in the plunged rod reading. For precise surveys the telescope can be plunged more that once.

Note: The numbers used in Table 6-4 and Table 6-5 are very similar. The reader would not be able to determine if the data was collected by double reading or plunging the telescope. The note keeper should indicate in the notes in the field book which method was used.

Also illustrated is the procedure that should be followed if the two readings are not the same. At station 203.8, Table 6-5, the foresight readings were 8.3 and 8.2. In this example the average is 8.25. If the average, 8.25, is subtracted from the instrument height, 106.3, the answer, 98.05, will have an additional decimal point. This indicates a false level of accuracy because all of the rod readings have an accuracy of a tenth of an inch, 0.1. The averaged number, 8.25, indicates the rod was read to two decimal points, 0.01, this is not true. The averaged rod reading should be reduced to one decimal place, 8.2, resulting in an elevation of 98.1. In this example the average was rounded to 8.2 following the rounding rule of leaving the last digit even when rounding off a 5.

Table 6-4 Data table for double reading

STA	BS	HI	FS	IFS	ELEV
0.0	5.00	955.00			950.00
118.75				2.92/2.92	952.08
146.53				3.02/3.02	951.98
203.47	6.46	960.11	1.35		953.65
333.33				2.19/2.19	957.92
400.00	1.53	958.82	2.81		957.30
TP2	3.03	956.92	4.93		953.89
0.0			6.92		950.00
Σ	16.1		16.1		

$$16.1 - 16.1 = 950.00 - 950.00$$
$$0.0 = 0.0$$
Notes Ok
$$0.1\sqrt{\frac{4010 \times 2}{5280}} = 0.04$$
$$0.0 < 0.04$$
Survey acceptable

Table 6-5 Data table for double reading

STA	BS	HI	FS	IFS	ELEV
BM	6.3/6.3	106.3			100.0
0.0				3.2/3.2	103.1
102.5				4.5/4.5	101.8
203.8	3.5/3.5	101.6	8.3/8.2		98.1
456.0				2.6/2.6	99.0
521.0	5.8/5.8	104.20	3.2/3.2		98.4
TP2	2.9/2.9	102.3	4.8/4.8		99.4
BM			2.2/2.2		100.1
Σ	18.5		18.4		

$$|18.5 - 18.4| = |100.00 - 100.10|$$
$$0.1 = 0.1$$
Notes Ok
$$0.1\sqrt{\frac{521.00 \times 2}{5280}} = 0.04$$
$$0.1 < 0.04$$
Survey acceptable

Selecting Intermediate Foresights

The purpose of intermediate foresights is to define the location and elevation of stations that are important to the design of the route. It is critical that all of the stations important to the survey are identified and measured, but marking and measuring unnecessary points is a waste of resources. The route must be evaluated with the intended purpose of the survey in mind. An inexperienced surveyor will find a hand level useful for evaluating the route. It can be used to determine changes in elevation and slope. Knowing these two factors helps in determining the placement of intermediate foresights and turning points.

One method of selecting intermediate foresights is called stationing. When the stationing method is used, a station is established at a uniform interval, every 100 feet for example, and then additional stations are included to locate the important features that fall between the standard stations. Another method is to study the route and just collect data on the important features. The question is "What additional features should be recorded?" The best answer is "It depends on the use of the data." Profiles are usually used to collect data for a route; therefore changes in elevation are important. The amount of change that is important is dependent on the use of the data. If the purpose of the survey is to plan the route for an overhead power line, small changes in elevation such as six inches would not need to be recorded, but six foot changes in elevation would be. A survey for an underground pipe drain would have different requirements. Elevation changes of six inches or less would be important. For an underground drain it would be important to identify the position and elevation of large rocks or rock ledges. The surveyor must learn how to determine the important features of the route without recording excessive data. Efficiency in selecting stations for profile leveling improves with experience.

Figure 6-1 and Figure 6-2 illustrate the effect the number of stations can make on the plot of the profile. Both graphs use the same data. The graph in Figure 6-1 was drawn including stations with six inches or more change in elevation. Figure 6-2 uses the same data, but it only includes changes in elevation of one foot or more. The difficult part of selecting intermediate foresights is that both graphs could be correct. It all depends on the use of the data.

Spot Elevations

Spot elevations are rod readings taken to record natural features such as rocks and trees, and man-made structures, such as sidewalks and buildings, that have the potential to affect the design or cost of the construction. Spot elevations are intermediate foresights and the appropriate methods must be used to insure they are located and their elevation recorded with the same level of precision and accuracy.

Station Identification

The second primary difference between differential and profile leveling is the technique used to identify the stations. Station identification is not an issue for differential surveys because the purpose of the survey is to determine the difference in elevation between two or more points. In differential surveys, any stations used between the starting and ending point are only used to extend the survey. These stations can be identified by any organized system of numbers, letters, or a combination of letters and numbers. It is important that the identification system is clear in the data table and agrees with the location of the stations in the sketch.

More attention must be given to identifying stations in profile surveys because profile surveys are usually plotted on an XY chart. To plot or chart the

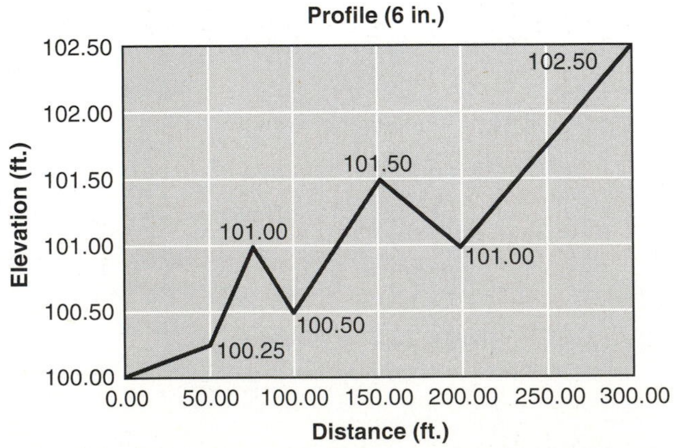

Figure 6-1 Profile recording 6 inch elevation changes.

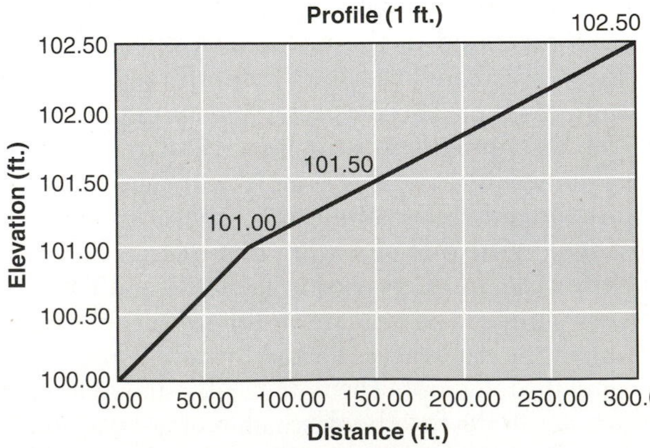

Figure 6-2 Profile recording 1 foot elevation changes.

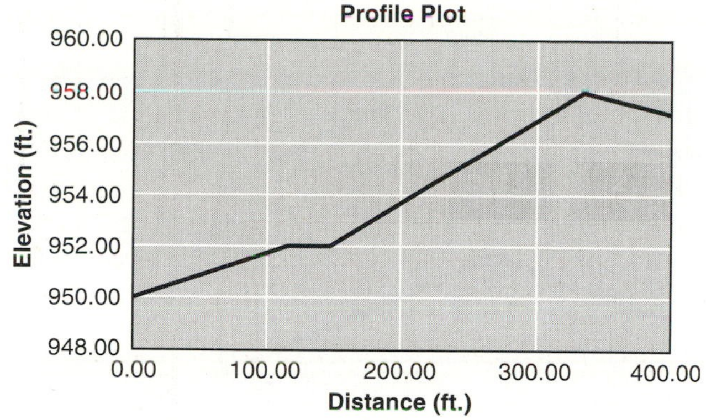

Figure 6-3 Example of plotted profile data.

data the distance of each station from the starting point is the value used on the X axis of the chart. A traditional method is to record the distance as the number of full chains plus the partial chain. Using this method 132 feet would be written as 1+32. This method is difficult to use with calculators and spreadsheets. When calculators and/or spreadsheets will be used, distances are recorded in decimal feet.

A two-dimensional graph requires data for both the vertical (Y) and horizontal (X) axis. When profile data is drawn on a graph, the elevation is used for the Y-axis and the distance from the starting point to the station is used for the X-axis. Graphs are very useful during design of the route to determine cuts and fills, maximum depths, minimum depths, slopes, etc. Figure 6-3 is an illustration of a profile graph.

DEFINING MAN-MADE AND NATURAL FEATURES

If the presence of man-made and natural structures affects the design or cost of the construction, they must be located and have their elevation recorded. One of the advantages of electronic surveying instruments is that some have the capability of capturing a digital picture of these structures and attaching the picture to the station information. The decision to include or exclude a structure must be based on the use of the data. Assume the purpose of the survey is to define the route for a sidewalk that will connect to two existing sidewalks. In this situation, the location and elevation of the existing sidewalks would be very important information. The existence of the same two sidewalks may be insignificant information if the purpose of the survey was to define the route for a new road and the route will be cleared by heavy equipment before construction begins. Surveys that require the location, size and elevation of man-made structures are more time-consuming because additional stations must be recorded to define these structures. The number of the stations that will be required will depend on the structure and the precision of the survey. The same concerns and procedures are true for natural features as ditches, hills, depressions, rocks, and trees.

To illustrate these points consider the number of stations that will be required to define a sidewalk. Using the criteria of location, size, and shape, a sidewalk will require at least two stations, one at each edge along the route. Determining the number of stations required for complex features, such as a road or ditch, is more complicated. This is illustrated in Figure 6-4.

For this simple ditch, at least five stations will be required. The size and cross section shape of the ditch will greatly influence the number of stations. Consider Figure 6-5.

In this example, at least nine stations will be required. These two illustrations demonstrate that a small change in the cross sectional shape of a ditch has had a dramatic effect on the number of stations that are required. In these two examples the number almost doubled. If a high-precision survey was required, the number of stations could easily double again.

The same consideration must be given to other man-made and natural structures. The surveyor must have a good understanding of the required precision and the use of the data to determine the features that should be defined, and the number of stations for each one.

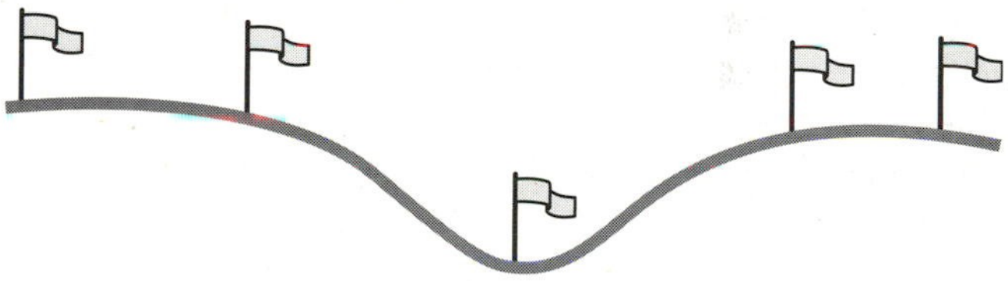

Figure 6-4 Number of stations for a simple ditch cross section.

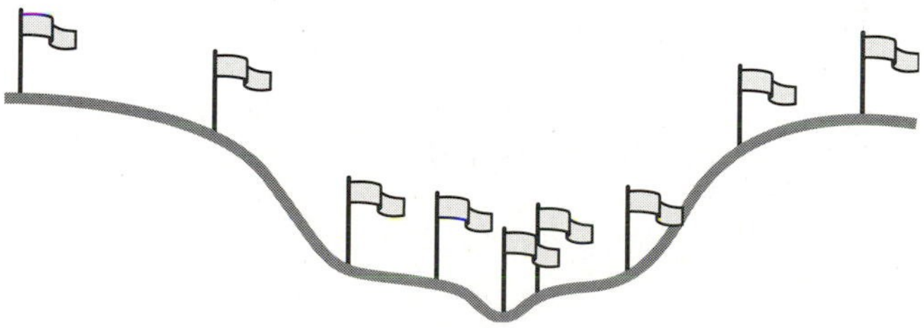

Figure 6-5 Number of stations for a more complex ditch cross section.

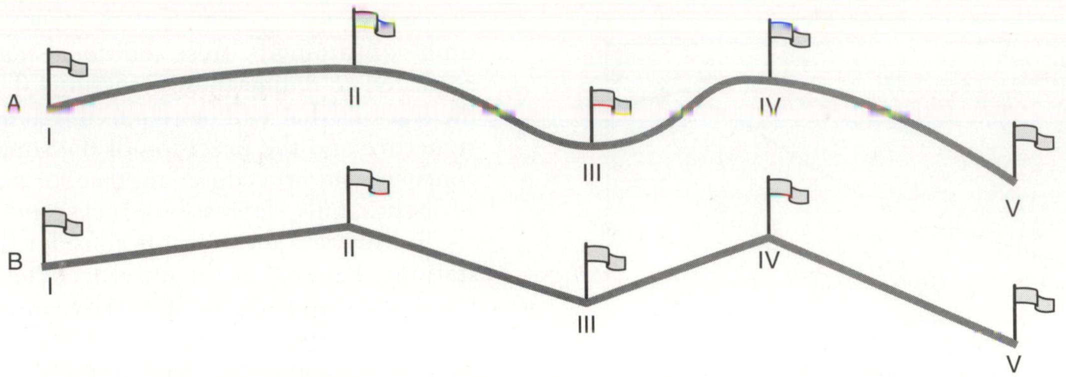

Figure 6-6 Effects of plotting data.

In addition, it is important to remember that when plotting or graphing profile data the lines are drawn dot to dot, not curved. The surface between any two points is treated as a plane. In Figure 6-6 if stations I, II, III, and IV on curve A are the only points the surveyor used to record the topography of the route, the plot of the data will look like line B. All features important to the profile must be identified and measured. Figure 6-6 shows that if the location and elevation of a structure are not recorded, they will not be included in the plot or any design work using the data.

PROFILE LEVELING

Before the instrument is removed from the case, someone must determine the precision required for the survey, the allowable error (K value), walk the route, and stake or flag every station. At this time the need for and location of turning points should also be determined. A hand level will aid in making these decisions. Organizing the route can require a lot of time and may seem unnecessary, but it is very important that the survey be well organized and proceeds in a logical manner. Otherwise there is a higher probability that data will be recorded incorrectly and an increased chance that other mistakes will be made.

After the stations have been identified, the distance from the starting point to each station is measured and recorded as the station identification. As in differential leveling the sights should be balanced, if possible. It is not practical to balance the intermediate foresights because this would require moving the instrument for every intermediate foresight. The sights for the backsights and true foresights should be balanced.

The first rod reading is taken at the benchmark or the first station of the route. The first station on the route may or may not be the benchmark, depending on the survey. This is explained in more detail in the next section.

Profile Benchmarks

The first step in a profile survey is the establishment of the benchmark. When a drainage survey is being completed, a culvert or other drainage structure that will not be affected by the construction can be used. If a structure is not available, a benchmark can be installed for the survey. In either case it is important to locate and describe the benchmark in the field notes.

There are two possible locations for the benchmark: along the route and to the side of the route. The selection of the benchmark location is determined by the survey. If the survey is for drainage purposes and a culvert, or other drainage structure, is part of the route and will not be disturbed during construction, then it can be used as a benchmark. If the survey is being conducted to construct a surface drain and none of the existing structures will be left undisturbed, then the benchmark must be established outside of the construction area. This decision is important because if the benchmark is damaged or destroyed anytime during the job, it cannot be used as a reference, and this can cause serious problems. An even worse scenario is if the height of the benchmark is accidentally changed and the change is undetected. This can lead to very costly mistakes. The location of the benchmark will also change how the data table is started and which values are included when plotting the profile.

Figure 6-7 is an example of profile leveling notes when the benchmark *was not* part of the route. When this data is plotted, the elevation of the benchmark would not be included in the profile because it does not represent an elevation that is part of the route.

In this situation, the backsight on BM1 would be the first row of the data table and the rod reading at station 0.0 would be an intermediate foresight. Station 0.0 would be the first point of the data plot, Table 6-6.

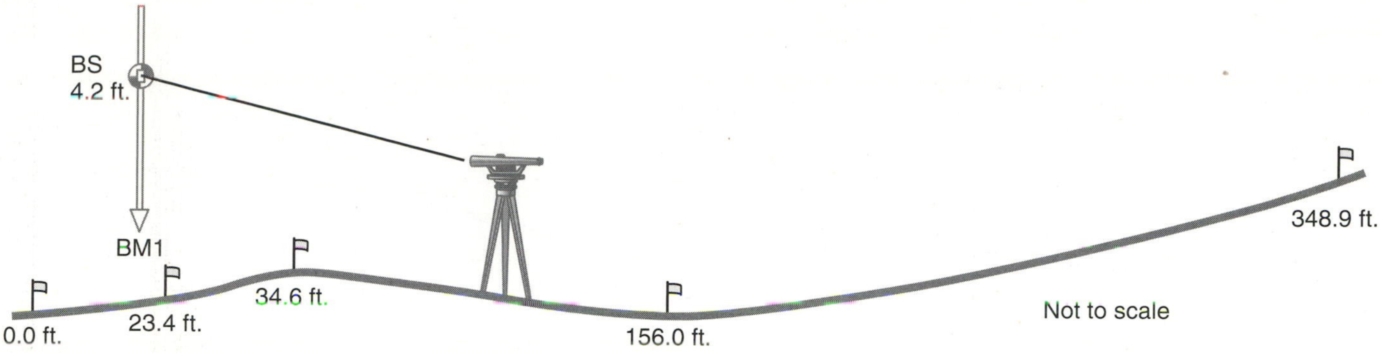

Figure 6-7 Profile with the benchmark off the route.

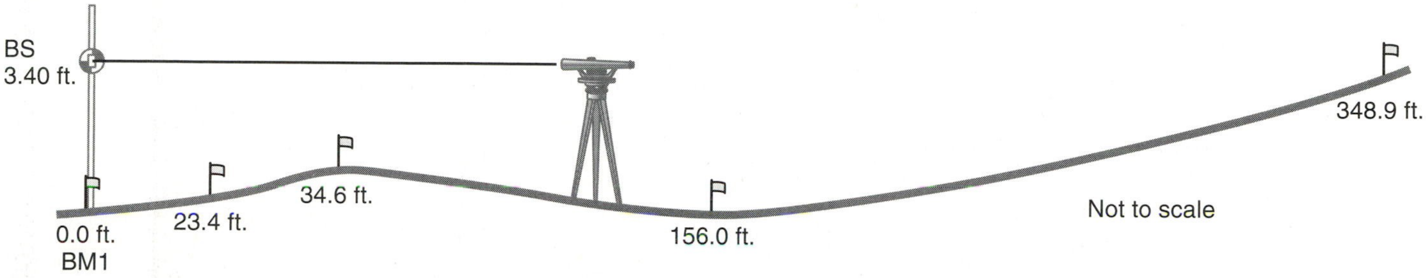

Figure 6-8 Profile with the benchmark as part of the route.

Table 6-6 Data table for benchmark off the route

STA	BS	HI	FS	IFS	ELEV
BM	4.2				
0.0				xxx	

Table 6-7 Data table for benchmark on the route

STA	BS	HI	FS	IFS	ELEV
0.0	3.40				
23.4				xxx	

Figure 6-8 is an example of profile survey data when the starting point (station 0.0) is also used as the benchmark.

In this situation, the survey is started with a backsight on station 0.0 and this rod reading is recorded on the first row of the data table, Table 6-7. The first intermediate foresight is on station 23.4. The elevation of station 0.0 would be the first point on the plot of the profile.

PROFILE EXAMPLE

Figure 6-9 is an illustration of a profile survey. In this illustration all of the data is presented at one time. When using this type of illustration, it is very helpful to visualize the process and add the measurements to the data table one number at a time, as if the survey were being conducted in the field.

In this example, the starting point (0.00) is also the benchmark.

Note: If stations 118.75, 146.53, and 333.33 were removed from the profile, the remaining stations would form a differential survey. These stations are the intermediate foresights. The illustration also shows that station 203.47 is also used as a turning point. We can also determine that station 203.47 is a turning point by analyzing the data. In the data table the row for station 203.47 has a BS value of 6.46. Backsights are rod readings on a point of known elevation, usually either a benchmark or a turning point. Station 203.47 is not a benchmark; therefore, it must be a turning point. Note that Table 6-8 also illustrates a survey that used double reading for the intermediate foresights.

The checks for error are the same for profile surveys as for differential. Surveying back to the benchmark closes the loop, the note check is completed, and the allowable error of closure is calculated. Figure 6-10 is the illustration of the closing loop for the example profile.

Table 6-8 Data table example profile

STA	BS	HI	FS	IFS	ELEV
0.0	5.00	955.00			950.00
118.75				2.92/2.92	952.08
146.53				3.02/3.02	951.98
203.47	6.46	960.11	1.35		953.65
333.33				2.19/2.19	957.92
400.00			2.81		957.30

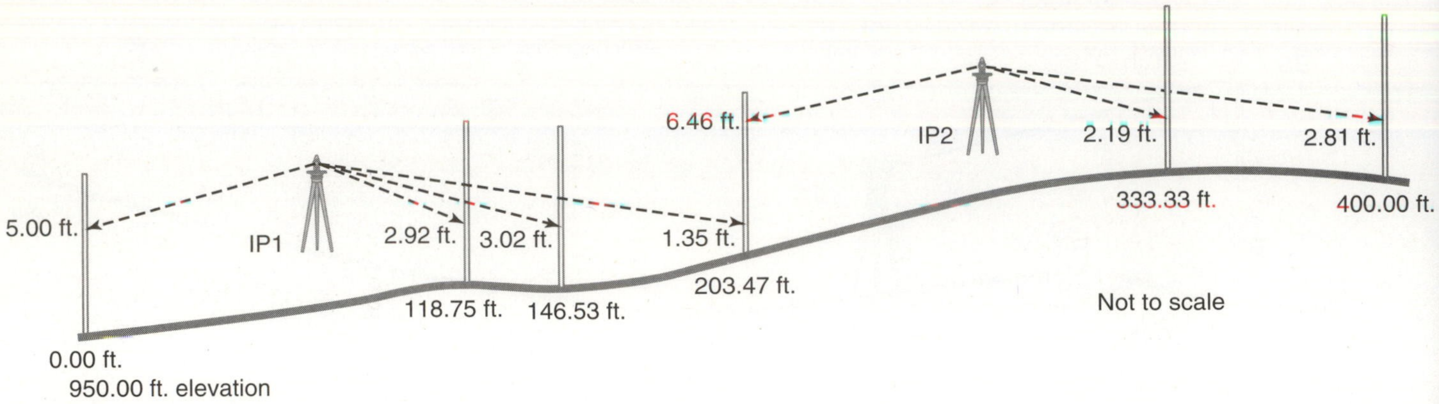

Figure 6-9 Profile example.

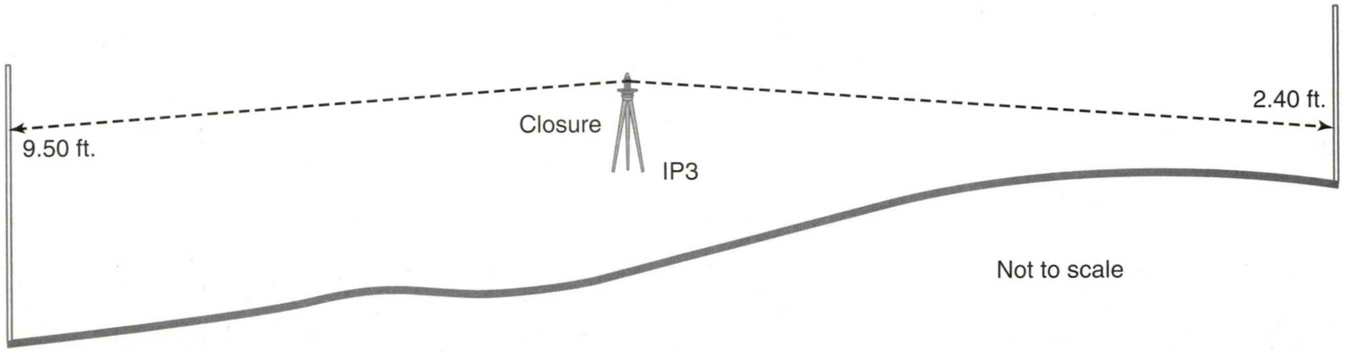

Figure 6-10 Closing loop for example profile.

Table 6-9 Data table for example problem with error checks

STA	BS	HI	FS	IFS	ELEV
0.0	5.00	955.00			950.00
118.75				2.92/2.92	952.08
146.53				3.02/3.02	951.98
203.47	6.46	960.11	1.35		953.65
333.33				2.19/2.19	957.92
400.00	2.40	959.97	2.81		957.30
0.0			9.50		950.20
Σ	13.86		13.66		

$$|13.86 - 13.86| = |950.00 - 950.20|$$

$$0.20 = 0.20$$

Notes Ok

$$AE = 0.1\sqrt{\frac{400.00 \times 2}{5280}}$$

$$= 0.02$$

$$0.20 > 0.02$$

Unacceptable

In this example, an alternative route was used for closure that did not require a turning point. This is an acceptable practice as long as none of the principles of surveying are violated. Once the closing loop is completed, the data table can be completed.

In this example for a K factor of 0.1, the actual error of closure is greater than the allowable error of closure. This means the data is not usable and the survey must be repeated. This can be a very costly situation. Using an instrument and traditional methods, a survey of only 400 feet could have involved several people for a day. All that effort was wasted because proper care was not taken to manage the errors.

PLOTTING PROFILE DATA

An experienced surveyor can look at a data table and interpret the results, but not all people have this ability. A common practice is to plot/draw the data on a graph. This provides a visual representation of the data. The proposed route can also be added to the graph.

Turning points can pose a problem. If the turning point is a stake or some type of structure that does not represent the elevation of the earth at that point, it should not be used as part of the profile when it is plotted. If the object used as a turning point has the same elevation as the surface, then it can be used as a point in the profile. For this reason the position of the turning points should be noted in the sketch and the information should describe the structure that was used for the turning point.

The following table is the profile data for a proposed drainage ditch. The purpose of the example is to design a drainage ditch with a 1% slope and a starting elevation 2.0 feet lower than the elevation of the benchmark.

Plotting the Profile

Table 6-10 contains sufficient information to design the ditch, but plotting this data gives a better visual reference for interpreting the drainage data, Figure 6-11.

The visual representation of the data in Figure 6-11 makes is much easier to see the variability of the topography and from the plot it should be evident that the variability in the topography may make it very difficult to install the drainage ditch. The feasibility of the design is evident when the proposed drainage ditch is added to the graph, Figure 6-12.

Table 6-10 Data table for drainage example

STA	BS	HI	FS	IFS	ELEV
BM	8.6	948.6			940.0
0.0		948.6		10.6	938.0
245.4		948.6		6.3	942.3
367.3	3.5	950.1	2.0		946.6
445.6		950.1		7.8	942.3
492.5		950.1		7.8	942.3
512.6		950.1		7.0	943.1
534.2	4.6	948.5	6.2		943.9
BM			8.5		940.0
Σ	16.7		16.7		0.0
Diff		0			
			0 = 0		

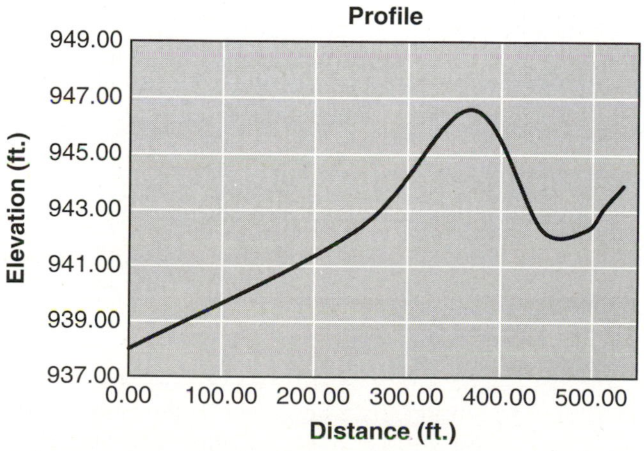

Figure 6-11 Plot of drain example profile.

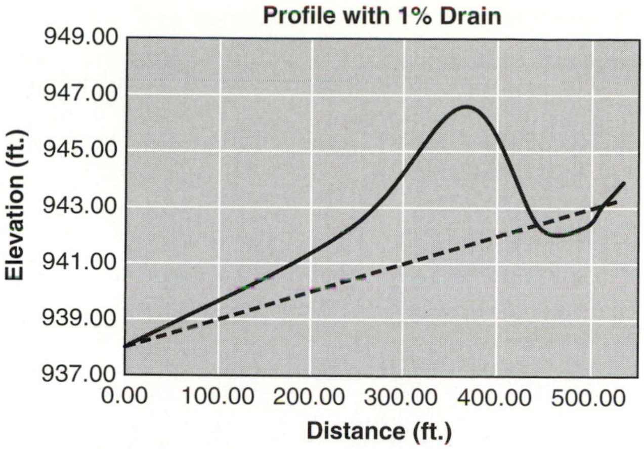

Figure 6-12 Profile with ditch at 1% slope.

Table 6-11 Calculations for drain

STA	STA Elev.	Drain Elev.
0	938	938.0
245.4	942.3	940.5
367.3	946.6	941.7
445.6	942.3	942.5
492.5	942.3	942.9
512.6	943.1	943.1
534.2	943.9	943.3

By studying Figure 6-12 it should be evident that the intended design is not feasible because the ditch will be above the ground at the 445.6 foot mark. Table 6-11 contains the data for plotting the drain. For a drain with a positive slope, the starting point for the drain is the lowest elevation; the drain elevations are determined by the equation:

$$\text{Drain Elev.} = \text{Starting elevation} + \left(\text{Station Distance} \times \frac{\%\ \text{slope}}{100}\right)$$

For a drain with a negative slope, the starting point for the drain is the highest elevation; the drain elevations are determined by the equation:

$$\text{Drain Elev.} = \text{Starting elevation} - \left(\text{Station Distance} \times \frac{\%\ \text{slope}}{100}\right)$$

For this example the drain starts at the lowest elevation; therefore at station 245.4 the drain elevation is:

$$\begin{aligned}\text{Drain Elev.} &= 938.0 + \left(\text{Station Distance} \times \frac{\%\ \text{slope}}{100}\right)\\ &= 938.0 + \left(245.4 \times \frac{0.01}{100}\right)\\ &= 938.0 + (245.4 \times 0.01)\\ &= 938.0 + 2.454\\ &= 940.5\end{aligned}$$

The remaining drain elevations are calculated using the same process.

This example demonstrates one of the advantages of computers because a spreadsheet can be designed to change the percent slope of the ditch and, through the charting function, see the immediate result.

An alternative design would be to change the slope of the ditch to 0.5%, Figure 6-13.

Figure 6-13 shows that a ditch slope of 0.5% is acceptable. Table 6-12 contains the ditch elevation for 0.5% slope.

The ditch elevation for station 245.4 is determined by:

$$\text{Drain Elev.} = 938.0 + \left(\text{Station Distance} \times \frac{\%\ \text{slope}}{100}\right)$$

$$= 938.0 + \left(245.4 \times \frac{0.005}{100}\right)$$

$$= 938.0 + (245.4 \times 0.005)$$

$$= 938.0 + 1.227$$

$$= 939.227 \text{ or } 939.2$$

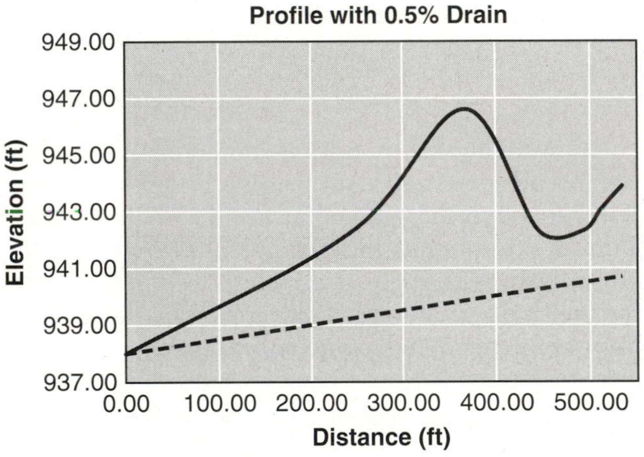

Figure 6-13 Profile with ditch at 0.5%.

Table 6-12 Data for ditch with 0.5 slope

STA	Profile Elev.	Drain Elev.
0	938	938.0
245.4	942.3	939.2
367.3	946.6	939.8
445.6	942.3	940.2
492.5	942.3	940.5
512.6	943.1	940.6
534.2	943.9	940.7

The data and graphs can be used to answer many questions about the construction of the ditch. Such as, "What will be the maximum depth of the ditch?" or, "What will be the minimum depth of the ditch?" can be answered.

Note that the scales for the X and Y axes are not equal. It is a common practice to exaggerate the vertical scale so changes in elevation are magnified.

With the use of spreadsheets it is easy to run other scenarios, such as the profile of the ditch if a slope of 0.75% is used. If a computer and spreadsheet are not available, the graphs can be completed by hand. The difference is that the calculations must be completed using a calculator, and the graph must be drawn by hand. See Appendix 1 for further instructions on drawing a graph by hand.

Profiles with Angles

In all of the examples presented in this chapter, the profiles have been in a straight line. Profile surveys are not limited to straight lines. When an angle is encountered along a route, it should be measured and noted in the field book. The angle may, or may not, be noted on the plot of the profile. It depends on the importance of the angle to the route. The types of angles used in surveying and their measurement are included in Chapter 7.

Summary

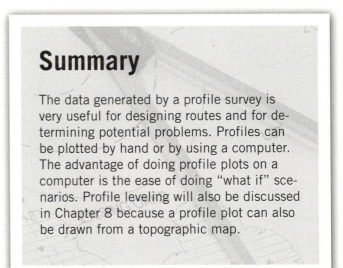

The data generated by a profile survey is very useful for designing routes and for determining potential problems. Profiles can be plotted by hand or by using a computer. The advantage of doing profile plots on a computer is the ease of doing "what if" scenarios. Profile leveling will also be discussed in Chapter 8 because a profile plot can also be drawn from a topographic map.

Student Activities

1. Complete the notes for profile leveling survey in the illustration.

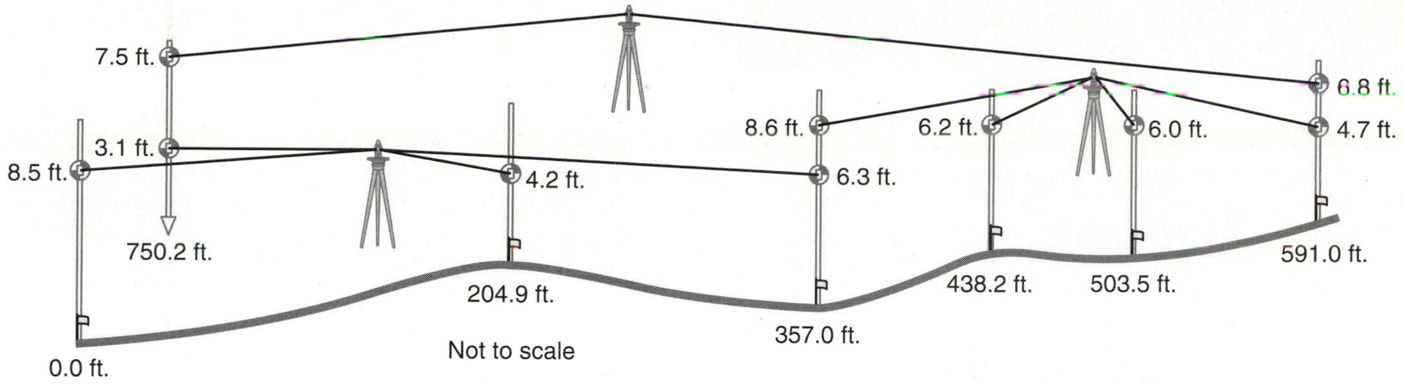

2. Draw or plot the profile. Leave room on the vertical axis for a point 6 inches below the elevation of 0.0.
3. Draw a drain on the profile. The drain will start at 6.0 inches below the elevation of 0.0 and have a slope of +0.5.

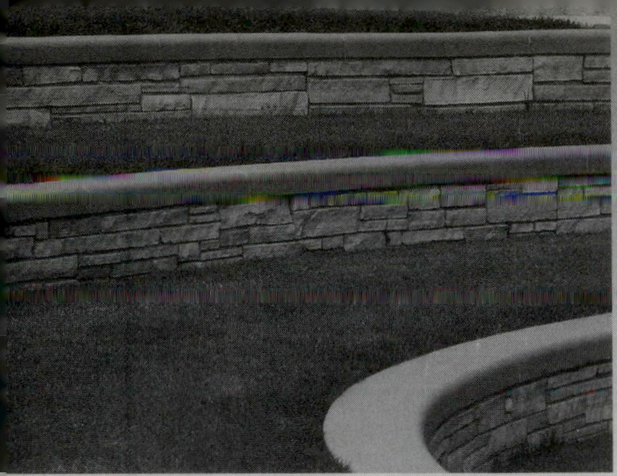

CHAPTER 7

Angles

Objectives

After reading this chapter the reader should be able to:

- Understand the difference between DD and DMS.
- Convert angles between DD and DMS.
- Explain the common indirect methods for layout angles.
- Layout an angle using each of the indirect methods.
- Know the difference between deflection and interior angles.
- Know the steps required to measure a horizontal angle using an instrument.
- Read a Vernier scale.
- Understand the difference between an azimuth and a bearing.
- Convert between azimuths and bearings.

Terms To Know

deflection angle
interior angle
chord
3-4-5 method
tape-sine method
sine
cosine
tangent
clockwise
counterclockwise
azimuth
bearing
least count
vernier
decimal degrees (DD)
degrees
minutes
seconds
DMS
quadrant
magnetic north
geographical north

INTRODUCTION

Angles are an integral part of site analysis and land measurement. Examples of their use include:

- Defining property boundaries
- Planning and laying out flowerbeds
- Designing fences and sidewalks
- Locating man-made and natural features
- Mapping
- Building retaining walls

This chapter will discuss the principles of angles and illustrate several of their uses.

ANGLE

An angle defines the rate at which two lines or planes diverge from a common point, as shown in Figure 7-1.

All angles have three parts. In surveying, the parts of an angle are called the backsight, the vertex, and the foresight. The backsight can be on a baseline, reference line, or a point that is used as zero angle, as shown in Figure 7-2.

The vertex is the point where the divergence of the lines starts. The foresight is the line or point that the instrument is rotated to.

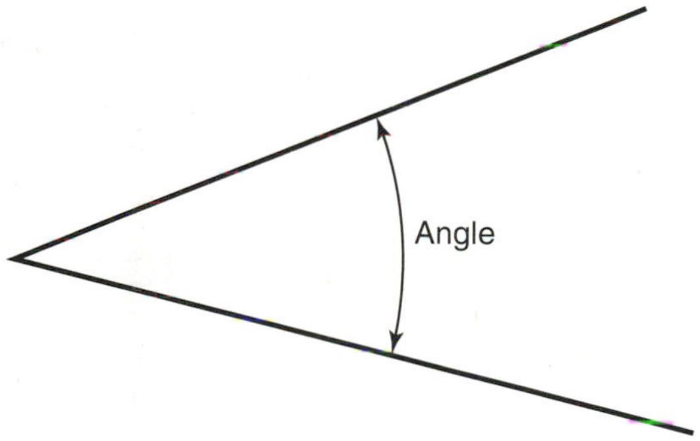

Figure 7-1 Angle.

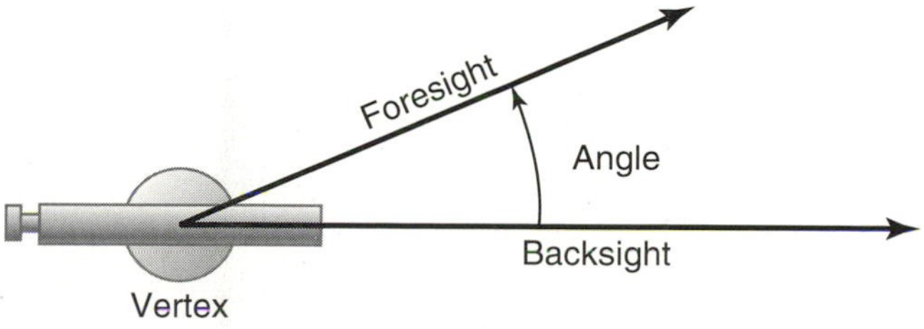

Figure 7-2 Parts of an angle.

DD AND DMS

Several different systems are used for recording angles. Two will be used in this text. They are decimal degrees (DD) and degrees-minutes-seconds (DMS). When DD are used, fractions of a degree are expressed as a decimal. When angles are measured using DMS, partial degrees are expressed as minutes (') and partial minutes are expressed as seconds ("). In the DMS system one degree equals 60 minutes and one minute equals 60 seconds. Electronic equipment can usually be set to use either form, but mechanical equipment typically use DMS. Angles in DD are usually preferred because they are easier to use in calculations and with computers.

Adding and Subtracting with DMS

One disadvantage of the DMS system becomes apparent when angles are used in calculations. For example, add the angles 20° 45' 27" and 30° 24' 35".

$$\begin{array}{r} 20^\circ\ 45'\ 27'' \\ +\ 30^\circ\ 24'\ 35'' \\ \hline 50^\circ\ 69'\ 62'' \end{array}$$

It is incorrect to have 69' and 62"; therefore, the number must be reduced. The first step is to subtract 60 from the seconds until the remainder is less than 60. In this example, only one subtraction is required leaving 2". This means one minute must be added to 69 changing it to 70. Next, 60 is subtracted from the minutes until the remainder is less than 60. In this example, only one subtraction is required, leaving 10'. This means one degree must be added to the degree column. The result is 51° 10' 2".

The same problem completed using DD is:

$$\begin{array}{r} 20.76^\circ \\ +\ 30.41^\circ \\ \hline 51.17^\circ \end{array}$$

This illustrates that DD is much easier to use in mathematical problems. When several angles are added together using DMS, large numbers may result. These angles must be reduced. The angle 34° 125' 215" would be reduced to 36° 8' 35".

Subtraction is also more difficult than addition because of the need to carry units of 60. Subtract 40° 18' 50" from 120° 15' 45". The problem starts by trying to subtract 50" from 45". Surveying does not use negative angles. The first step is to transfer numbers from the minutes to seconds. Because there are 60 seconds in one minute, the minutes are reduced by one and the seconds are increased by 60. This same thing occurs with degrees and minutes. Solution:

$$\begin{array}{rrrr} & 120^\circ & 15' & 45'' \\ - & 40^\circ & 18' & 50'' \end{array}$$

Converted and completed

$$\begin{array}{rrrr} & 119^\circ & 74' & 105'' \\ - & 40^\circ & 18' & 50'' \\ \hline & 79^\circ & 56' & 55'' \end{array}$$

Converting between DD and DMS

Another useful math function when using angles is being able to convert angles from DD to DMS and from DMS to DD. Scientific calculators are programmed to complete this function, but if one is not available, the conversion can be accomplished manually.

Entering DD into a calculator is usually not a problem because calculators use the decimal system. The problem arises when entering DMS into the calculator. The angle minutes and seconds must be entered in the correct syntax for the calculator. The operator must know the appropriate method for the calculator they are using.

Manual Conversion from DD to DMS

The first step manually converting from DD to DMS is to multiply the decimal part of the angle times 60. This results in the number of minutes. If the product has decimal digits, the second step is to multiply the decimal part of the minutes by 60 to determine the number of seconds. If this product has a decimal, the last step is to round off any decimal seconds to a whole number. For example, the DD angle of 120.43° converted to DMS is:

$$0.43 \times 60 = 25.8'$$

$$0.8 \times 60 = 48''$$

$$120.43^\circ = 120^\circ\ 25'\ 48''$$

and 45.873° converted to DMS is:

$$45^\circ$$

$$0.873 \times 60 = 52.38'$$

$$0.38 \times 60 = 22.8 \text{ or } 23''$$

$$45^\circ\ 52'\ 23''$$

Converting from DMS to DD Manually

Angles in DMS can also be converted to DD. This is accomplished by adding the angle plus the number of minutes divided by 60 plus the number of seconds divided by 3600. The DMS angle of 45° 32' 21" converted to DD is:

$$45^\circ + \frac{32'}{60} + \frac{21''}{3600}$$

$$= 45 + 0.5333 + 0.005833$$

$$= 45.53916\ldots \text{ or } 45.5^\circ$$

It is important to note that if the same angle is converted by different methods—manually, calculator, or computer spreadsheet the answers may be slightly different. This occurs because the different processes round numbers differently and determine significant figures differently.

HORIZONTAL AND VERTICAL ANGLES

An angle may occur in either a horizontal or a vertical plane. Angles in a horizontal plane are called **horizontal angles**, and angles in a vertical plane are called **vertical angles**. Complex angles occur when both planes are included.

Horizontal and vertical angles are reviewed in this section. A more detailed description is contained in Chapter 1.

Horizontal Angle

A horizontal angle is formed by the intersection of two vertical planes, Figure 7-3. The values for horizontal angles range from 0° to 360°.

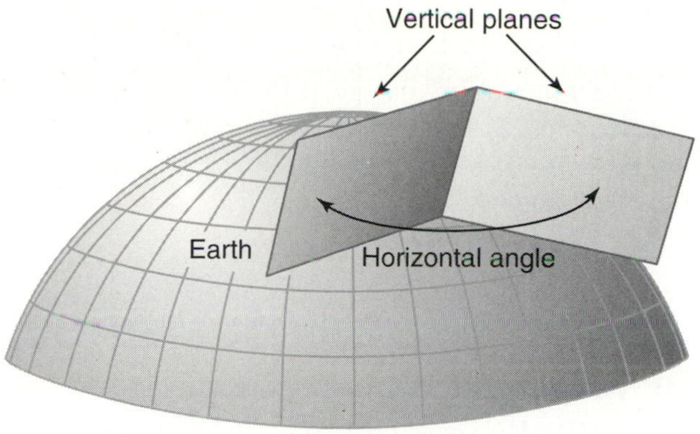

Figure 7-3 Horizontal angle.

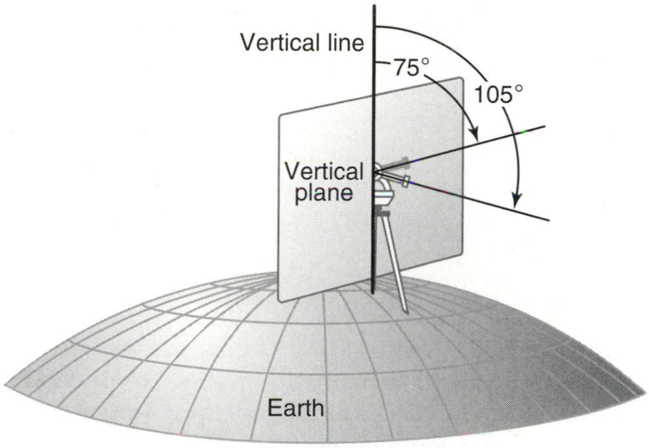

Figure 7-4 Zenith zero vertical angle.

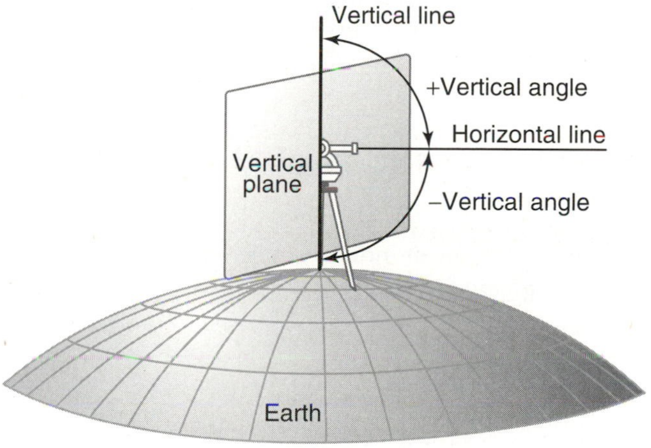

Figure 7-5 Horizontal zero vertical angle.

Vertical Angle

Vertical angles are measured from a vertical line or vertical plane. The reference line is a vertical line, zenith angle, or a horizontal line, horizontal zero angle. Zenith angles range in values from 0° to 180°. When zenith angles are used, 0° is directly overhead, Figure 7-4.

Horizontal zero vertical angles range in value from 0° to 90° and from 0° to minus 90° (−90°), as shown in Figure 7-5.

Deflection Angle

A **deflection angle** is the amount of deviation that has occurred from the direction of travel. The term deflection angle is usually associated with surveys that are used for a route, such as a utility line, property boundary, or a road.

Both angles in Figure 7-6 are deflection angles. Deflection angles must be identified as either left or right. In this example 34.65° is a right deflection angle and −37.73° is a left deflection angle.

INTERIOR AND EXTERIOR ANGLES

An **interior angle** is an angle that measures the angle between two adjacent sides of a polygon. For standard polygons interior angles are usually less than 180 degrees. All four angles in Figure 7-7 are interior angles.

Exterior angles also measure the angle between two adjacent sides of a polygon, but for standard polygons they are usually greater than 180 degrees, Figure 7-8.

Sometimes interior angles are less than 90 degrees and exterior angles are greater than 90 degrees. Study Figure 7-9. Note that the interior angle 3 is greater than 90 degrees. Switching interior and exterior angles like angles 3 and C is a common mistake when doing calculations with interior and exterior angles.

It is also important to note that for every angle there is an opposite angle, 360 degrees minus the angle. For every interior angle there is an exterior

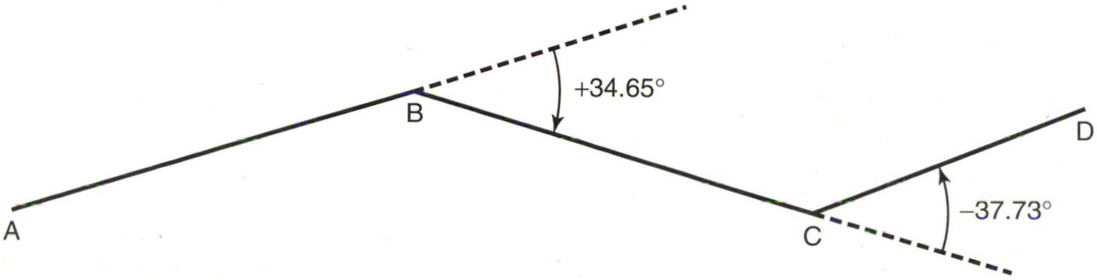

Figure 7-6 Deflection angles.

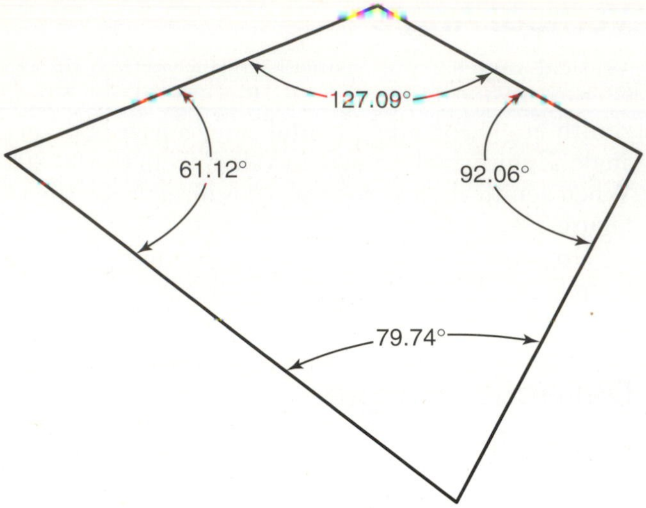

Figure 7-7 Interior angles.

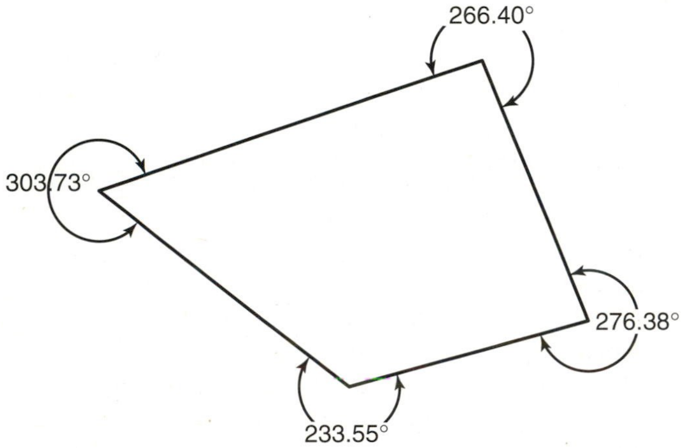

Figure 7-8 Exterior angles.

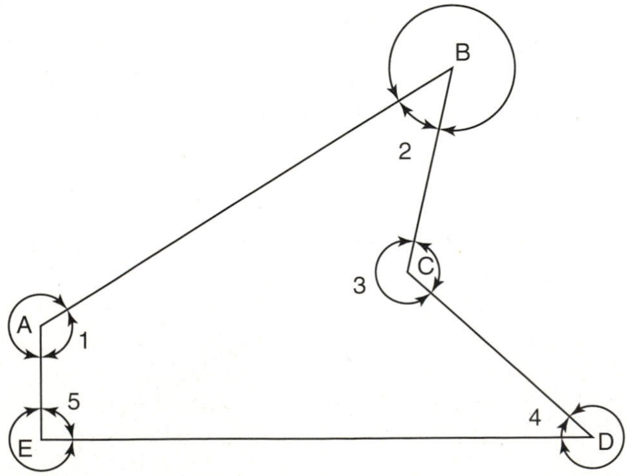

Figure 7-9 Angles for complex polygon.

angle and for every exterior angle there is an interior angle. For example, the 233.55 degree exterior angle in Figure 7-8 has an interior angle of 360 – 233.55 or 126.45 degrees. Accidentally using the opposite angle is a common source of error when surveying angles.

INDIRECT AND DIRECT METHODS OF MEASURING ANGLES

Angles can be measured using both indirect and direct methods. Indirect methods use the principles of arcs and chords, measurements, and calculations to measure and lay out angles. Several indirect methods for horizontal angles will be discussed in the following sections, but these indirect methods are not practical for measuring vertical angles. The selection of an indirect method for measuring horizontal angles is dependent on the uses of the data, the site, the skill of the surveyor, and the equipment that is available.

Both horizontal and vertical angles can be measured with an instrument, as long as it has a scale calibrated for reading angles. This is called the direct method of measuring angles. Angle scales used on mechanical surveying instruments will also include a Vernier scale. Vernier scales are a mechanical means of increasing the physical size of the smallest unit on the main scale. This provides another level of precision. They are discussed in more detail after the next section on indirect methods.

Indirect Methods of Measuring Angles

The three common indirect methods for establishing and measuring angles are:

- Chord
- 3-4-5
- Tape-sine

Chord Method for Laying Out a 90 Degree Angle

The term **chord** refers to a part of a circle. A chord is a line that connects two points on the perimeter of a circle, as shown in Figure 7-10. A chord that passes through the center of a circle is called the diameter.

The chord method is limited to establishing a 90 degree angle. This method is very simple and can be completed with two pieces of string, shoelaces, or any objects that can be used to establish two distances. The accuracy of the chord method depends on the care of the surveyor and the characteristics of the topography. Figure 7-11 illustrates the chord method.

Procedure for Chord Method

The first step in laying out an angle with the chord method is to locate and mark the vertex of the angle (point B) on the baseline, as shown in Figure 7-11. Next, points A and C are established. The distance between B and A, and B and C is not critical. The only requirement is that they must be the same, BA = BC. The next step is to attach one end of a second length to either station A or C and scribe an arc in the estimated location of station D. The length used for this distance is not critical either, it must be longer that the distance AB and BC. The last step is to attach one end of the same length to the other station, C or A, and establish an arc that crosses the previous one. The point where the two arcs intersect is station D. A line drawn from station B through station D will form a 90 degree angle with the baseline. This line can be extended by sighting or using a string line or chain. This method will work with any combination of distances as long as the two requirements are met, BA = BC and the distance AD and CD must be greater than the distance BA.

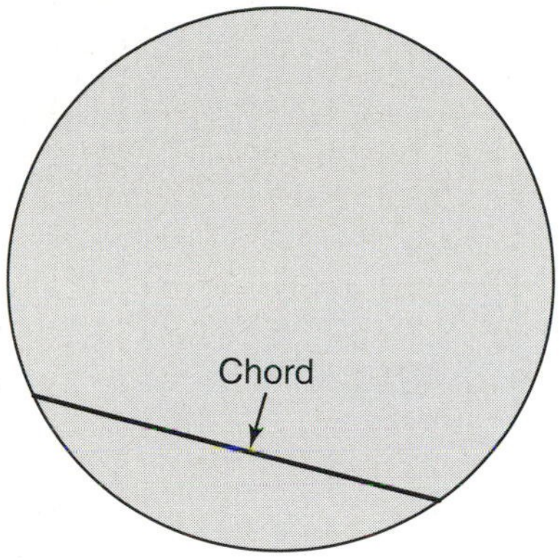

Figure 7-10 Chord.

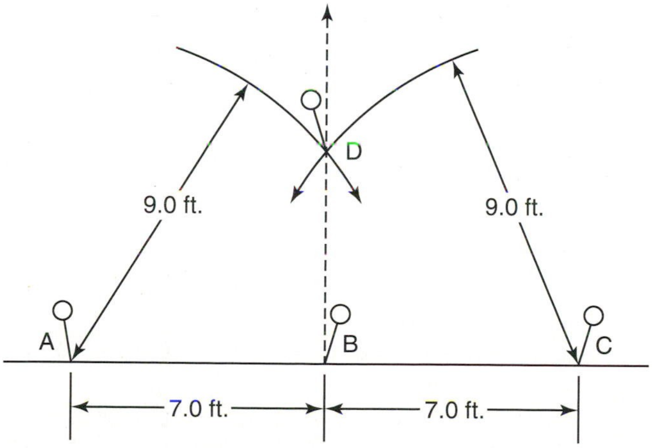

Figure 7-11 Chord method.

A problem may arise when establishing the arcs. If the area is a tilled field, it is easy to scribe the arcs on the ground with a stick. If the surveyor is faced with a more difficult surface, alternatives such as several flags, sticks, or chalk can be used to define the arcs.

Experience will show there is a practical limit for the minimum and maximum distances. The accuracy of the 90 degree angle will increase as the length of the distances used increases, but the errors will be more difficult to manage as the lengths increase.

The chord method is very useful for establishing a perpendicular line (90°). The biggest advantage of the chord method is its simplicity. No surveying instruments or other type of equipment is required. It has two disadvantages. 1) The baseline must extend past the corner being established. If the vertex of the angle is against a building or some other obstacle, this method cannot be used. 2) The chord method can only be used to layout 90-degree angles, and cannot be used to measure an existing angle.

String Chord Method for Establishing a 90 Degree Angle

A variation of the chord method can be used if a piece of string is available. In this method, any length of string can be used, 10 to 15 feet works well. This method is simple and will produce excellent results with proper attention to details. It is illustrated in Figure 7-12.

The first step is to mark the location of the angle vertex along the baseline. Next, hold the ends of the string together and locate the midpoint of the string. The midpoint must be identified with a durable mark. Tying a knot at this point works well. The second step is to fold the string in half again. This will result in a length one fourth of the total length. Use this length

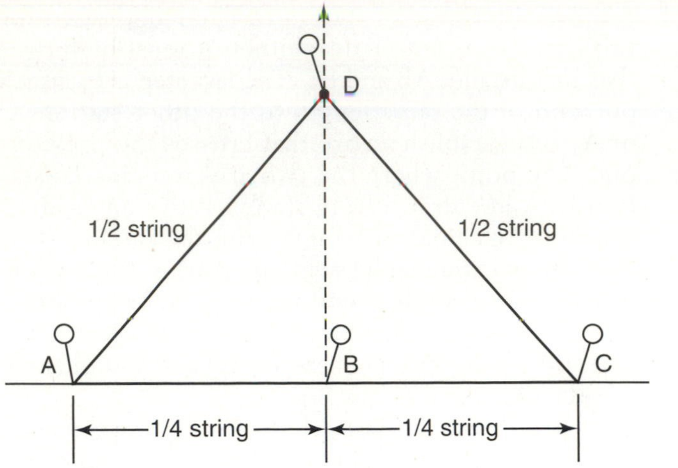

Figure 7-12 String chord method.

to layout and mark points A and C. The last step is to secure one end of the string at point A and the other end at point C. Pull the string away from point B until both sides have equal tension and drive a stake or flag where the mark or knot is in the middle of the string. A line between this point and point B will form a 90-degree angle with the baseline ABC. This line can be extended by sighting or using a string line. This method is very simple, but an acceptable level of accuracy for most construction work can be achieved if the surveyor is careful. There is a large chance for error if the distances from A to D and C to D are not the same.

The string chord method has the same advantages as the previous method plus one. With this method there is no need to establish arcs and determine where they cross. Only four stations must be marked. The string chord method has the same disadvantages as the chord method.

3-4-5 Method for Laying Out a 90 Degree Angle

The **3-4-5 method** is based on the principles of a right triangle. A right triangle with side lengths of 3, 4, and 5 will include one 90 degree angle. Other dimensions can be used, as long as they are the same multiple, for example 6-8-10 or 12-16-20. In this method, one or two measuring tapes must be used. If one tape is used, three people will be required, one for each corner. One person can complete the 3-4-5 method with two tapes.

The major limitation of the 3-4-5 method is that it is limited to laying out 90-degree angles. An advantage over the chord method is that it can be used in less space because the baseline does not need to be extended past the vertex of the angle.

Procedures for 3-4-5 with One Tape When one tape is used, it is important to add at least a two-foot loop at each corner. All metal tapes and some other types of tapes will be damaged if bent sharply. In this procedure, a five-foot loop was used at each corner, as shown in Figure 7-13.

The first step is to locate and mark station A, the vertex of the angle. One person extends the tape in the direction of station B. When the 4-foot mark is reached on the tape, a 5-foot loop is made and the tape is extended toward station C. The person at station B holds the 4-foot and 9-foot mark on the tape together. At station C, a person holds the 14-foot mark (9 + 5) on the tape and forms a 5-foot loop by holding the 14-foot and 19-foot mark together. The tape returns to the starting station. At station A, the person holds the 0-foot and the 21-foot mark (19 + 3 = 21) together, as shown in Figure 7-13.

The last step is taking the slack out of the tape. The person at station A holds the 0- and 22-foot mark on the vertex of the angle. The person at station B aligns

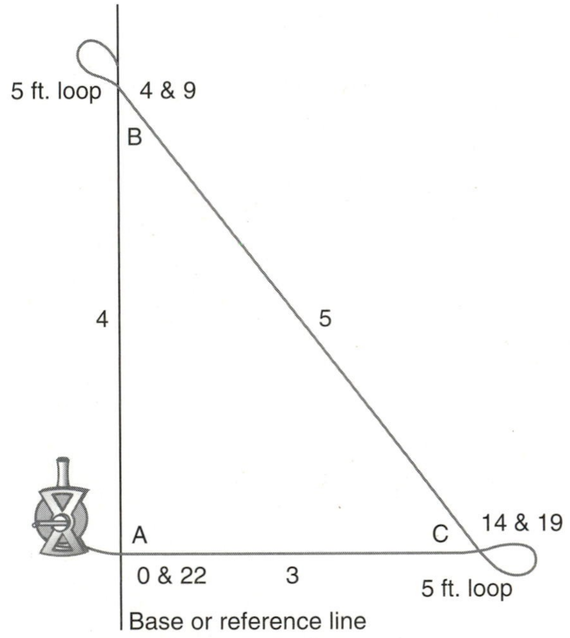

Figure 7-13 3-4-5 Method using one tape.

TIP

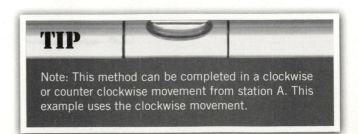

Note: This method can be completed in a clockwise or counter clockwise movement from station A. This example uses the clockwise movement.

their marks on the baseline and takes out the slack in the chain. The person at station C moves around until both sides have equal tension. When both sides have the same tension, corner C is marked. A line drawn from station A to station C will form a 90° angle with line AB. The line can be extended by sighting or by extending the chain from station A through station D.

Procedures for 3-4-5 with Two Tapes Laying out a 90 degree angle with two tapes starts with locating the vertex of the angle along the baseline. The distance AB along the baseline is measured and marked. See Figure 7-14. One tape is attached to station A and the second tape is attached to station B. The person then moves toward station C and roams around until the correct combination of 3 or 4 is held together on the two tapes. For example, if 6 feet is used for the baseline, then station C is located when the 8-foot mark on the tape attached to station A is held over the 10 foot mark of the tape attached to station B, as shown in Figure 7-14.

When both tapes have equal tension, station C is marked. A line connecting stations A and C will form a 90 degree angle with the baseline AB. The line can be extended by holding one end of the chain on station A and extending it through station C.

Tape Sine Method for Laying Out and Measuring Angles

The third indirect method for angles is call tape-sine. The **tape-sine method** uses a tape measure and the sine trigonometric function to establish and measure angles. This method is easy to use as long as the individual understands trig functions and has a calculator equipped with these functions. The tape-sine method has two distinct advantages over the other indirect methods, it can be used for any angle between 0° and 90°, and it can be used to establish a new angle or measure an existing angle. To understand this method it is important to have a basic understanding of trigonometric functions.

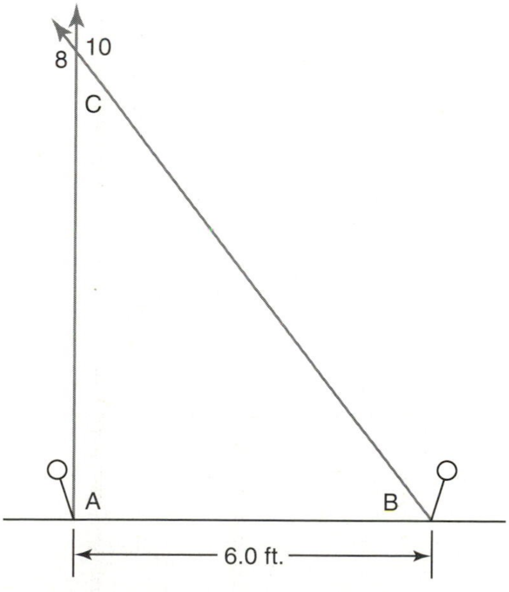

Figure 7-14 3-4-5 Method with two tapes.

Trigonometric Functions Trig functions can be used whenever the shape of the object forms a right triangle. A right triangle is a triangle that has one 90° angle. An understanding of trig functions is based on understanding two principles of right triangles. Two right triangles of different size that have the same angle will also have the same ratios between the lengths of any two sides. Secondly, there are six possible ratio combinations of any two sides of a right triangle. Figure 7-15 illustrates the first principle. The second principle is explained in the next section.

$$\frac{0.50\text{ in.}}{1.54\text{ in.}} = 0.32 \qquad \frac{0.80\text{ in.}}{2.47\text{ in.}} = 0.32$$

These two equations show that as long as the angles are the same, the ratio of the lengths of any two sides remains the same. The tape-sine method would not be useable unless this principle of right triangles was true.

The second principle of right triangles is the number of possible combinations of any two sides of a right triangle. In Figure 7-16 the sides of a triangle have been labeled a, b, and c.

Studying the triangle in Figure 7-16, should reveal six ratio combinations of any two sides. These are, a/b, a/c, c/b, b/a, c/a, and b/c. Mathematicians have assigned a name for each of these ratios. Each ratio and

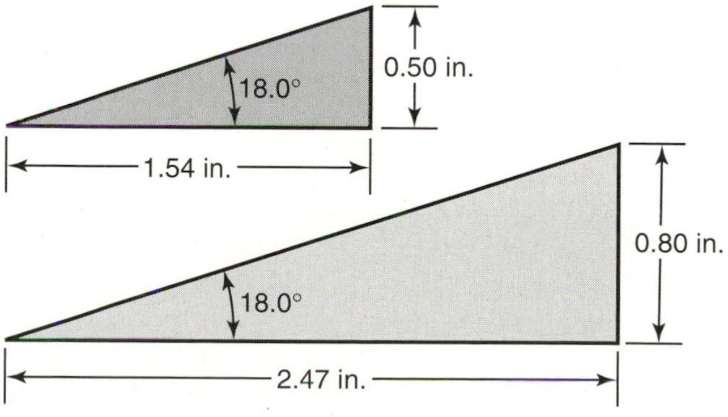

Figure 7-15 Ratio of right triangle sides.

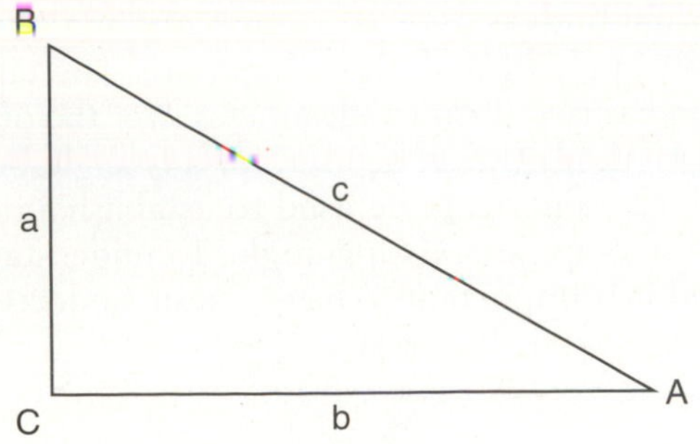

Figure 7-16 Right triangle labels.

the corresponding angle can be expressed as an equation to show the relationship of the three parts. These equations are called functions. Only three of the six functions will be included in this text. These three functions are **sine** (sin), **cosine** (cos), and **tangent** (tan):

$$\sin \phi = \frac{a}{c}$$

$$\cos \phi = \frac{b}{c}$$

$$\tan \phi = \frac{a}{b}$$

Note: the Greek letter phi, ϕ, is used to represent the term "angle".

These functions are easier to remember if terms are used for the sides instead of labels. The terms are relative to the angle being used. An easy way to remember the names of the sides is to visualize looking through the angle, as shown in Figure 7-17.

When terms are used, the three trig functions are called:

$$\sin = \frac{\text{opposite}}{\text{hypotenuse}}$$

$$\cos = \frac{\text{adjacent}}{\text{hypotenuse}}$$

$$\tan = \frac{\text{opposite}}{\text{adjacent}}$$

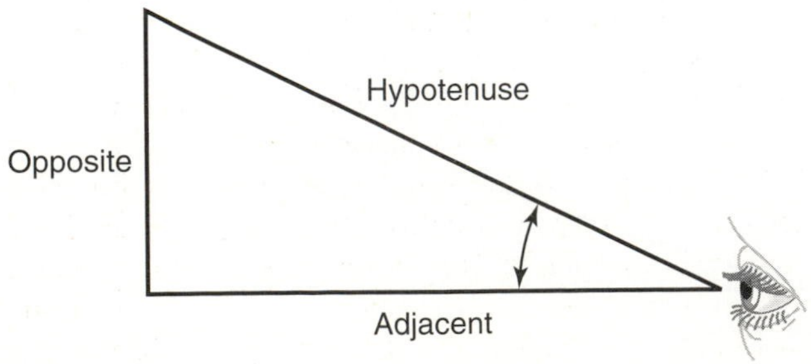

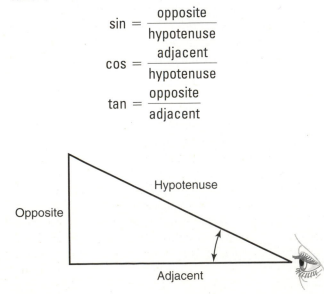

Figure 7-17 Sides of a right triangle.

Trig functions are equations with three variables, therefore, if any two of the variables are known the equations can be rearranged to solve for the third variable. One additional point must be clear when using trig functions. It is important to remember when one of the trig function buttons on a calculator is activated, the result will be the ratio of the angle. For example, enter 25 and use the cosine button. The calculator produces the number 0.906307787. This is the ratio of the length of the opposite side divided by the length of the adjacent side for all triangles with an angle of 25 degrees. When you know the ratio of the sides, then the appropriate inverse trig function must be used to determine the angle. For example, if the cosine ratio for the lengths of two sides is 0.8090, then the angle is 36 degrees.

Selecting the Right Trig Functions A common problem when using trig functions is determining which one should be used. The selection of the appropriate trig function is determined by which one of the three variables is unknown and which two are known. Study Figure 7-18.

Illustration A of Figure 7-18 shows which function should be used when the angle is unknown and one of the three sides is the unknown. In Illustration A the sine ratio is used because the sides that are known are the opposite and the hypotenuse.

$$\sin \theta = \frac{\text{opposite}}{\text{hypotenuse}}$$

In Illustration B, the cosine is used because the length of the adjacent and hypotenuse side is known.

$$\cos \theta = \frac{\text{adjacent}}{\text{hypotenuse}}$$

Illustration C shows the situation when the length of the opposite and the adjacent sides are known. The tangent function is used in this situation.

$$\tan \theta = \frac{\text{opposite}}{\text{adjacent}}$$

The same process is used to determine which trig function should be used when the angle is known and one of the sides is the unknown.

In Illustration A, of Figure 7-19, using the sine function is appropriate because the known side is the hypotenuse and the unknown is the opposite side.

$$\text{opposite} = \sin \theta \times \text{hypotenuse}$$

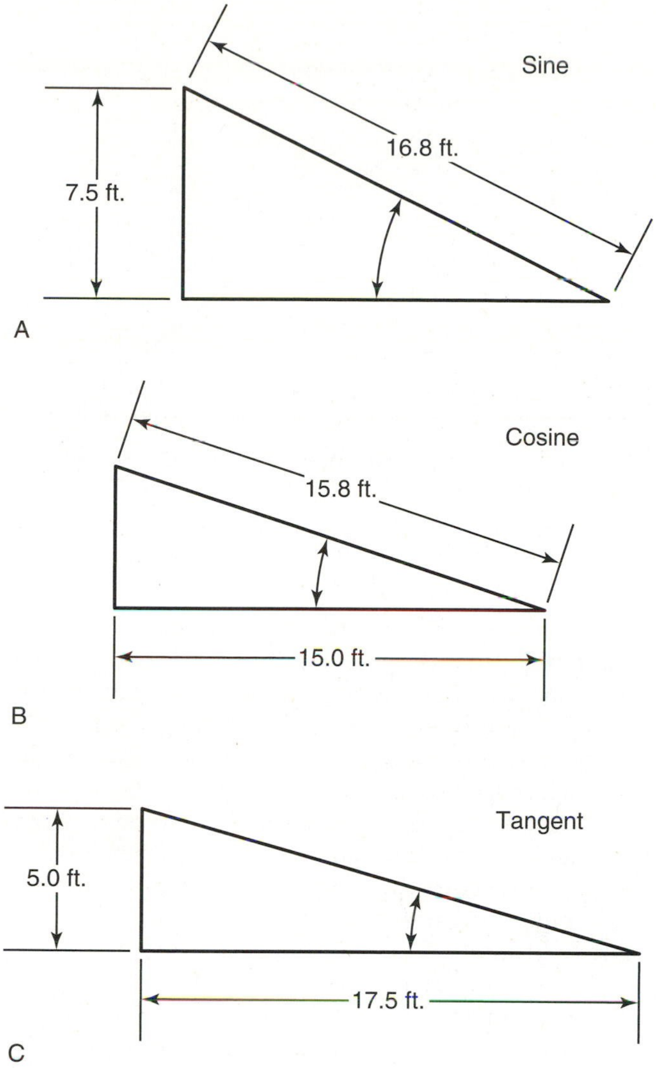

Figure 7-18 Selecting the appropriate trig function when the angle is unknown.

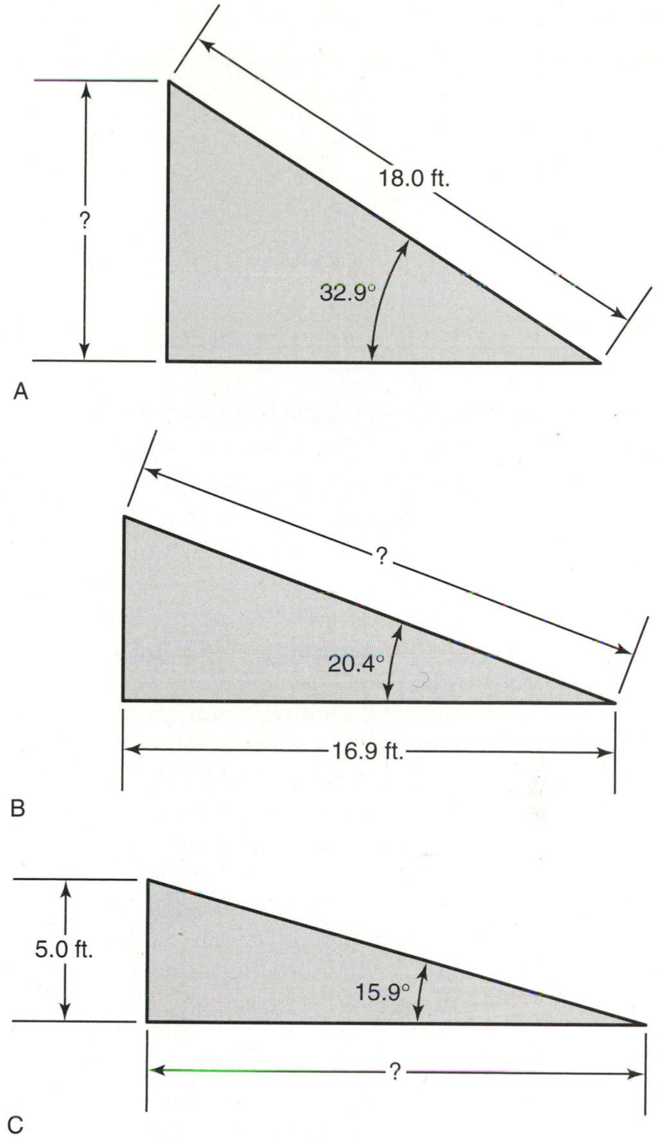

Figure 7-19 Selecting the appropriate trig function when the angle is known.

Illustration B, of Figure 7-19, must be solved using the cosine function because the adjacent and the hypotenuse are used.

$$\text{hypotenuse} = \cos\theta \times \text{adjacent}$$

The length of the unknown side in Illustration C of Figure 7-19 is determined using the tangent function because the opposite and adjacent sides are used.

$$\text{adjacent} = \tan\theta \times \text{opposite}$$

Procedure for Laying Out An Angle Using Tape-sine The procedure for laying out an angle using the tape-sine method is a combination of measuring, marking, and performing calculations. For this example, the desired result is a cross fence that is at a 40 degree angle with the existing fence.

The first step is to remember that it takes two points to establish a line. In this situation, one station (A) is established by deciding where the cross fence will attach to the existing fence. The second step is to establish a second station (C) such that a line passing from point A to point C forms a 40 degree angle with the baseline (existing fence). The tape-sine method is used to locate this second station.

In Figure 7-20, station A is the start of the cross fence. It takes two points to form a line and the location of the second station is unknown. Four possible positions for station C are shown in the above figure for illustrative purposes, but once station A is marked there is only one line that will form a 40 degree angle with the baseline. It is impossible to just select a point

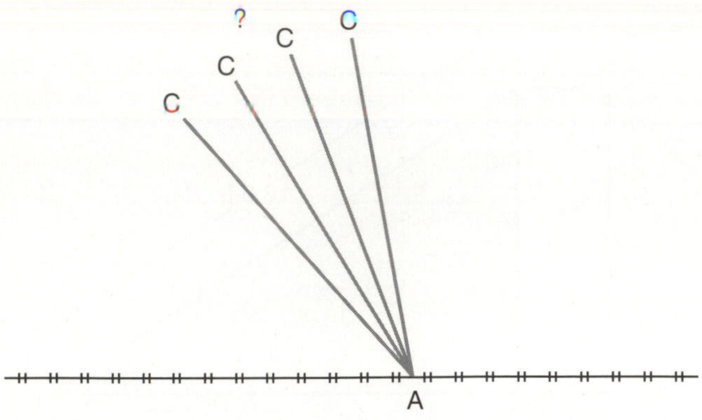

Figure 7-20 Possible locations of second point.

and have the correct angle. At least one point must be established on the 40 degree angle line to establish the location of the line. To establish station C another principle of math is used. When a point must be established relative to an existing point, either three distances must be known or an angle and a distance must be known. The tape-sine method uses three distances. This method is called trilateration.

The first distance is established by measuring from station A along the baseline. This distance is not critical; it should be appropriate for the terrain and the purpose of the angle. As the distance increases the accuracy of the angle will increase, but increasing the distance also increases the chances for making an error in measuring. The second distance is the same as the distance along the baseline because the distances AB and AC are the radius of the same circle, as shown in Figure 7-21.

In this example, a baseline distance of 50.0 feet is used. We know that an arc turned away from the baseline with a radius equal to the distance from station A to station B will contain station C. At this point two of the three distances required for trilateration are known. The third distance is the distance between B and C. This distance is determined by using the sine trig function.

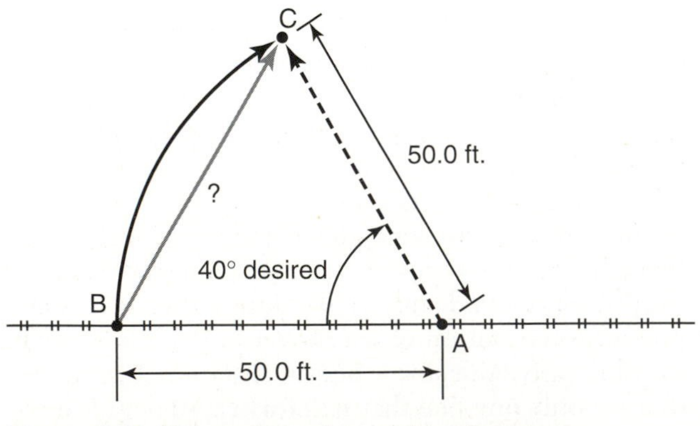

Figure 7-21 Illustration of tape sine distances.

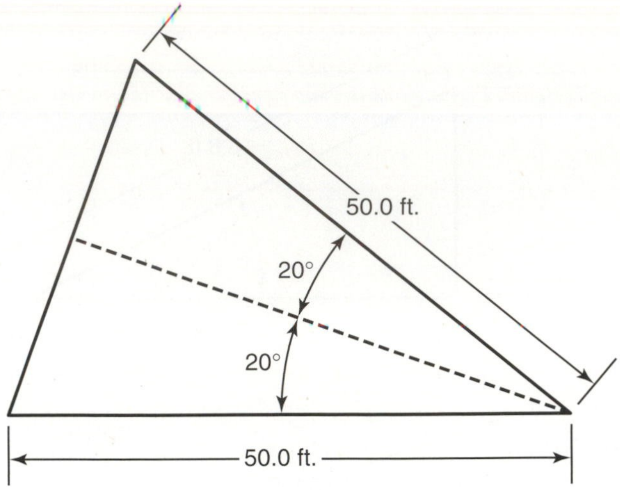

Figure 7-22 Forming two right triangles.

The first response is that it is not possible to calculate this distance using trig functions because trig functions only work with right triangles, but the triangle formed by A, B, and C is an isosceles triangle but an isosceles triangles can be bisected to create two right triangles, as shown in Figure 7-22.

Note: Additional information on isosceles triangles can be found in Chapter 11. Solve for half of the unknown distance between station B and station C and double this answer to find the distance between stations B and C.

The desired result is a cross fence that forms a 40 degree angle with the existing fence. Dividing the desired angle in half results in two right triangles with angles of 20 degrees. Solving for the length of the opposite side of either triangle will result in half of the distance between station B and station C. Using the sine trig function, half the distance is:

$$\sin \phi = \frac{\text{opposite}}{\text{hypotenuse}}$$

$$\begin{aligned}\text{opposite} &= \sin \phi \times \text{hypotenuse}\\ &= \sin 20° \times 50.0 \text{ ft.}\\ &= 0.3420 \times 50.0 \text{ ft.}\\ &= 17.1 \text{ ft.}\end{aligned}$$

The total distance is:

$$17.1 \text{ ft.} \times 2 = 34.2$$

With this information, the position of station C can be determined. This can be accomplished by marking the arc formed by the radius of 50.0 feet and then measuring along the arc until the chord distance is 34.2 feet or by establishing two arcs. One arc will

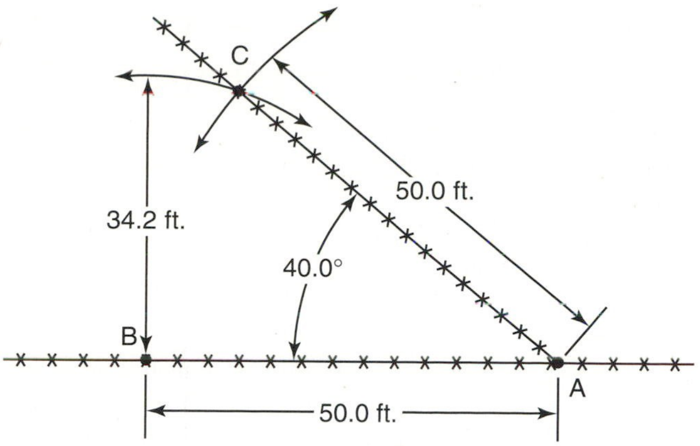

Figure 7-23 Establishing point C.

have a radius of 50.0 feet and be centered at station A and the second arc will have a radius of 34.2 feet and be centered at station B. Establishing a line from station A through station C will layout out the cross fence and a 40 degree angle to the baseline, as shown in Figure 7-23.

Procedure for Measuring an Existing Angle Using Tape-sine The tape-sine method can also be used to measure any angles between zero and 90 degrees. To measure an existing angle start by measuring and marking an equal distance along both sides (AC and AB) of the angle and then measure the distance BC. The angle is calculated by visually splitting the angle formed by stations A, B, and C into two right triangles. This will result in two triangles with equal length hypotenuse and equal length opposite sides that are half the distance between stations B and C. Solving for the angle of either of these triangles and multiplying the results by two will produce the desired angle. Refer to Figure 7-24. What is the angle formed by the retaining wall in the illustration?

The answer is 53.3 degrees. As in the previous example, before the Sine trig function can be used, a right

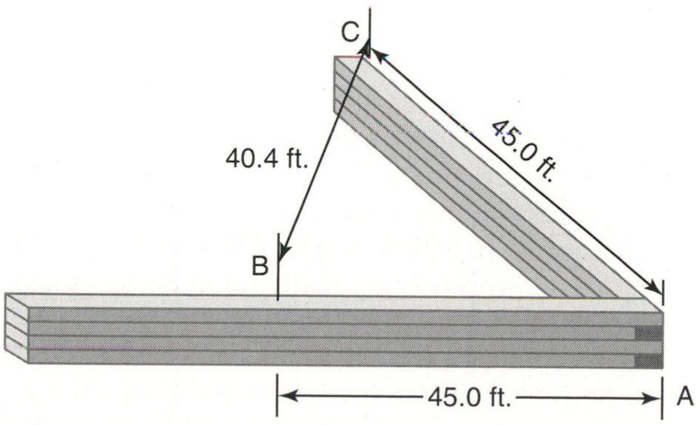

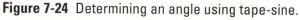

Figure 7-24 Determining an angle using tape-sine.

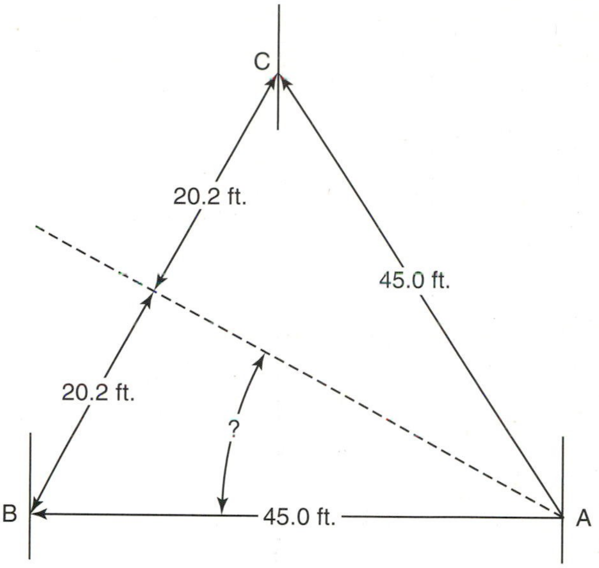

Figure 7-25 Forming two right triangles for retaining wall example.

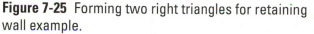

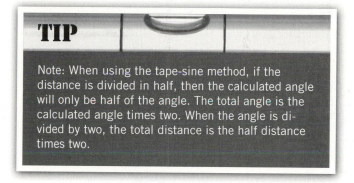

> **TIP**
>
> Note: When using the tape-sine method, if the distance is divided in half, then the calculated angle will only be half of the angle. The total angle is the calculated angle times two. When the angle is divided by two, the total distance is the half distance times two.

triangle must be formed. In this example, splitting the distance between station B and station C in half forms two right triangles, as shown in Figure 7-25.

Solve for the angle formed by either right triangle and multiply it by two to determine the angle of the retaining wall.

Tape-sine can be used to layout or measure any angle between 0° and 90°. Angles greater than 90 degrees can be laid out and measured if the angle can be split into angles that are less than 90 degrees.

Direct Methods of Measuring of Angles

Direct measuring of angles requires an instrument with angle scales. Most surveying instruments will measure horizontal angles. To measure vertical angles the telescope must be able to tilt up and down. This capability is limited to transits, theodolites, total stations, and a few other types of instruments. Another difference is the precision of the angle scales. The precision of dumpy and automatic levels is usually one degree or maybe 20 minutes. Transits, theodolites, and total stations are higher-precision instruments. Many transits are 5 second instruments and most theodolites are single second instruments. Total stations will measure decimal degrees to several decimal points. The steps in measuring an angle with an instrument include:

1. Center and level the instrument
2. Zero setting the angle scale
3. Aligning the telescope on the backsight
4. Turning the angle
5. Recording the angle

Note: the sequence of the 5 steps will not be the same for all instruments.

Each one of these steps will be discussed with further detail in the following sections.

Centering the Instrument

The first step is to insure the instrument is centered over the vertex of the angle. Instruments are centered over the vertex by suspending a plumb bob beneath the instrument or by using an optical or laser plummet. For accurate surveys, the plumb bob suspended beneath the instrument or the optical plummet must be aligned with a surveyor's nail in a stake or a surveyor's nail driven into an asphalt or concrete surface.

Example of Centering Error The potential for centering error is illustrated in Figure 7-26.

If the position of the instrument should be at A_1 but the instrument is set at A_2 instead, a distance of 5.0 inches, the result is an angle error of 1.0°. For some applications, an error of 1.0° would not affect the use of the data, but for other uses this would not be acceptable. For example, when completing the survey for a legal boundary the total error of all angle measurements must be less than the smallest unit of the angle scale on the instrument. This will vary depending on the instrument, but it can be as small as one second.

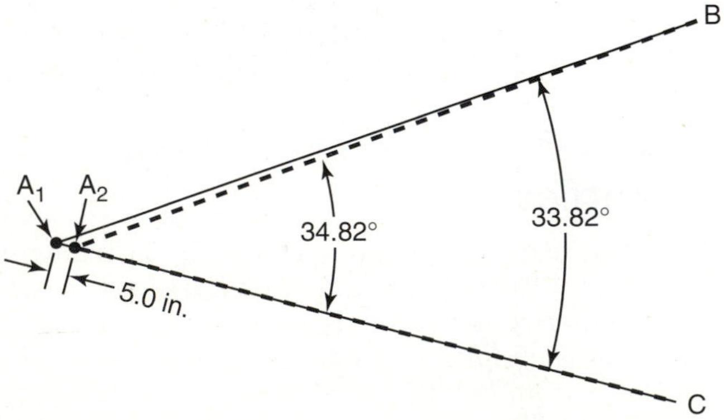

Figure 7-26 Instrument centering error.

Procedures for Centering an Instrument The procedures for centering an instrument over a point will vary depending on the instrument used. A common feature of instruments that are designed to measure horizontal angles is capability of a slight lateral movement after the instrument is mounted on the tripod and the leveling screws are loose. This is accomplished by either slightly loosening the mounting screw(s) or the leveling screws. The ideal condition for centering an instrument start with the tripod set with the head level and centered over the point. Centering the tripod is not difficult for an experienced surveyor. For the inexperienced surveyor, the process will be easier if they suspend the plumb bob through the center of the head of the tripod as it is set up. This should result in the tripod being close enough to center that the lateral movement of the instrument on the tripod head will be sufficient to center the instrument. The procedures for centering the instrument are slightly different depending on whether a plumb bob or an optical plummet is used. Both procedures will be explained in the following sections.

When a plumb bob is used, the instrument is mounted on the tripod, but the lateral adjustment is left slightly loose and the plumb bob is suspended beneath the instrument. The instrument is leveled and then shifted to the left or right until the plumb bob is centered over the point. Be sure to slide the instrument not rotate it. Rotating the instrument will move it out of level if the head of the tripod is not perfectly level. If the amount of lateral movement in the instrument is insufficient to center the plumb bob, the position of the tripod must be adjusted and the procedure repeated. When the instrument is centered, the lateral adjustment is secured and the instrument leveled again. At this step, the location of the plumb bob should be checked. If it has shifted off center, the instrument must be loosened, centered, tightened, and leveled until the instrument is level and the plumb bob is centered on the point. One disadvantage of plumb bobs is that they are affected by wind. In windy conditions a field notebook opened and held around the plumb bob is usually effective in blocking the wind so the plumb bob will hang straight.

Optical plummets are a set of lenses built into the instrument that allow the operator to look through an eyepiece on the side of the instrument and see the ground beneath the instrument. The optics include a crosshair or small circle that is used to center the instrument and may include a focusing adjustment. To center an instrument using an optical plummet the operator must center the tripod over the point and loosely attach the instrument. Laser plummets are used in the same way as optical plummets.

Next the instrument is leveled. This is an important step because the line-of-sight of the optical plummet is aligned with the vertical centerline of the instrument. If the instrument is not level, the centerline will not be a vertical angle. Centering the instrument in this condition will cause the crosshairs to shift off the mark when the instrument is leveled. This results in wasted effort because the instrument must be centered again. Some instruments will include a spirit level in the base of the instrument and another one on the instrument. The spirit level in the base is used to level the instrument the first time. The operator looks into the optical plummet eyepiece or turns the laser plummet on and determines which direction the instrument needs to be moved. When moving the instrument to center, it is important not to rotate the base of the instrument. Rotating the instrument base will move the telescope out of level and the steps will need to be started over.

The last step is to secure the instrument, level it with the spirit level mounted on the instrument and check the optical plummet. If the instrument is not centered, it is loosened and centered again. These steps are repeated until the instrument is secured on the tripod, level and centered.

Zero Setting the Instrument

After the instrument is centered and leveled the angle scale is set on zero or the desired starting angle. When a single angle is being measured, it is a common practice to set the angle scales on zero before measuring the angle, as shown in Figure 7-27. Other situations may require setting a starting angle different from zero. Regardless of the situation, the angle scale must be set on the desired angle. Different methods are used to accomplish this task. A transit has an upper and a lower plate. The upper plate movement is unlocked and rotated until the zero on the Vernier scale is aligned with the zero on the angle scale. The upper plate is locked and the lower plate is released to align the telescope on the backsight. The angle is turned by locking the lower plate, releasing the upper plate, and rotating the telescope to the foresight.

Most dumpy and automatic levels require the sequence to be reversed. These instruments are zero set by loosening the horizontal rotation lock and aligning the telescope on the backsight. The angle scale is loosened and rotated until the desired angle is set. Then the instrument is rotated to the foresight.

Total stations are similar to dumpy and automatic levels. The telescope is aligned on the backsight first and zero set by activating the appropriate buttons on the control panel. It is very important that the individual in charge of the instrument know how to operate the instrument. Individuals that attempt this procedure without knowing how to operate the instrument may damage the instrument and produce erroneous data.

Aligning the Instrument on the Backsight

As discussed in the previous section some instruments require that this step be completed before zero setting the angle scales. For those that do not, the telescope must be aligned with the backsight. To aid in this process most instruments have a tangent screw, or fine adjustment, associated with the movement lock. The tangent screw allows very slow movement of the telescope. When shooting longer

TIP

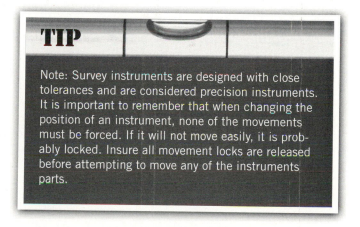

Note: Survey instruments are designed with close tolerances and are considered precision instruments. It is important to remember that when changing the position of an instrument, none of the movements must be forced. If it will not move easily, it is probably locked. Insure all movement locks are released before attempting to move any of the instruments parts.

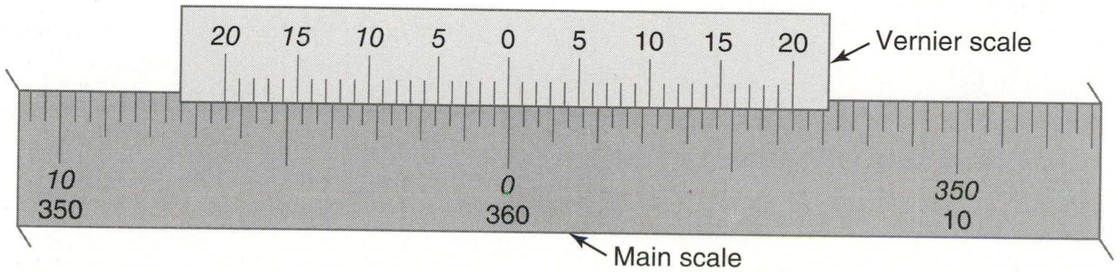

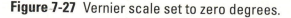

Figure 7-27 Vernier scale set to zero degrees.

TIP

Note: The instrument and method of aligning the instrument should be appropriate for the level of survey. For example, if the use of the data requires a transit or total station the backsight marker must have more precision that a wooden stake. A nail must be used in the stake and if a rod is used, the centerline of the telescope must align with the edge of the rod that is aligned with the nail in the stake. The opposite is also true. If an automatic level is the appropriate instrument for the survey, it is wasted resources to use a surveyor's nail in the stake and a plumb bob to transfer the location because some automatic levels have a precision of only one degree.

distances, very small movements of the telescope will result in wide swings of the crosshairs on the target. Using the tangent screw saves time and allows precise alignment of the telescope.

Aligning the telescope on the backsight or foresight is easy when using a transit or total station because the telescope can be depressed, rotated vertically down, and rotated horizontally until the crosshairs are centered on the surveyor's nail or pin. For instruments that cannot depress the telescope, the foresight position on the ground must be transferred up into the line-of-sight of the instrument. Range poles or rods can be used for low-precision surveys. A plumb bob can be used when more precision is required. When a range pole or rod is used, it is critical that they are plumb when the instrument is aligned on the pole or rod.

When less precision is acceptable, the marker could be a pin, flag, or stake. The reference point may be a previous station in the survey, a designated object, or north. Aligning the backsight to north will require the use of a compass. For accurate results, the compass must be built into the instrument. Many transits, total stations, and theolodites have this capability.

Turning the Angle

The last step is to rotate the instrument until the telescope is aligned with the second point (foresight). The steps required are unique for the type of instrument being used. It is important that the operator understands the controls for the instrument they are using. Releasing the wrong plate lock on a transit, for example, will require repeating the previous two steps. The alignment of the telescope on the foresight follows the same principles as aligning the telescope on the backsight. A range pole, rod, or plumb line should be used to establish a vertical line from the ground to the instrument line-of-sight if the telescope cannot be depressed.

Recording the Angle

Electronic instruments use an LED screen to display the angle measurement. The operator must be able to read the angle scale when a mechanical instrument is used. Mechanical instrument angle scales usually include a Vernier scale. To read a Vernier scale you must be able to determine the least count. The least count for an instrument is the smallest unit the instrument can measure. All instruments do not use the same level of precision for the Vernier scales.

Determining the Least Count A Vernier scale is a mechanical means of magnifying the last unit of measure on the main scale to provide a smaller least count. One limitation of mechanical scales is that the smallest unit of measure is determined by the thickness of the line used for the graduations. As the line is made thinner the least count can be increased, but the scale becomes harder to read. Older mechanical instruments require a small magnifying glass to read the Vernier scales efficiency.

The least count for a Vernier scale is determined using the same technique that is used to read a ruler or carpenters tape, as shown in Figure 7-28.

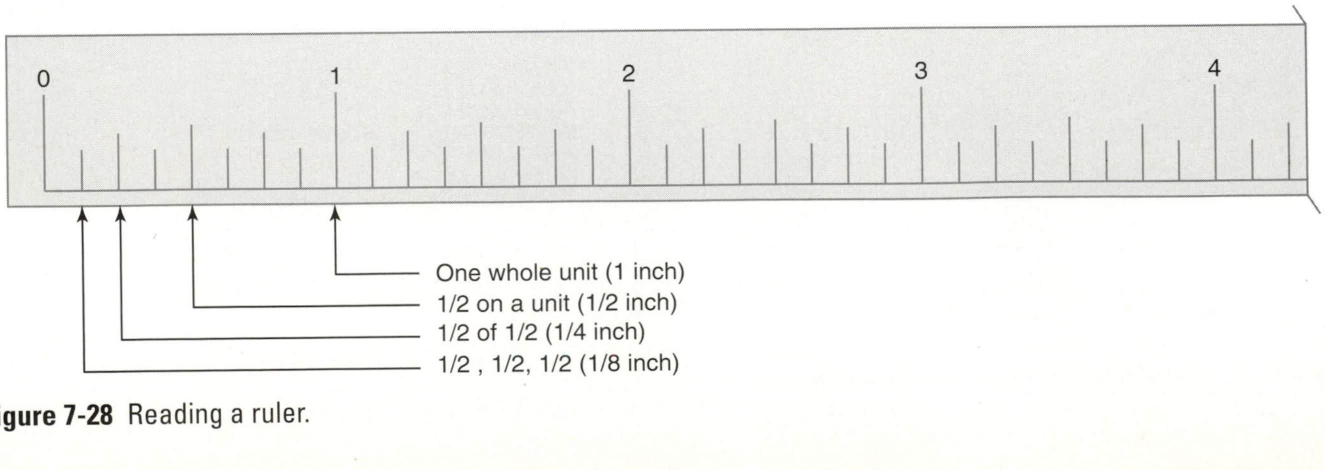

Figure 7-28 Reading a ruler.

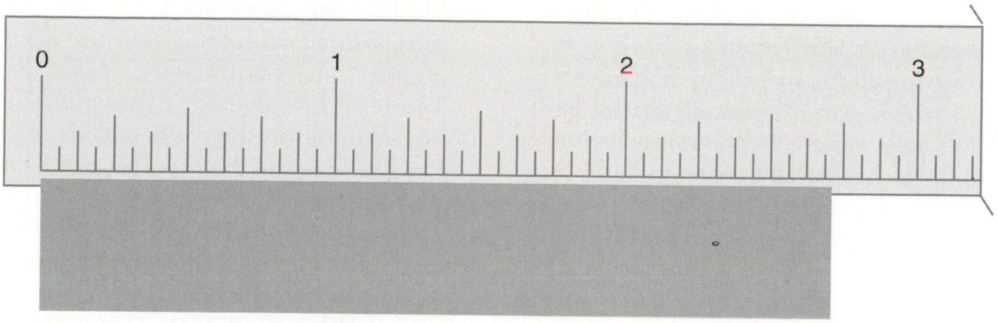

Figure 7-29 Measuring a block with a ruler.

A standard ruler is read by first determining the largest whole units (inch). On most rulers and angle scales, the largest whole units are usually labeled with a number. Figure 7-28 illustrates a segment of a standard ruler. For this ruler the numbers 1, 2, and 3 identify the inch marks. The next step is determining the fractions of an inch. This is accomplished by noting the relative lengths of the different lines and the number of lines of similar length. The ruler in Figure 7-28 has one long line between each inch. One line equals two spaces, this line is ½ unit or ½ inch. Between each ½ inch line is one line that is shorter than the ½ inch line, but longer than the other lines. One half of an inch divided by 2 results in ¼ of an inch. The process is completed until the smallest unit, least count, is identified. The least count for the ruler in Figure 7-28 is ⅛ of an inch.

Once the reader knows the whole unit and the smallest unit of measure (least count) for any scale, measurements can be read. For example, the block in Figure 7-29 is 2 and ¼ inches plus 3/16 or 2 and 11/16's of an inch.

The same principles are used when reading angle scales and Vernier scales.

Least Count of Vernier Angle Scales

The least count of an angle or Vernier scale is the smallest unit of measure. The first step in determining the least count is to determine the whole degree, and then the number of divisions for the whole degree. Figure 7-30 illustrates a segment of an angle scale.

Note that there are two lines identified as 30 and 40 degrees. Halfway between 30 and 40 degrees is one longer line. One line equals two spaces. Half the angle between 30 degrees and 40 degrees is 35 degrees. This line is 35 degrees. There are four of the next length lines between 30 and 35 degrees. Four lines equals five spaces, $(35 - 30)/5 = 1$. These lines are one degree each. Between each degree are two lines. Two lines equal three spaces. Each degree equals 60 minutes. Sixty divided by three equals 20. Each one of these lines equals 20 minutes. The least count of this scale is 20 minutes. This process can be used to determine the least count of any scale.

Compare Figure 7-30 and Figure 7-31. It may appear that Figure 7-30 and Figure 7-31 are the same, but this is not true. The scale in Figure 7-31 has one long line between 30 and 40 degrees. This line is 35 degrees. There are four lines of the same length between 30 and 35 degrees. Each one of these is one degree. Each degree is divided by one line. One line equals two spaces and one degree divided by two equals 30 minutes. The least count of this scale is 30 minutes.

The last step is determining the divisions on the Vernier.

Deciphering Verniers is manageable if you remember that the Vernier is a mechanical means of magnifying the last unit on the main scale. They have two characteristics. The Vernier scale divisions can be determined using the same method that is used for the angle scale. Secondly, the maximum value on the Vernier scale will be equal to the least count of the angle scale. An angle scale with a least count of 20 minutes will have a Vernier

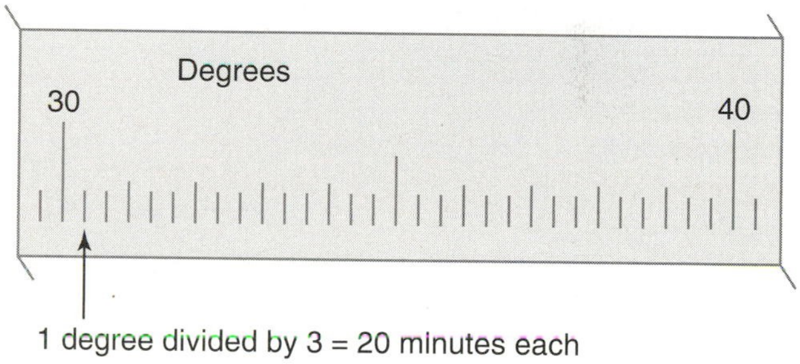

Figure 7-30 Twenty minute angle scale.

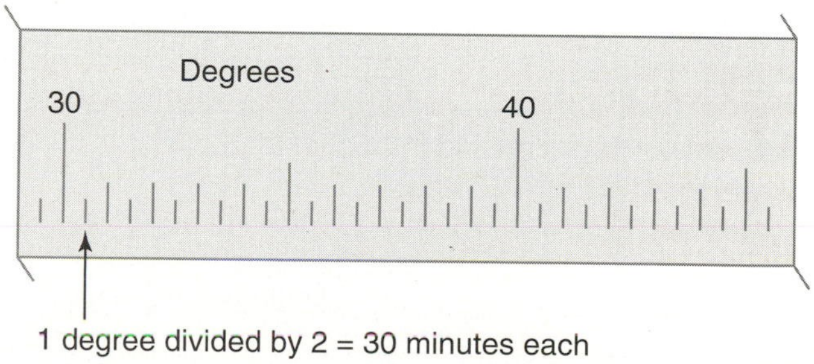

Figure 7-31 Thirty minute Vernier scale.

with a scale from 0 to 20 minutes, as shown in Figure 7-32. An angle scale with a least count of 30 minutes will have a Vernier that will measure 0 to 30 minutes. The Vernier in Figure 7-32 has 19 lines, 20 spaces, between the 0 and 20. This Vernier is a one minute Vernier and the least count of the instrument is one minute.

It is important to remember that the Vernier scale may also be subdivided. A Vernier scale that ranges from 0 to 20 minutes that has one line between each major division will have a least count of ½ a minute or 30 seconds, as shown in Figure 7-33.

The least count of the Vernier in Figure 7-33 is thirty seconds.

Figure 7-34 is another example of a sub-divided Vernier. The least count of this Vernier is 5 minutes.

Study the two examples in Figure 7-35. What is the least count of Illustration A?

Solution: In this example, the units are not identified. This illustrates that the scale can be read even if the units are not known. Note that several lines are numbered, 0 to 50, and that there are nine lines between each pair of numbers—nine lines equals 10 spaces. It is safe to assume that for whatever unit that is used, each line on the main scale has a value of one. The Vernier scale has a range of 0 to 10. This means that the last unit on the main scale is divided by 10. Therefore, the least count is $\frac{1}{10}$ or 0.1.

When reading an instrument with a Vernier, the starting point is the zero line of the Vernier. In Illustration A of Figure 7-35, the zero of the Vernier is in alignment with the 10 on the main scale. The reading for scale A is 10. For scale B, the zero line of the Vernier is not aligned with any line on the main scale. The reading on the main scale is between 16 and 17. Pick the lower of the two, 16, and then read the Vernier. The reading of the Vernier is the line that coincides with a line on the main scale. In Illustration B of Figure 7-35, the first line to the left of the line labeled with the 10 is the line of coincidence. The measurement for this scale is 16.9. When reading the Vernier scale, the value of the line on the main scale is not important, only the value of the line of coincidence on the Vernier scale is recorded.

The line of coincidence can be difficult to determine when reading Vernier scales on instruments. If it appears that two lines coincide, select one because remember an instrument precision is plus or minus one of the least count. If it appears three lines coincide, use the center one.

TIP

It is very difficult to produce drawings that accurately depict Vernier scales. It will be difficult to find the correct line of coincidence in the drawings.

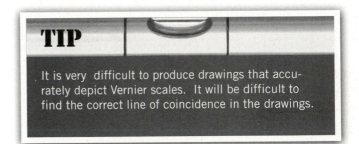

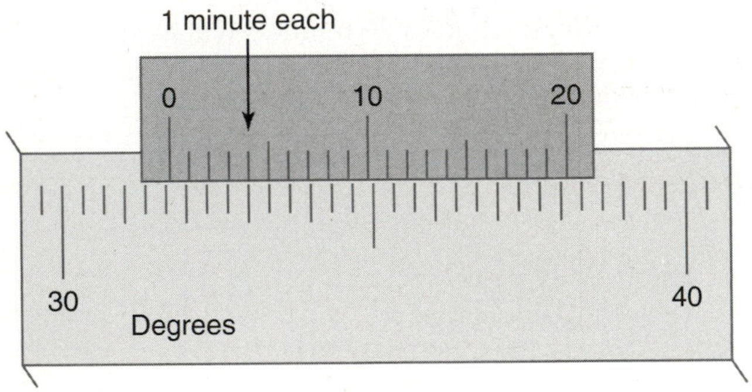

Figure 7-32 One minute Vernier scale.

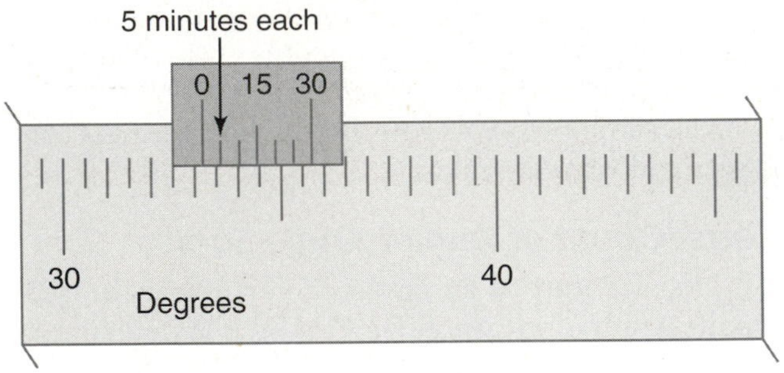

Figure 7-34 Five minute Vernier.

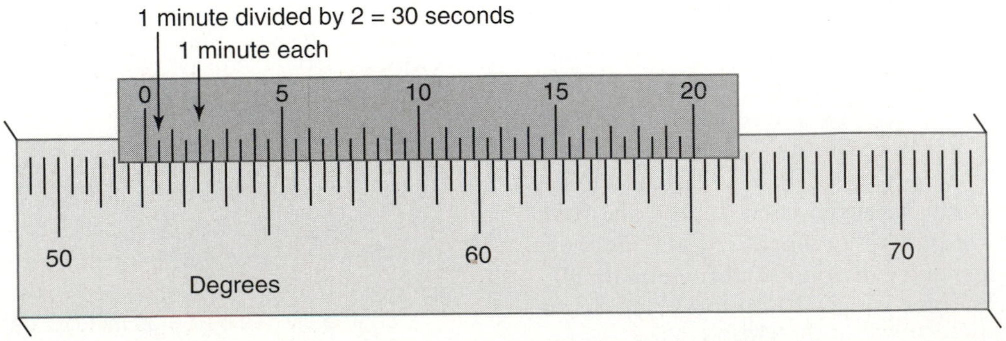

Figure 7-33 Thirty second Vernier.

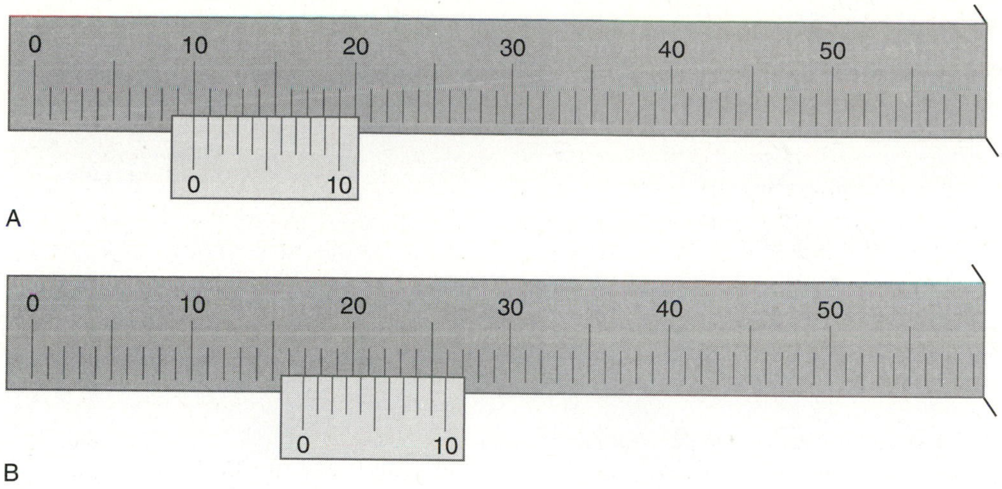

Figure 7-35 Two Vernier examples.

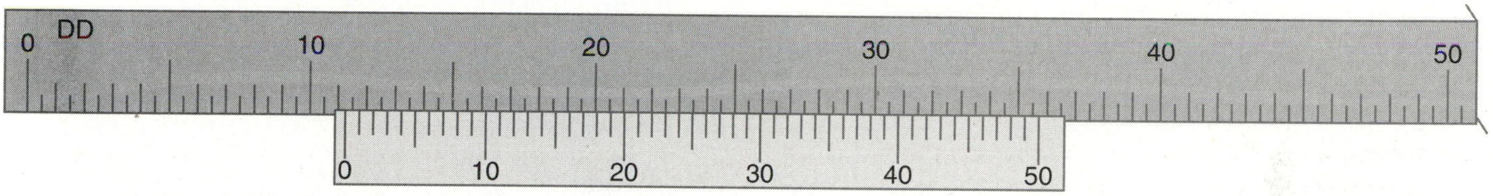

Figure 7-36 DD practice Vernier.

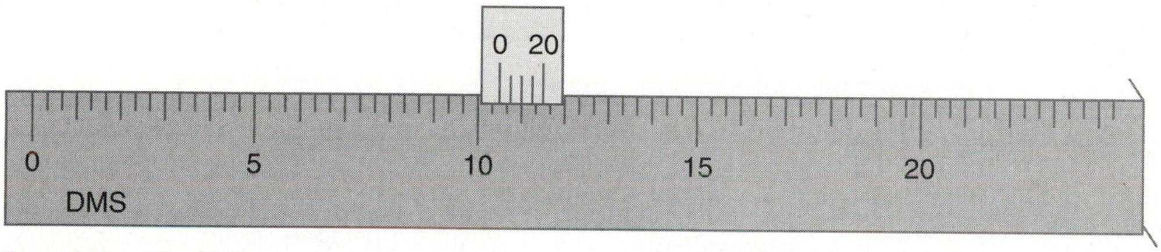

Figure 7-37 DMS practice Vernier.

Practice reading Verniers by recording the correct reading for the Verniers in the following illustration: The DD on scale A, in Figure 7-36, means the scale is in decimal degrees. The DMS for scale in Figure 7-36 means the units are in degrees (°), minutes ('), and seconds ("). One degree equals 60 minutes and one minute equals 60 seconds.

It is difficult to determine if the line of coincidence in Figure 7-36 is 22 or 23. If 22, then the correct answer is 11.22 degrees. Main scale = 11 degrees. The line of coincidence is 22 degrees: 11 + 0.21 = 11.21 degrees. If the line of coincidence is 23, then the correct answer is 11.23.

The reading for Figure 7-37 is 10 degrees + 20 minutes + 10 minutes = 10° 30'.

A variety of Vernier scales are used on direct reading mechanical instruments. They may be single, folded, or double and they may have different least counts. A single Vernier is the simplest, but it is not as useful because it can only measure angles directly in one direction of rotation. If the opposite direction is used, the angle that is read from the instrument must be subtracted from 360 to arrive at the correct angle. This increases the chances of error. To eliminate this problem, instruments use double angle scales and double Verniers.

Reading Double Angle Scales

Double angle scales are used to provide direct reading when turning angles in both directions. When reading a double angle scale, it is important to read the main scale in the direction of increasing values. Figure 7-38 is a segment of a double angle, double

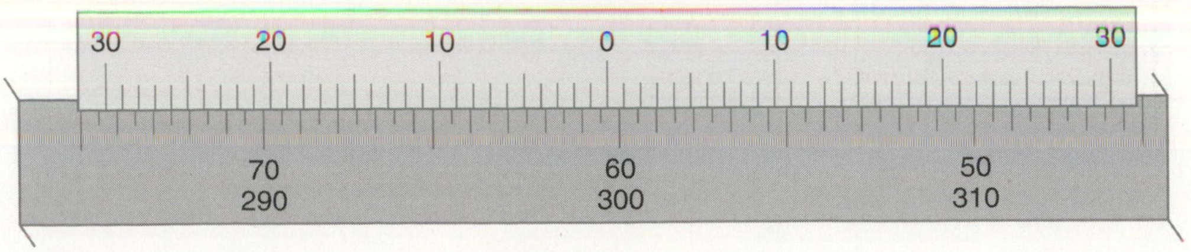

Figure 7-38 Double Vernier angle scale.

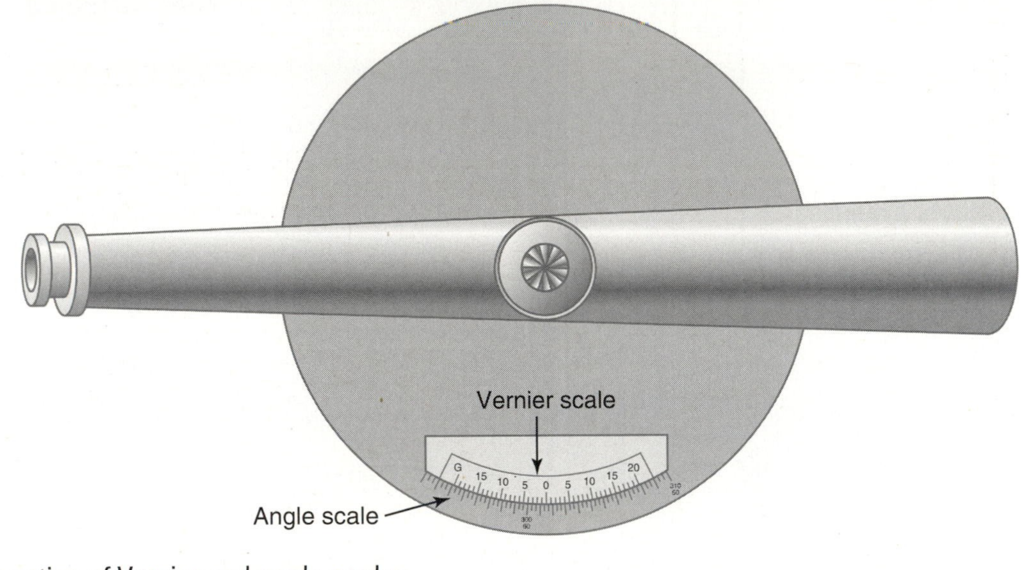

Figure 7-39 Location of Vernier and angle scales.

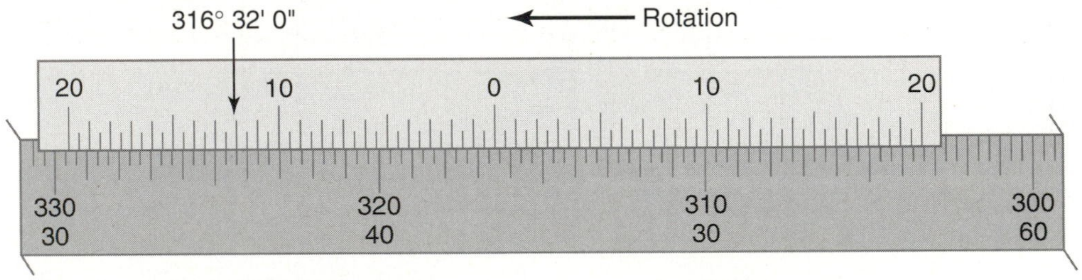

Figure 7-40 Reading the left Vernier for clockwise rotation, right turned angle.

Vernier scale. Note the numbers on the main scale are the opposite angle of each other, 70 and 290, for example.

A common error when reading this scale is reading the incorrect side of the Vernier. The correct side is determined by the direction of rotation of the instrument.

Note: It is important to remember that the angle scale is a complete circle and it is stationary when rotating the telescope of the instrument. The Vernier scale is a segment and it rotates with the telescope of the instrument.

When the instrument is rotated clockwise, the left half of the double Vernier scale is read, and when the instrument is rotated counterclockwise, the right half of the double Vernier scale is read. Another way to remember which half of the Vernier to read, is read the leading side, the side that leads in the direction of rotation. Selecting the correct Vernier is demonstrated in Figure 7-39.

In Figure 7-40 the correct reading is 316 degrees, 32 minutes, and 0 seconds. The instrument was turned to the right, rotated clockwise (CW), the increasing angle was the inside scale, and because the angle was right turned, the left half of the double Vernier scale was used.

Note: The answer included the seconds even though they were zero. This is the proper form because it tells the reader the least count for the instrument is in the seconds of a degree.

In Figure 7-41 the angle was turned to the left, so consequently the angle scale is rotating counterclockwise (CCW), the opposite direction of the clock hands, therefore the outside angle scale and the right half of the Vernier scale are used.

The correct reading is 43 degrees, 31 minutes, and 30 seconds.

Two rules are true when reading double Vernier angle scales:

1. The amount of time it takes to collect angle readings in the field will be reduced as the operator becomes familiar with the different types of scales that are used.
2. The amount of errors that occur when reading angles will be reduced as the operator practices reading double Vernier angle scales.

Practice Reading the Following Verniers

The first step is to determine the least count. The least count for the Vernier in Figure 7-42 is 30 seconds. The correct reading is 114° 19' 30".

It is important to notice that the angle for the Vernier in Figure 7-43 was left turned.

The least count is the same (30 seconds) and the correct answer is 245° 20' 0".

What are the least count and the reading for the angle scale in the following illustration, Figure 7-44?

Solution: Least count equals 30 seconds and the reading is 14° 43' 30".

What is the reading for the following Vernier, Figure 7-45?

The correct reading is 5° 34' 0".

AZIMUTHS AND BEARINGS

More than one system is used in surveying for expressing angles. The primary difference is in the reference point used for the angle. Two common systems are azimuths and bearings.

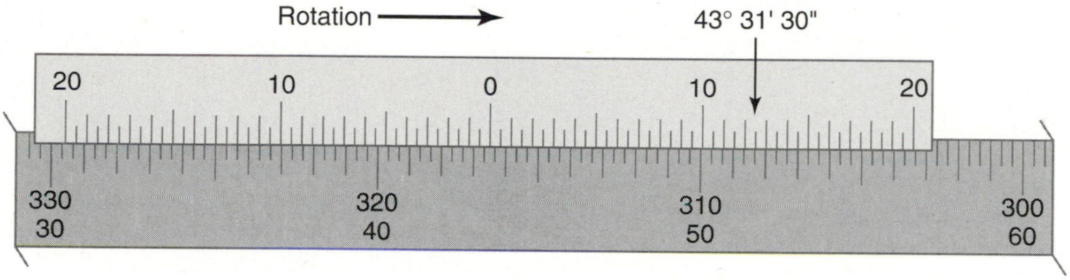

Figure 7-41 Reading the right Vernier for counterclockwise rotation, left turned angle.

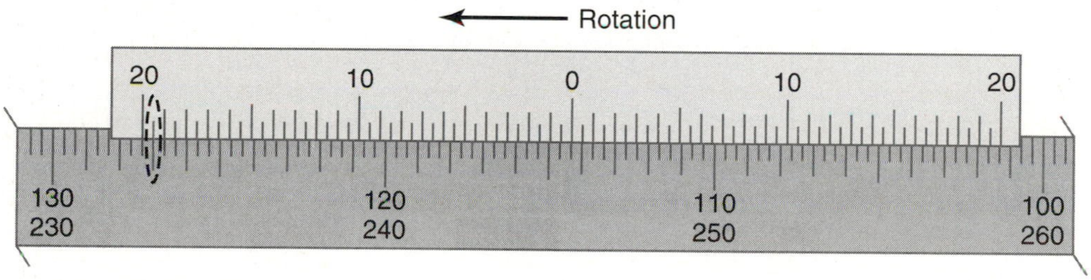

Figure 7-42 Practice Vernier, right turned.

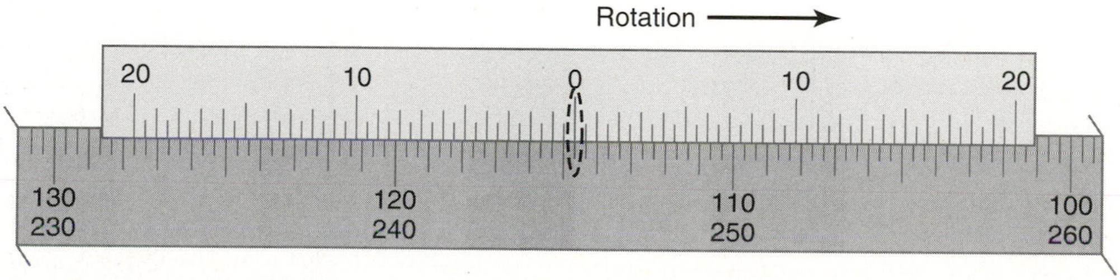

Figure 7-43 Practice Vernier, left turned.

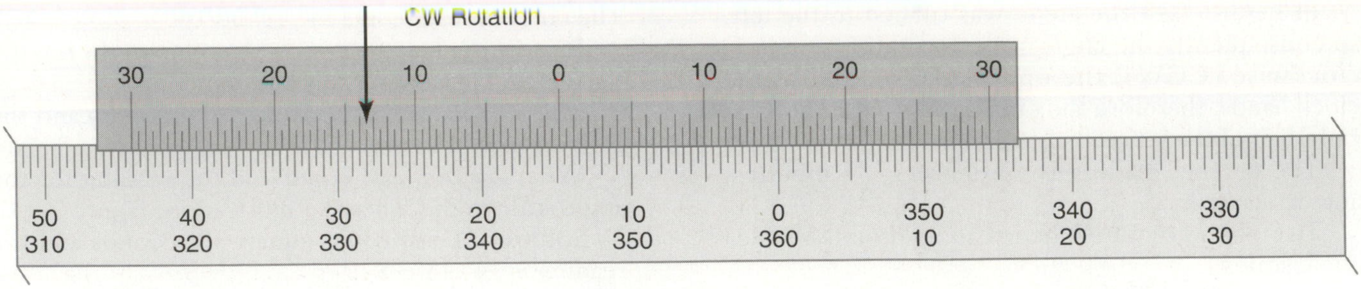

Figure 7-44 Practice Vernier, right turned.

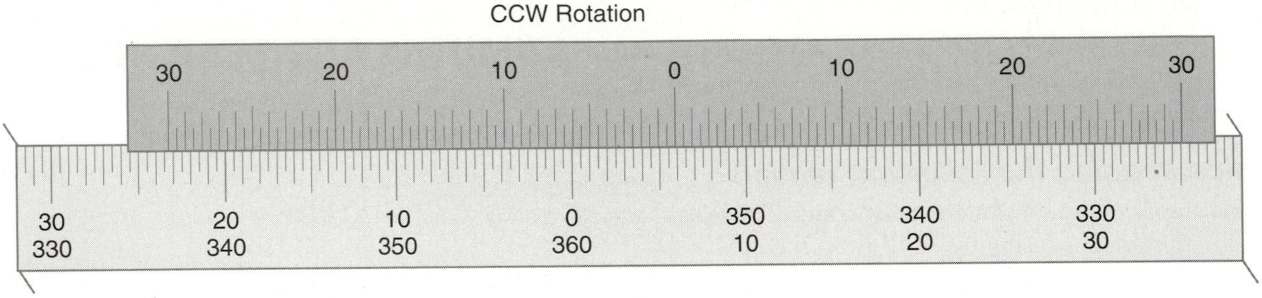

Figure 7-45 Practice Vernier, left turned.

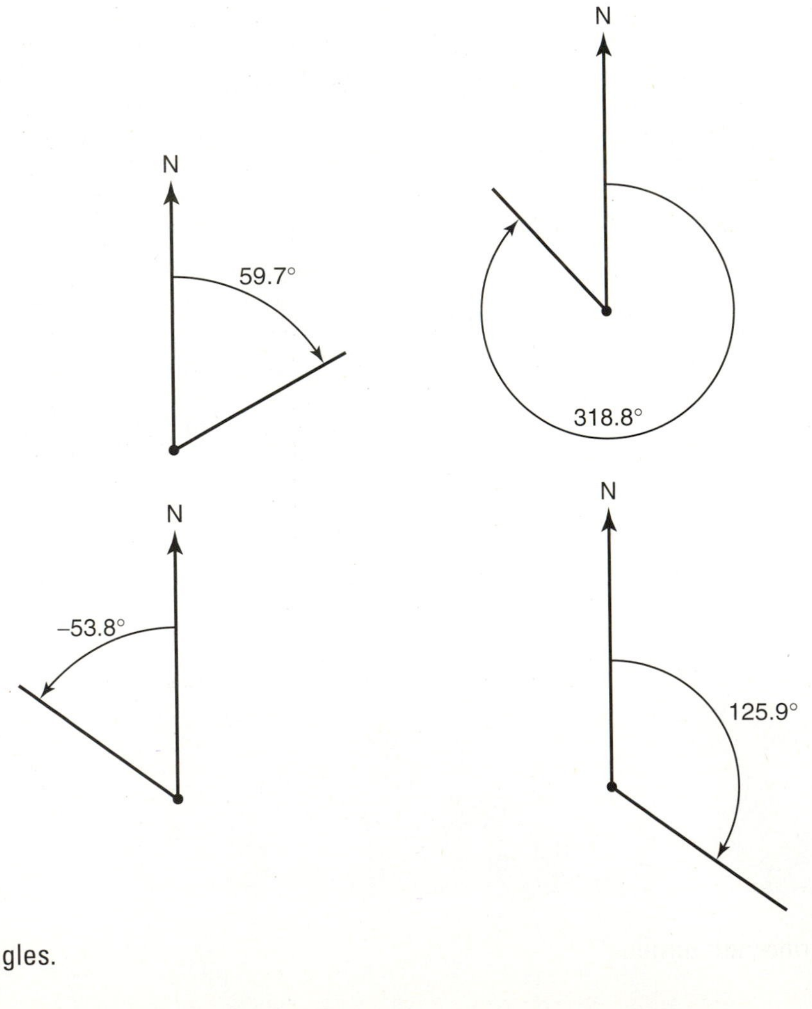

Figure 7-46 Azimuth angles.

AZIMUTHS

An **azimuth** is an angle measured from north. In mapmaking and surveying, different references for north are used. Geographical north and magnetic north are two of these. Geographical north is the physical North Pole. All lines of longitude pass through geographical north. Magnet north is the pole for the earth's magnetic field. The location of the magnetic north pole is not constant. Magnet north moves on a regular basis.

An azimuth angle will have one of these reference points and it will be a continuous angle turned either to the right (CW) or to the left (CCW). The direction the angle was turned must be included with each angle recorded. A common method for indicating the direction the angle was turned is to use a plus sign (+) to indicate turning to the right and a minus sign (–) for turning to the left. This is illustrated in Figure 7-46.

Note: Many individuals do not use the + in front of a right azimuth, they only include the – in front of a left turned azimuth.

Azimuth angles will range from 0° to 360°. One additional point to remember is that because azimuth angles use the full circle and are turned to the left and right, the same telescope position can have two different values. Study Figure 7-47.

In this example, the right turned angle of 64.2 degrees is the same as the left turned angle of –259.8 degrees. When recording azimuths from instruments, it is very important to include the direction of rotation with the angle reading.

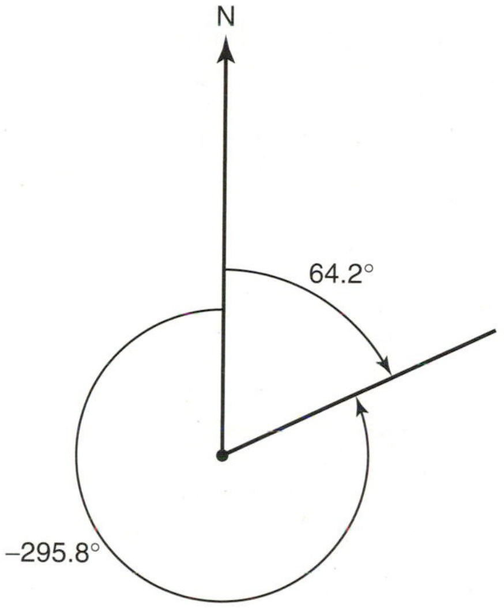

Figure 7-47 Plus and minus azimuths.

BEARINGS

Bearings are relative angles. They are used to express the direction of travel from one point to the next. The direction of travel is determined by dividing the circle into four quadrants. The centerlines of the quadrants are aligned with north and south, and east and west. The quadrants can be labeled northeast (NE), southeast (SE), southwest (SW), and northwest (NW), Illustration A or numbered as in Figure 7-48.

Using four quadrants restricts bearing angles to values between 0° and 90°. Bearing angles are always referenced to the north-south line. In the NW and NE quadrants zero degrees is at North. In the SW and SE quadrants zero degrees is at South. A bearing angle by itself could be in any one of the four quadrants. Therefore, it is important to always include the compass directions. Bearings will always start with either N or S and will always end with either E or W. Three examples of the correct way to record bearings are illustrated in Figure 7-49.

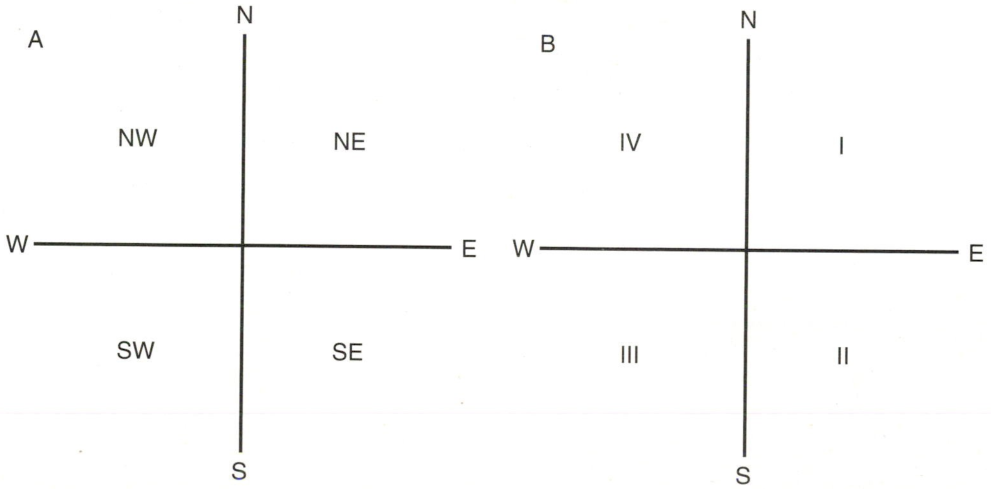

Figure 7-48 Bearing quadrants.

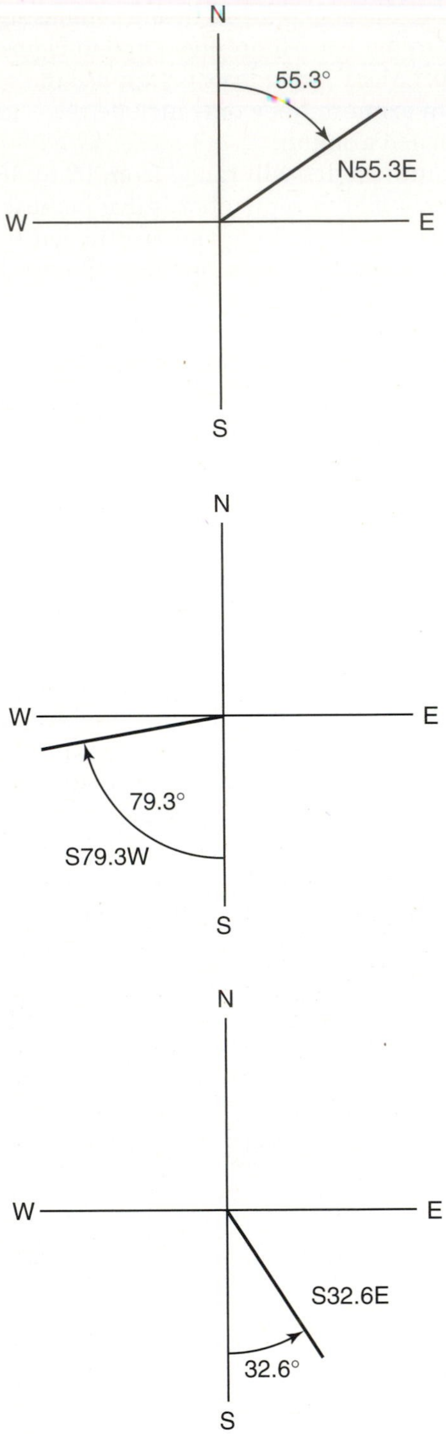

Figure 7-49 Examples of bearings.

CONVERTING AZIMUTHS AND BEARINGS

Occasionally it is necessary to convert from azimuths to bearings and from bearings to azimuths. When converting from azimuths to bearings, it is important to remember that azimuths are measured from north, but bearings are measured from a north-south line. It is also important to remember that there are 360° in a circle and 180° in a half circle.

Converting from azimuths to bearings can be accomplished with a calculator. For example, an azimuth of 42.3 degrees is a bearing of N42.3E because it is a right turned angle and being less than 90 degrees it is in the northeast quadrant. An azimuth of –95.6 degrees is a bearing of S84.4W because the angle was turned to the left and is greater than 90 degrees. This places the angle in the southwest quadrant. One hundred eighty degrees minus 95.6 degrees leaves 84.4 degrees. For the individual that is not comfortable with math, a diagram will make the conversion easier and reduce the chance of making mistakes. Figure 7-50 contains the diagram for converting an azimuth of 102.3 degrees to a bearing.

To use the graphical method, start by drawing the NESW grid and then add to the drawing the known angle. In this example: +102.3 degrees. Knowing bearings are referenced to the north-south line it is easy to see that the correct answer is determined by subtracting 102.3 degrees from 180 degrees.

An azimuth of 102.3° is a bearing of S77.7E.

Figure 7-51 is an example of the diagram that can be used to convert an azimuth of –285.8 to a bearing.

TIP

Note: it is important to remember that the (–) sign is used for two purposes. It is used to indicate the direction the angle was turned and it is used to indicate subtraction when doing math. Do not confuse the two uses when converting azimuths to bearings.

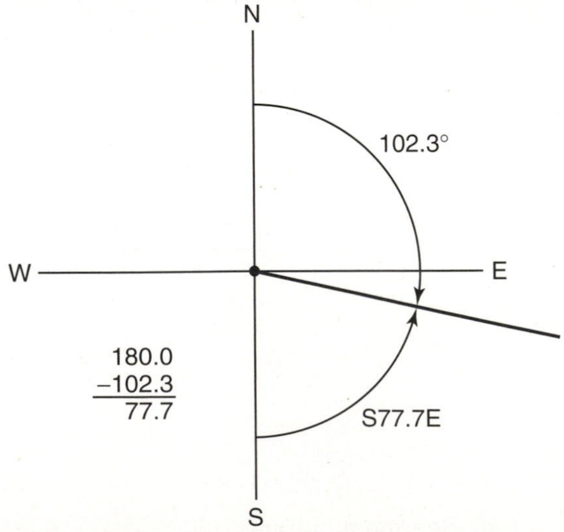

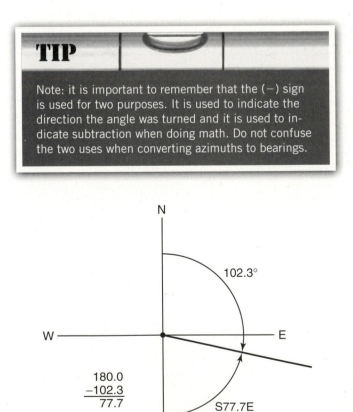

Figure 7-50 Converting an azimuth to a bearing.

Bearings are converted to azimuths using the same principles. For every bearing there are two possible azimuths, a plus azimuth (CW) and a minus azimuth (CCW). Figure 7-52 is the diagram used to convert the bearing S23.0E to a plus azimuth.

The conversion of the bearing S23.0E to a minus azimuth is illustrated in Figure 7-53.

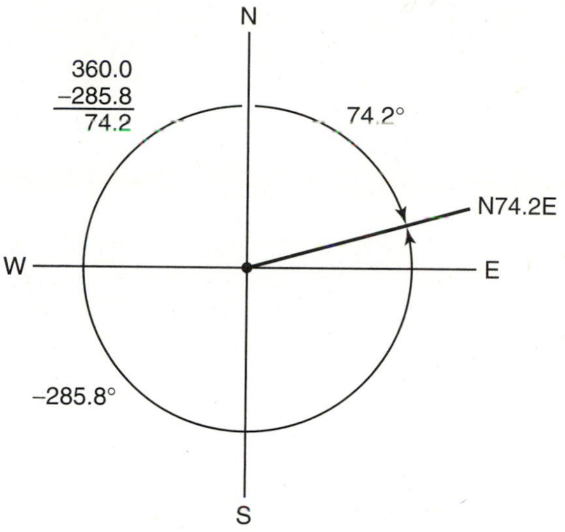

Figure 7-51 Converting an azimuth to a bearing.

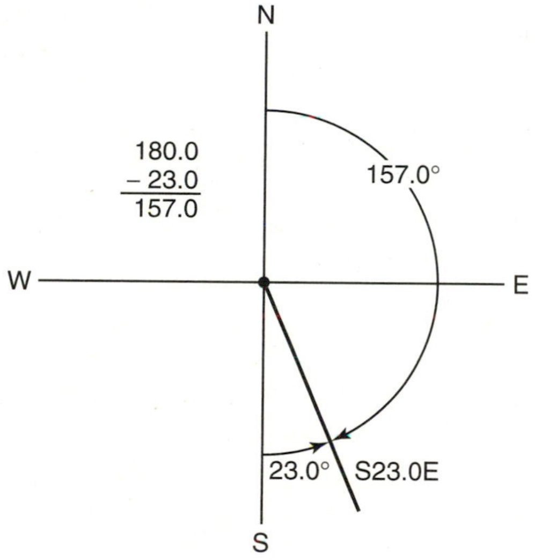

Figure 7-52 Converting a bearing to a plus azimuth.

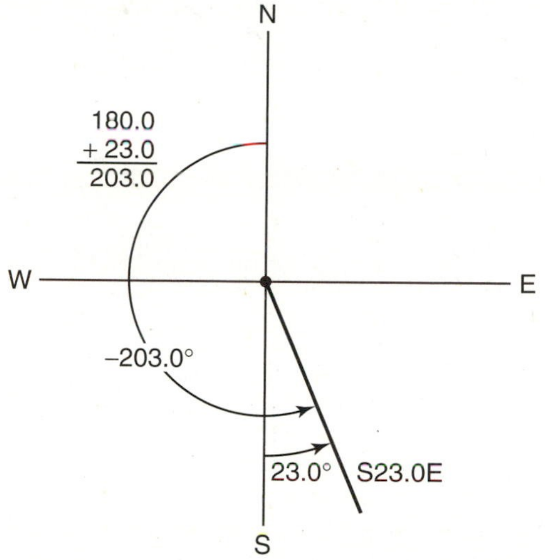

Figure 7-53 Converting a bearing to a minus azimuth.

Summary

This chapter explained the parts of an angle, indirect and direct ways they can be measured, and several applications of these principles. Both the indirect and direct methods of measuring angles are useful in land measurement and surveying for nonengineering purposes. Angles are used in many different applications when working with utilities, horticulture, agriculture, and general construction.

Student Activities

1. Explain the steps for laying out an angle using the 3-4-5 method.
2. Explain the steps for laying out an angle using the chord method.
3. Determine the + azimuths for the following bearings.

   N67.5E
   N34.8W
   S37.8E
   S55.8W

4. Determine the bearings for the following azimuths.

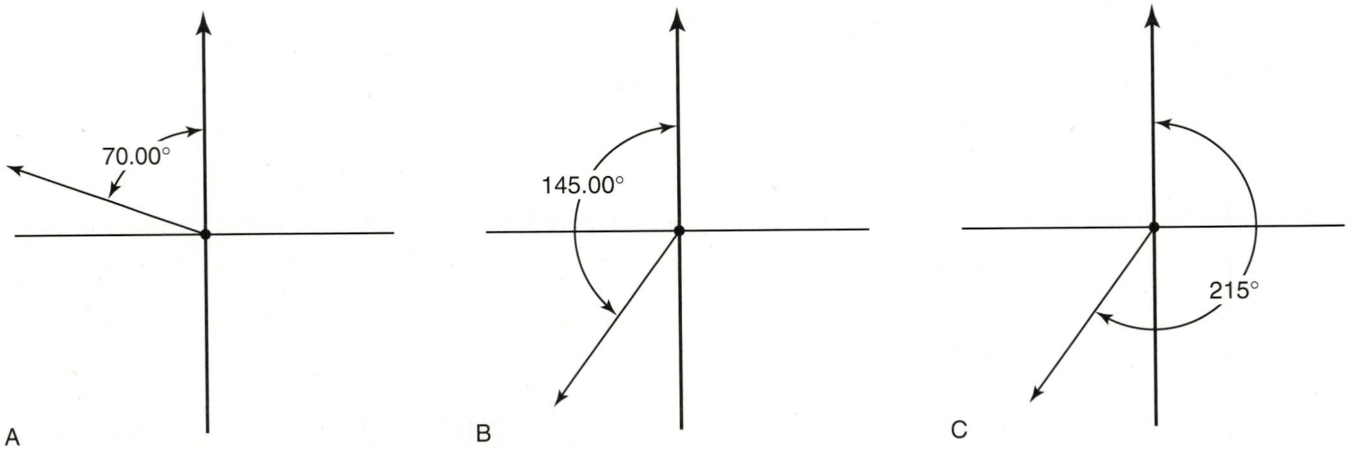

5. What are the opposite angles for the azimuths in the previous problem?

CHAPTER 8

Topographic Survey

Objectives

After reading this chapter the reader should be able to:

- Explain the features of a topographic map.
- Understand the importance of selecting the right locations for stations.
- Understand the difference between the two main types of topographical surveys.
- Draw topographic maps.
- Know the characteristics of contour lines.
- Know how to interpolate between two points.

Terms To Know

topographic map
contour lines
hachure marks
interpolating
USGS
scale
7.5 minute quadrangle
control point
proportional distance

INTRODUCTION

Topographical surveying is used to collect the information required to draw topographic maps. A **topographic map** is drawn with distances to scale, as a road map, but it also includes information about the surface of the earth. The information for the surface will include the relief (elevation) and the position of the natural and man-made features. The relief is described using lines of constant elevation called **contour lines**. The use of contour lines is the distinguishing characteristic of topographic maps. The use of contour lines produces a three-dimensional map on a two-dimensional surface. The natural and man-made features are represented by a system of symbols and may include their elevations. This chapter will discuss uses of topographic maps, methods of collecting data for topographic maps, and the procedures for drawing topographic maps.

TOPOGRAPHIC MAPS

Topographic maps are very useful for planning construction and land use because they are three-dimensional. With a topographic map, the reader can determine the elevation of any point on the map, the distance between any two points, and the slope between any two points. A topographical map can also be used to draw a profile of any route across the map. Topographic maps are the preferred source of information for construction, landscape design, and community planning.

Maps for a small area, town or state, are usually drawn to a scale. Map scale is defined as the ratio of the distance on the map compared to the corresponding distance on the ground. If a map had a scale of 1:100, every inch, foot, etc. on the map would represent 100 units on the ground. Another method is to specify the units of the scale. For example 1 inch = 100 feet. Maps for large areas, countries, or the world, are projected. More information on map projections is included in Chapter 10.

The United States government has produced topographic maps of the entire country. A common size of map is called a 7.5 minute quadrangle, because they cover 7.5 minutes of latitude and longitude. The scale that is used determines the amount of detail provided by the map. For the United States Geological Survey (**USGS**) maps the common scales are 1:24,000, 1:100,000, and 1:250,000. The 1:24,000 scale is the most popular. It shows sufficient detail to meet the needs of hikers, campers, hunters, and large scale planning.

Maps for construction and planning of typical sites are usually produced for each job. These maps are used for planning roads, sidewalks, retaining walls, and other types of construction. A scale must be selected that provides the desired level of detail, without making the dimensions of the map too large to be useable. For example, if the construction area measures 800 feet × 1500 feet then a map with a scale of 1 inch = 100 feet would measure 8 inches by 15 inches. A useable size, but an object that was 10 feet square would only be 1/10 inch square on the map. If the scale was changed to 1 inch = 10 feet, then the map would measure 80 inches by 150 inches—not practical. More information is presented on map scales in a later section in this chapter.

Topographic Map Symbols

Topographic maps use symbols to represent different man-made and natural features. The symbols show the location and an estimate of the size of the features. Symbols are used for features such as lakes, roads, buildings, towers, benchmarks, and bogs. Figure 8-1 is an example of the symbols commonly used by the USGS. Other government and private agencies also use symbols. These may not be the same as the USGS symbols, but they will be similar.

Contour Lines

Contour lines are used to show the relief of the earth's surface. A contour line is an imaginary line that connects points of equal elevation. The points may be on dry land or on the land surface underwater. Before contour lines can be drawn, the survey crew must record the location and elevation of numerous points within the area being surveyed. In addition, the survey party should determine the contour interval that will be used. The decision on contour spacing is another judgment based on the volume of the data and the use of the data. Many USGS maps use a

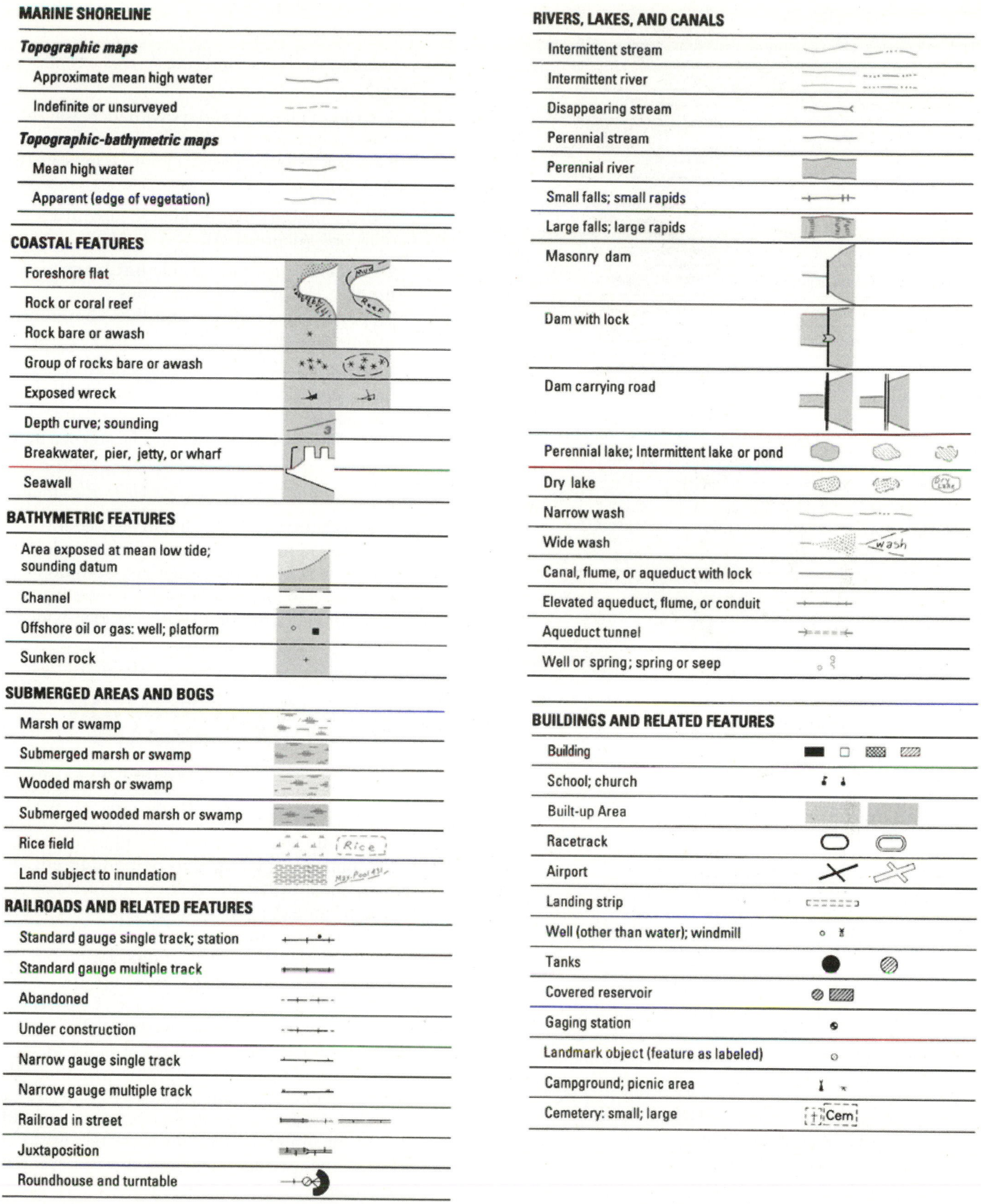
MARINE SHORELINE
Topographic maps
Approximate mean high water
Indefinite or unsurveyed
Topographic-bathymetric maps
Mean high water
Apparent (edge of vegetation)
COASTAL FEATURES
Foreshore flat
Rock or coral reef
Mud
Reef
Rock bare or awash
Group of rocks bare or awash
Exposed wreck
Depth curve; sounding
3
Breakwater, pier, jetty, or wharf
Seawall
BATHYMETRIC FEATURES
Area exposed at mean low tide; sounding datum
Channel
Offshore oil or gas: well; platform
Sunken rock
SUBMERGED AREAS AND BOGS
Marsh or swamp
Submerged marsh or swamp
Wooded marsh or swamp
Submerged wooded marsh or swamp
Rice field
Rice
Land subject to inundation
RAILROADS AND RELATED FEATURES
Standard gauge single track; station
Standard gauge multiple track
Abandoned
Under construction
Narrow gauge single track
Narrow gauge multiple track
Railroad in street
Juxtaposition
Roundhouse and turntable
RIVERS, LAKES, AND CANALS
Intermittent stream
Intermittent river
Disappearing stream
Perennial stream
Perennial river
Small falls; small rapids
Large falls; large rapids
Masonry dam
Dam with lock
Dam carrying road
Perennial lake; Intermittent lake or pond
Dry lake
Narrow wash
Wide wash
wash
Canal, flume, or aqueduct with lock
Elevated aqueduct, flume, or conduit
Aqueduct tunnel
Well or spring; spring or seep
BUILDINGS AND RELATED FEATURES
Building
School; church
Built-up Area
Racetrack
Airport
Landing strip
Well (other than water); windmill
Tanks
Covered reservoir
Gaging station
Landmark object (feature as labeled)
Campground; picnic area
Cemetery: small; large
Cem

Figure 8-1 USGS topographic map symbols.

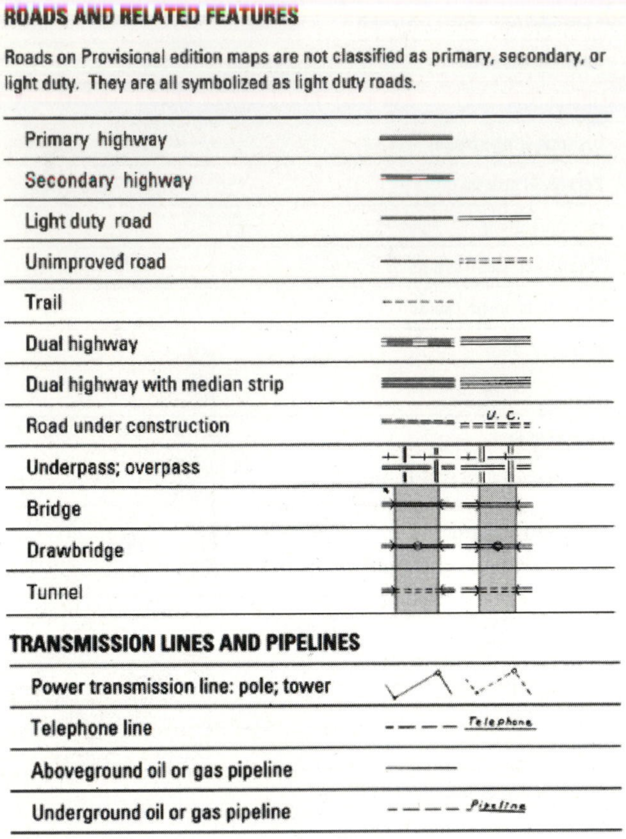

Figure 8-1 *(continued)*

contour interval of 200 feet; this is not very useful for construction or landscaping. A contour interval of two or five feet is very common for construction topographic maps, but there are situations where an interval as small as six inches or as large as ten feet would be appropriate. This is discussed more in the following section on collecting data.

Characteristics of Contour Lines

Government agencies and other groups that draw contour lines have established several guidelines for individuals producing topographic maps.

1. Contour lines are continuous. Some contour lines may close within the map, but others will not. In this case, they will start at a boundary line and end at a boundary line.
2. Contour lines are relatively parallel unless one of two conditions exists. They will meet at a vertical cliff and they will overlap at a cave or overhang. If contour lines overlap, the lower elevation contour should be dashed for the duration of the overlap.
3. A series of V shapes indicates a valley and the Vs will point to higher elevation.
4. A series U shape indicates a ridge. The Us will point to lower elevation.
5. Evenly spaced lines indicate an area of uniform slope.
6. A series of closed contours with increasing elevation indicates a hill and a series of closed contours with decreasing elevation indicates a depression.
7. Closed contours may be identified with a + (hill) or – (depression).
8. Closed contours may include **hachure marks**. Hachures are short lines perpendicular to the contour line. They point to lower elevation.
9. The distance between contour lines indicates the steepness of the slope. The greater the distance between two contours the flatter the slope. The closer the lines are the steeper the slope.
10. Contour lines are perpendicular to the slope.
11. A different type of line should be used for contours of major elevations, for example, at 100-, 50-, and 10-foot intervals. Common practice is to identify the major elevations lines, or every fifth line, with a bolder, wider line.

COLLECTING INFORMATION FOR TOPOGRAPHIC MAPS

The scale of the map and the contour interval that will be used influence the equipment and procedures used to collect the data. Before the instruments are set up someone must decide on the type of topographic survey that will be conducted, and the contour interval. Knowing the contour interval is important because it is misleading to produce a map with small contour intervals when the differences in the stations' elevations were much larger. Drawing a map with two-foot intervals when the differences in station elevations were 10 feet or more can result in bad designs and costly mistakes. This can be a problem with computer generated maps because most topographic mapping programs will allow the operator to select the desired contour interval. The operator can produce any number of maps with different contour intervals from the same data set. This can result in topographic maps that do not represent the topography.

The selection of the contour interval will also influence the amount of data that must be collected. Figure 8-2 illustrates the number of stations that are

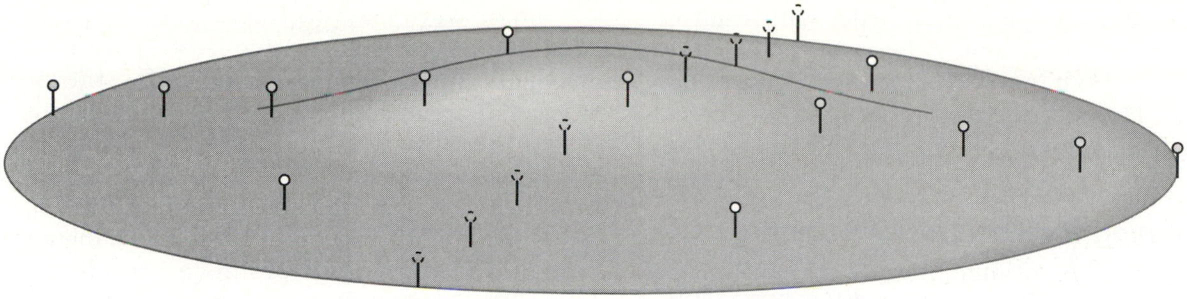

Figure 8-2 Stations required to map a hill.

required to define a hill. It can also be used to illustrate the effect of contour interval. As discussed in the previous section, reducing the contour interval increases the number of stations that should be measured. For example, in Figure 8-2 the number of stations would increase dramatically if the interval were reduced by half. It should be noted that it is better to err on the side of collecting too much data. It is easy to throw out data that is not needed. To collect missing date requires another survey.

The choice to produce a topographic map by hand or by computer hinges on several factors. First, to draw a topographic map by computer you must have the computer, the program, and the ability to use both. Second, maps are usually drawn on paper larger than 8.5 by 11.0 inches. This requires a different printer or a plotter. Third, drawing a map by hand will require more time and some drawing ability. A hand drawn topographic map represents many hours of work. Computer drawn maps are worth their cost for large or complicated areas, but hand drawn maps are still useful for small areas and for individuals that do not have a computer or computer program and have the time to hand draw maps. A discussion of topographic drawing computer programs is beyond the scope of this text. Change the sentence to "Before selecting topographic software the reader should consult a professional surveyor and/or read product reviews." This chapter will discuss how to draw simple maps by hand.

Map Scales

All maps are drawn to scale because it is not practical to have a map the same size as the area. Various scales are used for different agencies and purposes, but they are always integers and usually multiples of 5 or 10. A common way of expressing the scale of a map is to use the notation of a number, usually one, a colon, and a second number. The scale on a map could be written as 1:10. On a map with this scale, each foot on the map represents 10 feet on the ground. The common USGS scale of 1:24,000 means that each foot on the map represents 24,000 feet on the ground or each inch on the map represents 24,000 divided by 12 or 2000 feet. It is important for the reader to study the map scale to insure they are interpreting the distances correctly.

The mapmaker must decide the appropriate size and scale for the size of the area surveyed and the amount of detail that is required. A map scale that uses multiples of ten reduces the complexity of the math when hand drawing a map.

Number of Stations

One aspect of topographic surveys that is constant across all methods is the care that must be taken when selecting the stations. The first response is to take all you think you need and then a few more. It is important to remember that three items of information must be measured and recorded for each station, two for the location and one for the elevation. An excessive number of stations will greatly increase the time required to record the data and the volume of data. If an insufficient number of stations are recorded, the user of the map may not have sufficient information to make accurate decisions. The optimum number of stations is the minimum required to define the topography, and the additional sites, for the use of the data.

The intended use of the data is a very important factor to consider. Assume the purpose of the survey is to develop a map that will have a large scale and will primarily be used for general planning. In this situation, the number of stations can be less because the precision of the map will be low. The opposite is true if the purpose of the survey is to develop a map that will be used to produce a drainage plan for an area that has very little slope. In this case it may be necessary to record a large number of stations. Recording elevation changes as small as six inches, coupled with needing a small scale, will result in a large number of stations. A large area using these criteria will result in a very large data set and a large size map. These criteria are used for this situation because it is important to measure small changes in elevation.

Determining the Number of Stations

Determining the optimum number of stations is an important aspect of topographical surveying. It is possible that the person drawing the map has not seen the site. The only information they have is what the surveyors have recorded. They will assume all of the man-made and natural features have been documented and that the surface between any two stations is a straight line. A depression or tree that was left out of the data will not appear on the map. This kind of mistake can lead to very expensive errors during the design and construction phase of a project. It is the responsibility of the survey team to record the appropriate data. The ability to select the right stations is more of an art than a science. For this reason it is usually done by the most experienced member of the team.

In Chapter 6 , Figures 6-4 & 6-5 were used to illustrate the number of stations required to define the profile of a ditch. Greater care must be taken when the surveying team records surface features such as ditches, hills, and depressions for topographic maps because topographical maps are three-dimensional.

Figure 8-3 is a example of a simplified ditch. In this example three cross sections and one additional point were used to record the shape of the ditch. The simple ditch in Figure 8-3 would require at least 20 to 25 stations to define the ditch and a small area around it in three dimensions. To draw or plot in three dimensions you must be able to locate each station and know the elevation for each station. Locating each station will require at least two dimensions, or an angle and a distance (X and Y). The third piece of information is the elevation of the station (Z).

Not only must the team decide on the number of stations needed to define a natural or man-made feature such as a ditch, they must also remember that the length and shape of many features are not constant. As the shape and elevation of the feature changes, additional information must be collected to define the changes. Figure 8-2 and Figure 8-3 illustrate that when small changes in elevation are important or a complex structure is being surveyed 20 or more stations may be required. It is also possible that the presence of this hill is of no consequence for the construction being planned. In this case, it would be ignored. The determining factor is always the use of the data.

In summary, selecting the stations is a very important part of topographical surveying. The surveying team must balance the uses of the data against the amount of resources required to collect the data. Even so, it is better to err on the side of collecting too much data than not enough. Someone must reconnoiter the site with an understanding of the purpose for the data, to determine the location of the stations.

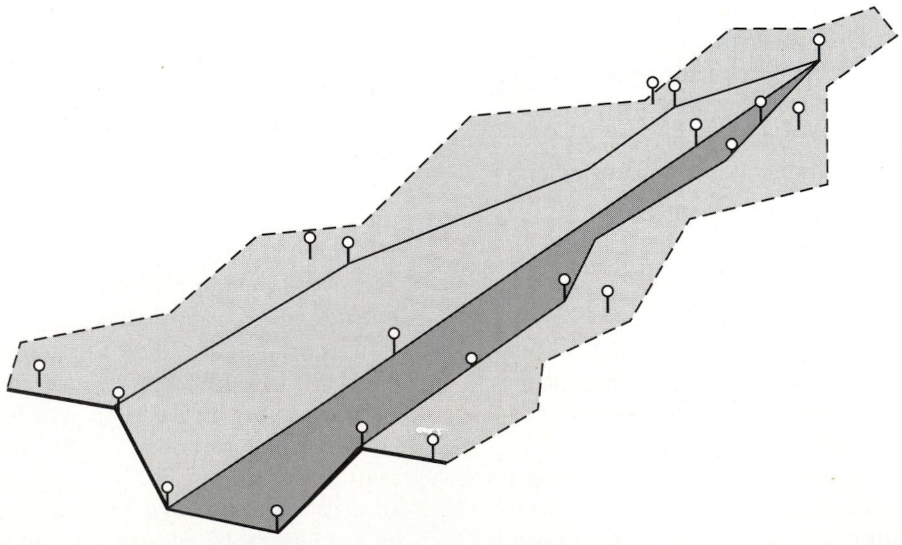

Figure 8-3 Number of stations for a ditch.

Instrument Site

Selecting the optimum site for the instrument can have a dramatic effect on the time required to complete the survey. The ideal site is one that will allow the instrument to see all of the stations for the survey. The stations will include:

- All of the corners on the property,
- The stations required to define the topography,
- The stations that are used to locate and establish the elevation of the man-made and natural features of the property.

For a small area, relatively flat and devoid of trees or structures, site selection will be easy. As the area being surveyed becomes larger, the topography more varied, and the number of trees and other features increases, site selection becomes more complicated. If all of the stations can not be seen from one instrument station, then one or more turning points will be required. The use of turning points will increase the time required to complete the survey and increase the opportunity for errors. In addition, the location of the instrument for each turning point must be carefully documented when the angle and distance method is used. This will be discussed in greater detail in a later section.

When turning points are used, the note keeper must insure that sufficient information is recorded to determine whether or not the turning point can be used as one of the stations for the map. When the top of a stake is used as a turning point, that elevation cannot be used as a station in the map because the elevation of the top of the stake does not represent the elevation of the earth at that point. When a mark on a sidewalk, utility access cover rim, etc. is used as a turning point, the elevation of the turning point can be included in the topographic map. This point can be included because the elevation of the turning point represents the elevation of the earth's surface. It is very important to record the structure used for the turning point in the notes for the survey. The structure used and the location of the turning point is part of the sketch. It is important to include this information on the sketch to prevent someone from using the data incorrectly. A well drawn and documented sketch is a requirement of good survey notes.

Control Point

A **control point** is a station in the survey that is used as a reference, similar to a turning point or a benchmark. Control points may be established around a large structure or a large area to reduce the distance that must be traveled to check the elevation of a point. In this use they are very similar to a benchmark. Control points can also be used when multiple instrument setups are required to complete a survey. They provide a means for checking for errors that can occur when turning points are used. In this use, a control point would be a station that is used with more than one instrument setup.

The structure used for a control point should be similar to a turning point and the same rules apply for including the elevation in the topographic map. The use of a control point serves as a check for error when using turning points in an angle and distance topographic survey and a check for instrument height when using the grid method. These two methods will be explained in detail in a later section of this chapter.

In Figure 8-4, station H is a control point. At instrument position one (IP1), the location and elevation of the turning point were recorded to extend the survey. In addition, the location and elevation of station H were also recorded. Station H may be a point that will be included in the topographic map, or it may not be included. The determining factor is the type of structure that was used to mark the location of the station. See the previous section on turning points. When the instrument is set up at IP2, a backsight is taken on the TP to establish the height of the instrument and a foresight is recorded on the control point (H). The elevation of station H from IP2 is compared with the elevation of station H from IP1. If the elevations are not the same, an error has occurred.

Control points also help blend more than one set of data. When a large area, or an area with a large number of stations, is surveyed it may not be possible to complete the survey in one day. Using a common benchmark is standard practice to combine data, but including at least one control point will help insure the two data sets can be combined.

INTERPOLATING

Interpolating is a mathematical process that determines the location of a contour line between two stations. The location is determined by calculating the proportional distance. When drawing maps by hand this is an arduous process because of the number of calculations that must be completed. Setting up a spreadsheet will greatly reduce the amount of time required to draw a topographic map by hand.

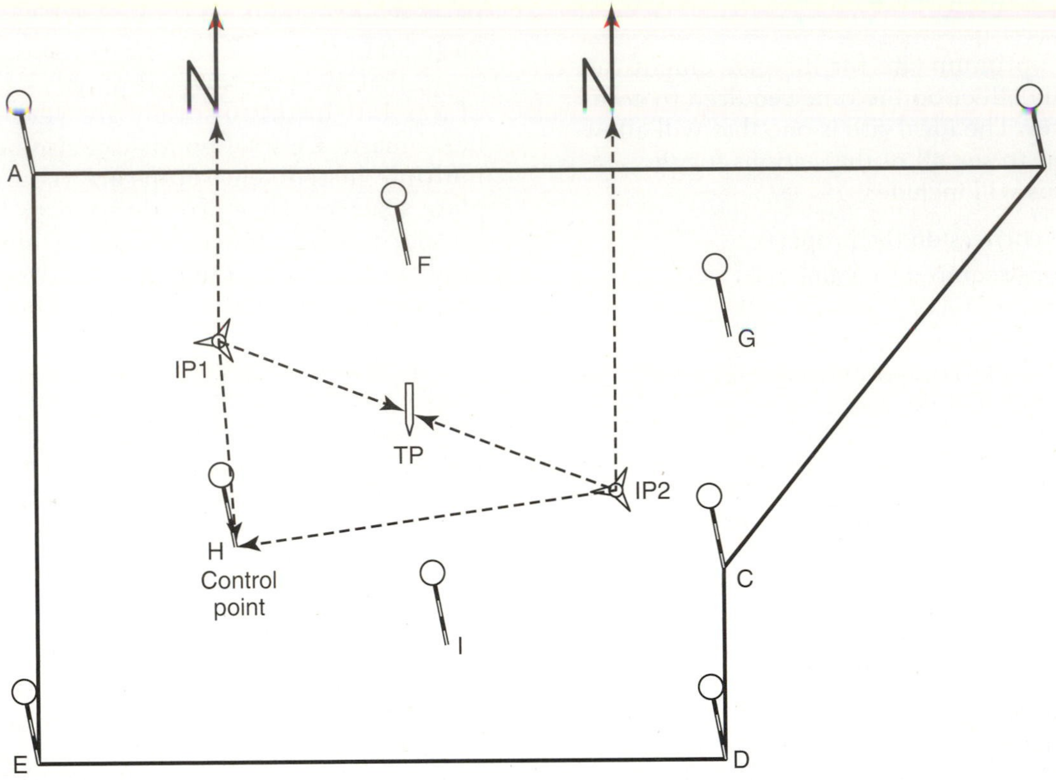

Figure 8-4 Control point.

Interpolating for Grid

Assume a 100-foot contour will pass between two stations. One station is at an elevation of 99 feet and the second station elevation is 101 feet. In this example, the 100 foot contour would be an equal distance from both stations and so it would be drawn halfway between the two stations. Unfortunately, topographic interpolating is not always this easy. Determining the location of a 100 foot elevation contour between stations with elevations of 103.5 feet and 97.2 feet is more difficult and would encourage the use of a calculator or a computer spread sheet.

For those situations that require a calculator or spreadsheet, a formula can be used. The formula determines the proportional distance for the contour line from the highest elevation.

$$\text{Proportion} = \frac{\text{High elevation} - \text{Contour elevation}}{\text{High elevation} - \text{Low elevation}}$$

The proportion determined by the formula, times the distance between the two stations on the map, determines the distance between the station with the highest elevation and the contour line.

Interpolating Example

Determine the location of a 98.0 foot contour line on a topographic by grid map when the two stations are 101.0 feet and 96.0 feet elevation. The map is drawn with a scale that results in one inch between stations.

$$\text{Proportion} = \frac{101.0 - 98.0}{101.0 - 96.0} = 0.60$$

If the contour grid spacing is one inch, then contour line crosses an imaginary line between the two stations a distance of 1.0 inch $\times$ 0.6 = 0.6 inches from the station with the highest elevation, Figure 8-5.

The same process is used when the distance between the grid stations is not equal to one inch, and for interpolating the diagonals for a grid survey. A map with a one inch grid will have 1.4 inches between the diagonal stations, Figure 8-6.

$$\begin{aligned}\text{Diagonal distance} &= \sqrt{a^2 + b^2}\\ &= \sqrt{1^2 + 1^2}\\ &= \sqrt{2}\\ &= 1.4\end{aligned}$$

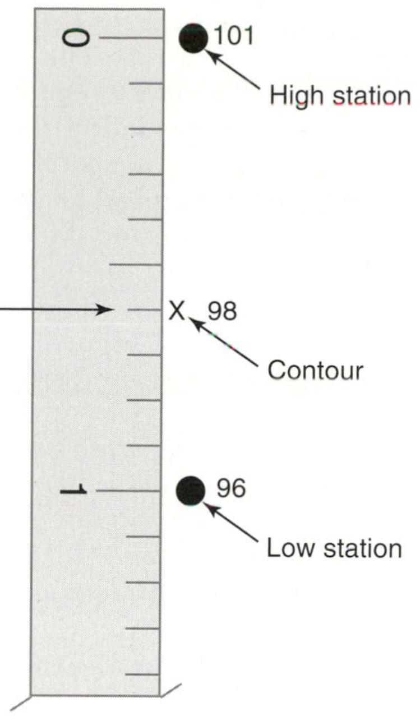

Figure 8-5 Locating contour line between stations.

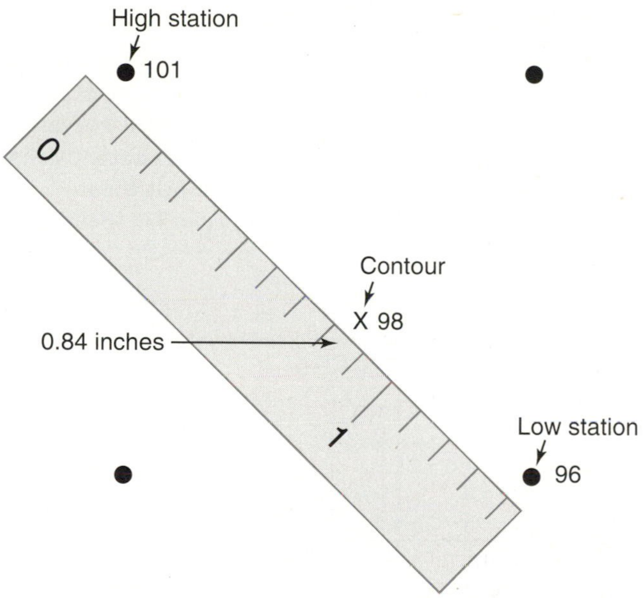

Figure 8-6 Locating contour line between diagonal stations.

The location of the contour line is determined by multiplying the proportion times the distance between the two stations on the map. Assuming the same station elevation and contour:

$$\text{Contour distance} = 1.4 \times 0.6$$
$$= 0.84$$

The contour line crosses an imaginary line between the two diagonal stations at 0.84 inches from the station with the highest elevation, Figure 8-6.

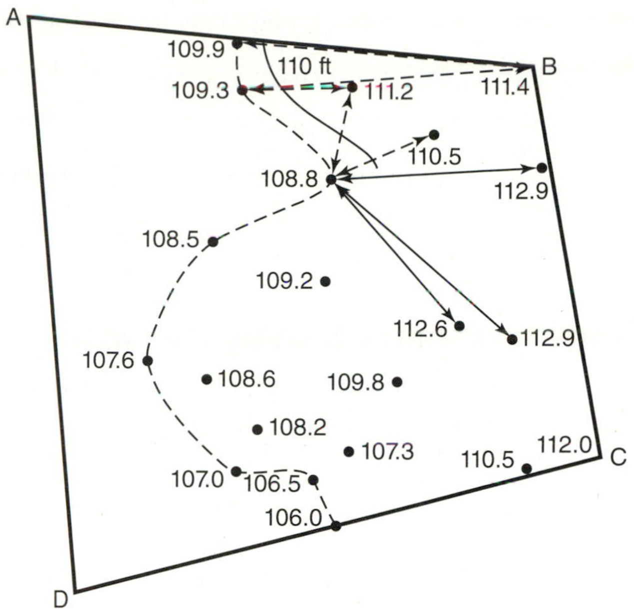

Figure 8-7 Interpolating for angle and distance map.

This process is repeated for each pair of stations and the grid diagonals that the contour passes between until the contour line is completed or it intersects a boundary.

Interpolating for Angle and Distance Maps

Interpolating for angle and distance surveys uses the same formula. The primary difference is the distance between stations is more variable and the map drawer must be careful that they do not miss some stations. In Figure 8-7 the 110 foot contour has been partially drawn. The dashed lines with double arrows indicate the stations that were used to locate the portion of the contour line that has been drawn on the map. The solid lines with double arrows indicate the stations that would be used to locate the next three points on the contour line.

The map drawer must carefully analyze the map to insure all possible combinations of stations have been considered. Some of the stations may be several inches apart on the map.

TYPES OF TOPOGRAPHIC SURVEYS

The data for topographic maps can be collected using several different methods. A popular method, especially for a large area, is aerial photography. Using special cameras and computers, surveyors can fly over an area and develop topographic maps from the photographs. This method is time-consuming and expensive, but for difficult terrain that will require a large amount of human resources, it can be more economical.

Two popular types of land based topographic surveys are grid, and angle and distance. These two method will be explained in this chapter.

Topographical Survey by Grid

The first step in conducting a grid topographic survey is to identify and locate the boundary corners. This is an example of where a traverse survey could be used. Information on traverses is included in the next chapter. After the boundaries are located, a grid is established, Figure 8-8. Elevations are recorded at each point the lines cross and at each intersection of a grid line with the property line. A common practice for identifying the stations is to use the rows and columns identified by letters and numbers. The grid identification system used is not important as long as it results in a unique identification for each station. For example, the identification for the station marked by a circle in Figure 8-8 is C3 and the one marked by a diamond is J8.

When establishing the grid, it should be aligned with the longest boundary. This will reduce the number of non-typical grids. For the illustration in Figure 8-8, stations J2 through J9 and A11 through C11 are non-typical grids. The distances between the non-typical station and the nearest station on the grid must be recorded. Otherwise, these stations will be located incorrectly on the topographical map. When non-typical grids are used and not identified, the person using the data would assume all of the stations were on the standard grid and the boundary of the area would not be drawn correctly. Figure 8-8 and Figure 8-9 illustrate this situation. Figure 8-8 is the grid of the area being surveyed. Figure 8-9 is how the map would be drawn if the mapmaker didn't know that the data contained non-typical grids. Figure 8-9 would be a major mistake.

The grid is the reference for locating all stations and features. In the grid method the location of the instrument is not important when drawing the map, but it is good practice to identify the location(s) of the instrument in the sketch that accompanies the data in the field book.

Spot Stations

Spot stations are stations that are important to the survey, but are not part of the grid. These stations may be taken to record the location and elevation of additional man-made or natural structures that are important to the survey, but are not located at a grid intersection. Spot elevations can be located by

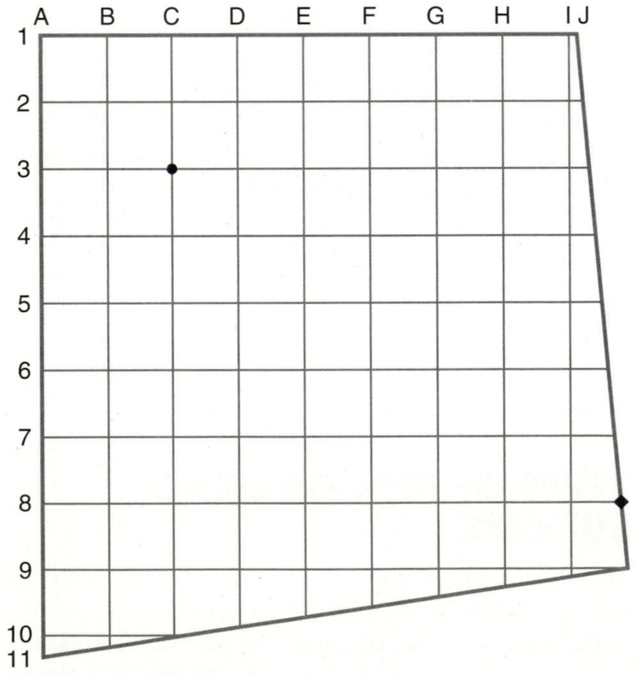

Figure 8-8 Example of survey grid.

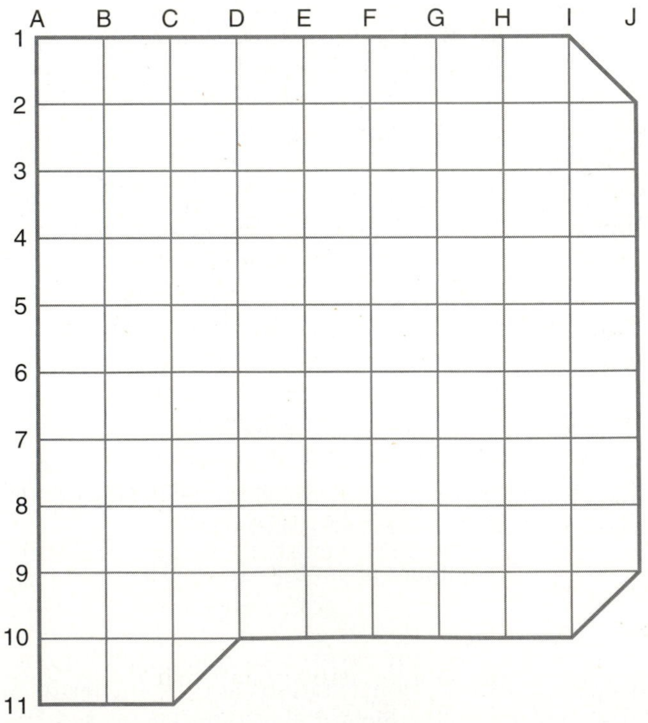

Figure 8-9 Incorrect grid.

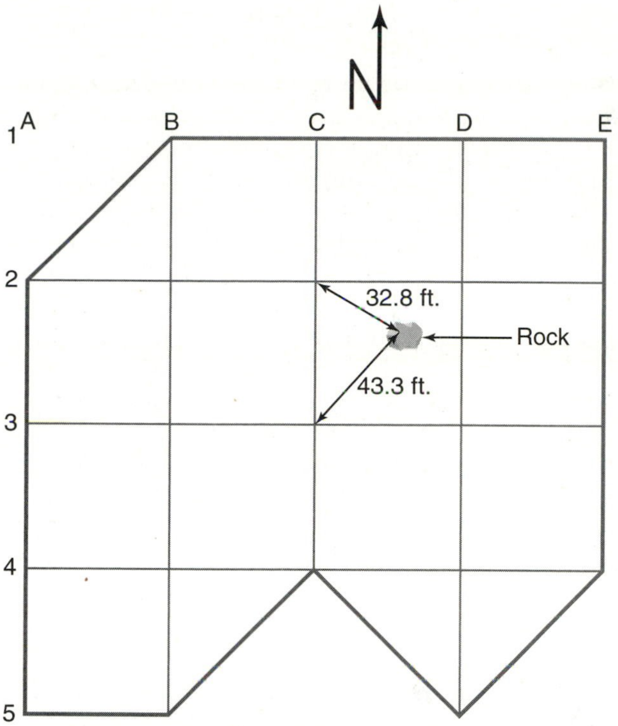

Figure 8-10 Spot station in a grid.

trilateration from two grid intersections. The rock in the illustration in Figure 8-10 is located 32.8 feet SE from station C2 and 43.3 NE feet from station C3.

Trilateration requires three dimensions. The third dimension is the grid spacing, which was previously defined by the positions of C2 and C3. The data table could include the spot elevations with the grid elevations. It would also be appropriate to include a separate table of spot elevations. The surveyor must also include the information that was used to locate the spot elevation in the field book.

Data Table

Topographic data can be organized in several different data tables. The style of table is not important. What is important is the quality of the data. The stations must be identified, and the data must be clear, complete, and easy to read and understand. Figure 8-21 in Page 136 is an example of a notebook page for a topographic uniform grid survey. The data table for the example in Figure 8-10 would have the appropriate number of rows and tiers, but there should not be a place to record the elevation for A1, C5, and E5. These stations do not exist on the parcel of land being surveyed.

Grid Spacing

A very important decision when using the grid method is determining the size of grid. Too small and an excessive amount of resources are used collecting

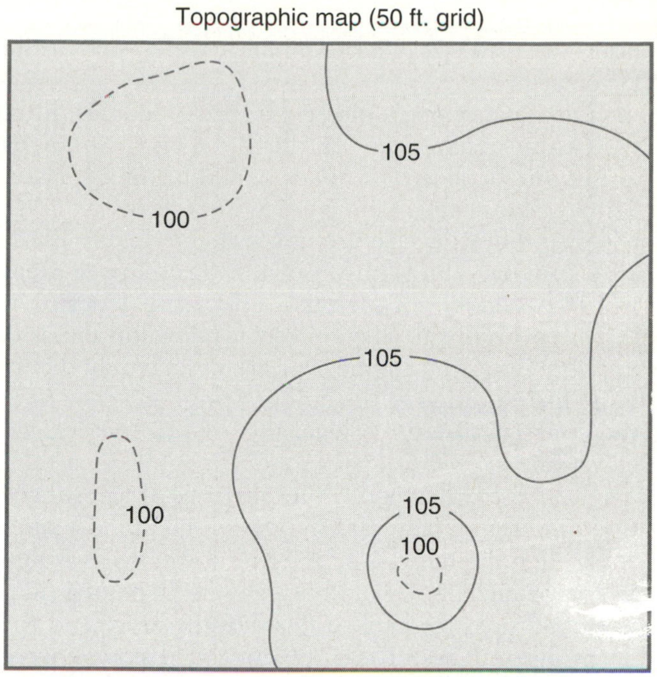

Figure 8-11 Topographic map with 50 foot grid.

data; too large and important information will be missed. The determining factors are the steepness of the slopes and the contour interval that will be used. Steep slopes and a small contour interval would require a small grid. A smooth surface and a large contour interval indicate a large grid should be used. This point is illustrated by studying Figure 8-11 and Figure 8-12. Both figures use the same set of data. The difference is that in Figure 8-12 every other row

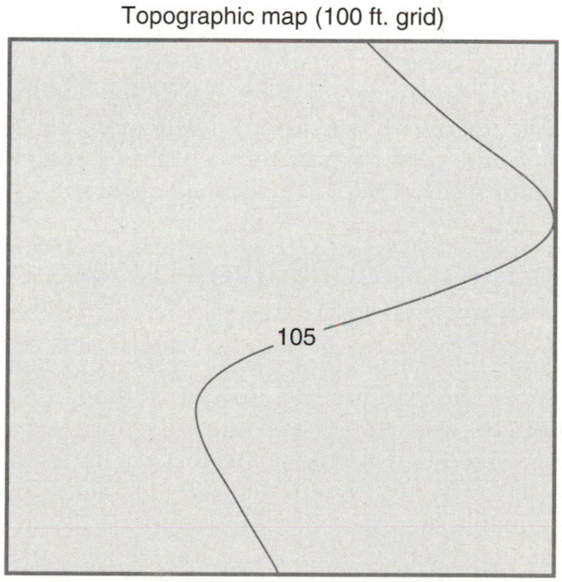

Figure 8-12 Topographic map with 100 foot grid.

and column was left out to simulate the effect of doubling the grid size. There is a clear difference in the two maps.

The issue of appropriate grid size becomes more confusing because under the right circumstances either one of these two maps would be appropriate. The smaller grid size in Figure 8-11 reveals four areas of lower elevation that are not visible in Figure 8-12. Not knowing the location of the three lower areas would probably not seriously affect construction if the purpose of the survey were to develop the area for a building. The engineer or architect would determine the elevation of the building floor and then have the construction crew level the area to the desired elevation.

If on the other hand, the purpose of the survey was to design a park and the design included a sidewalk, then the location of the four lower areas would be important. They would change the slope of a sidewalk and possibly cause the sidewalk to exceed the desired standards. If the 100-foot grid survey were all that was available, the designer would not know the three depressions even existed. A small area, 10 acres or less, with very little variability could be surveyed with a grid spacing of 50 or 100 feet. If the surface is more variable, it may be necessary to use a 25-foot or smaller grid.

It is always important to remember the grid size determines the number of stations that will be recorded. Ten acres with a 50-foot grid will result in 196 stations. If the grid is reduced to 25 feet, the number of stations increases to 729. Increasing the number of stations dramatically increases the time and other resources that will be required to collect and manage the data. For large areas it is customary to use a grid spacing of 100 feet or more. For example, 100 acres surveyed with a 25 foot grid results in 7056 stations. It is not good business for a surveying team to spend the amount of time required to record 7000+ stations unless the use of the data dictates this requirement. The client or the personal preferences of the head surveyor may ultimately decide the selection of grid size.

Topographical Survey by Angle and Distance

Locating the stations by angle and distance is an alternative to the grid method. In this method all of the stations are spot elevations because the only elevations that are recorded are the points that are important to the survey. When conducting an angle and distance topographic survey, numbers or letters usually identify the stations, and the stations are located by measuring the angle and distance from the location of the instrument. Therefore, it is very critical that the location of the instrument is measured and recorded. A common practice is to locate the instrument by measuring the distance from two or more of the boundary corners or two or more features within the survey area. When two or more features are used, the features must be located before the instrument can be located.

An angle must be measured from a point or line. Before stations in an angle and distance topographic survey can be identified, the survey crew must determine what will be used as the backsight for the angles. Historically, transits were the instrument of choice for angle and distance surveys because they measure angles accurately and most styles of transits included a compass in the base. The compass could be used to orient the instrument to magnetic or geographical north. North could be used as the backsight for all angles. An alternative is to use two or more easily identifiable, permanent landmarks as the backsight for the angles. Any two corners of the area being surveyed are the best. When landmarks such as a utility pole or tower are used, it is important to insure that they are identified in the field notes and their location must be carefully measured. If you cannot place the landmark used to locate the instrument, you cannot locate any of the other stations.

As with the grid method, someone must determine the number and location of stations. The station selection must provide adequate data to define all of the man-made and natural features that are important for the use of the data.

One of the requirements of the angle and distance method is measuring the distance from the instrument to each of the stations. The selection of the method for measuring distance will greatly affect the time required to collect the data. Chaining the distances would be acceptable for a small area, but it would be a time-consuming process for a large area. If a transit or dumpy level is used, distances can be measured by stadia. This would only be acceptable for low-precision surveys because of the limitation of distances to a precision of 0.1 feet. This is a situation where a total station or GPS unit would be the instrument of choice. A total station produces very accurate and precise data for distances and angles. Another plus is that higher end units can be operated by a single person. A GPS unit could be set up to provide location information on a grid system, which makes it much easier to draw the map. Collecting data in a grid system is discussed in Chapter 10.

Angle and Distance Example

The following example will illustrate the process for completing an angle and distance topographic survey. Figure 8-13 is an illustration of an area of land encompassing about 1.9 acres.

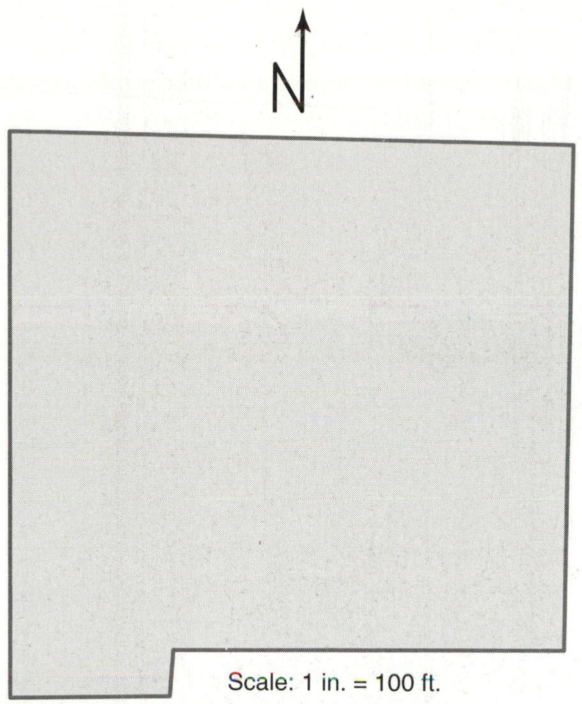

Figure 8-13 Boundary of angle and distance topographic example.

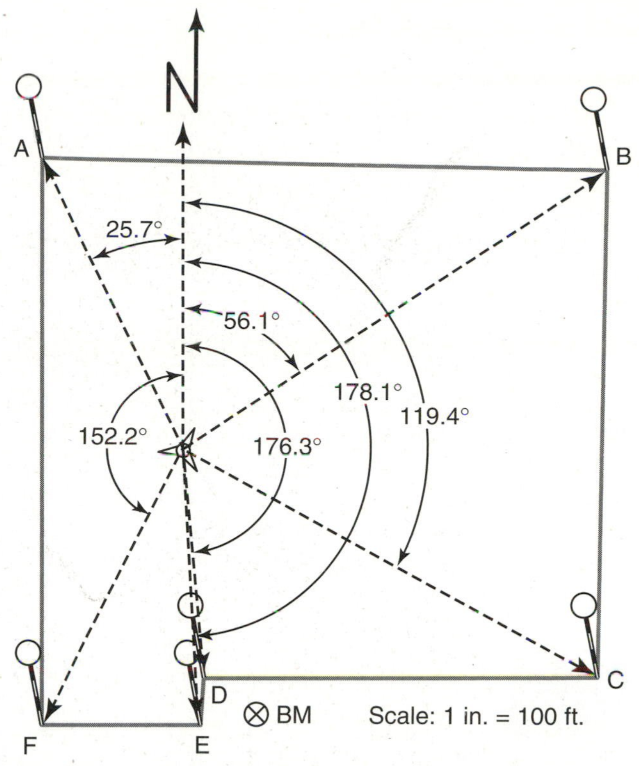

Figure 8-14 Angles for sample topographic problem.

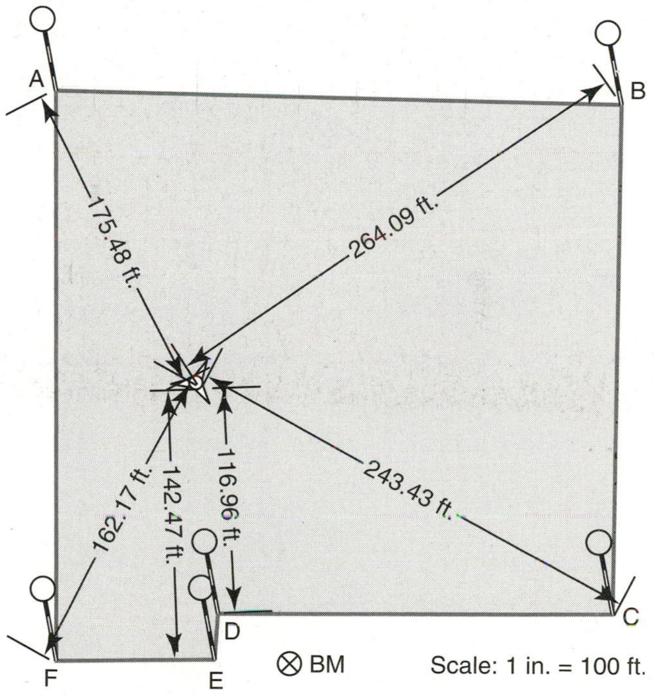

Figure 8-15 Distances for sample topographic problem.

The first step is to set up the instrument and record the corners of the property and their elevations. It is worth the time to study the site by checking elevations with a hand level, pacing some distances, etc. to determine the best location of the instrument.

Measuring Angle and Distance The process of determining the angles and distances for the corners from the instrument position is illustrated in Figure 8-14 and Figure 8-15. In this example, a compass was used to establish the backsight for all of the angles.

The sequence for recording data will depend on the methods used. The angles and distances can be recorded simultaneously if the distance is measured by stadia or EDM. If not, then two separate operations will be required to record the angles and distances. Several options also exist for recording the elevations. If a good boundary map is available, the elevations for the six corners could be recorded in a separate operation, using a laser for example, or they could be recorded when the angles or the distances are measured. An advantage of using a total station or GPS is that all of the data can be recorded simultaneously. Using equipment that requires the survey to be broken down into several operations will increase the amount of time required to complete the survey.

Regardless of the methods used, the survey crew must insure that all of the necessary data are recorded and that it is accurate. The data must be recorded in an appropriate table.

Table 8-1 contains the data for the topographic survey that was completed on the land in Figure 8-13.

Table 8-1 contains the data for the six corners of the land parcel. A topographic map could be produced with this data, but it would only be accurate if the topography of the 1.9 acres was very uniform. Any hills, ditches, etc. within the boundaries of the property would not be evident because there is no data to locate them and define their shape. Any person reading the data table would make the assumption that no other features or structures exist on the property. Figure 8-16 shows what the topographic map would look like using just the data for the corners.

This type of data should be suspect. Just the addition of one point with an elevation different from the elevations on the map would make a dramatic difference. Assume an additional point was recorded in the middle of the property and it had an elevation of 99.8 feet. Figure 8-17 illustrates the results of adding this one station.

Table 8-1 Angle and distance data for topographic example

Station	Angle (°)	Distance (ft.)
BM	172.0	177.21
A	−25.7	175.48
B	56.1	264.09
C	119.4	243.43
D	176.3	116.96
E	178.1	142.47
F	−152.2	162.17

Station	BS	HI	FS	IFS	Elevation
BM	5.2	105.2			100.0
A				3.1	102.1
B				3.2	102.0
C				4.2	101.0
D				5.6	99.6
E				5.7	99.5
F				4.8	100.4
BM			5.2		100.0
Σ	5.2		5.2		0.0
		0.0 = 0.0			

$$AE = K\sqrt{M}$$
$$= 0.1\sqrt{177.21}$$
$$= 1.3$$
$$0.0 > 1.3$$

Note: Two of the angles are recorded as (−) angles. Refer to Chapter 7 if you need to review why this symbol is used.

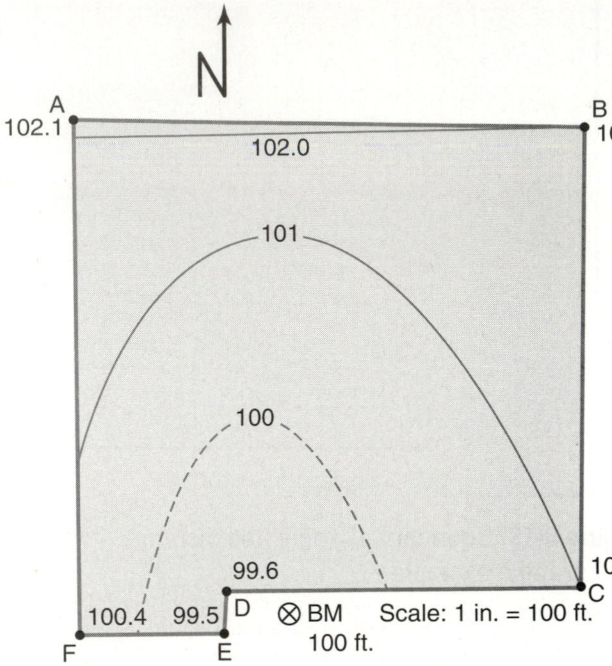

Figure 8-16 Topographic map of sample problem.

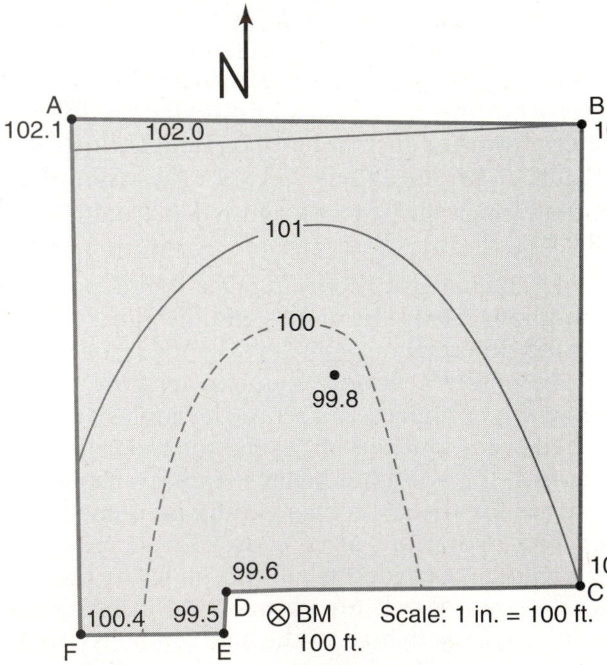

Figure 8-17 Angle and distance sample problem with one additional station.

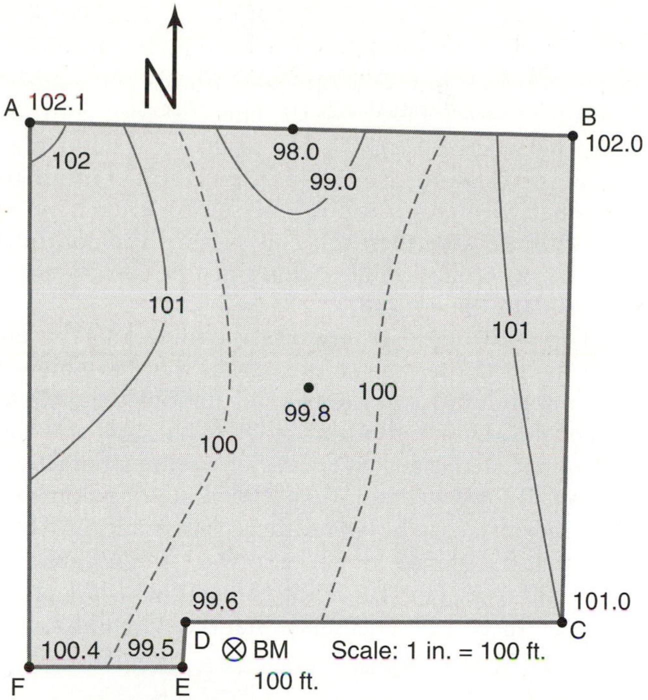

Figure 8-18 Angle and distance sample problem with two additional stations.

With the addition of this one point, the appearance of the map changes. The greater the difference is between the new data and the existing data, the more dramatic the change will be in the topographic map. To emphasize this point one step further, assume an additional point is included halfway between stations A and B and that the elevation of this station is 98.0 feet. Figure 8-18 illustrates the effect of adding this station to the data set. This causes a dramatic change.

Comparing Figure 8-16, Figure 8-17, and Figure 8-18 illustrates the dramatic effect additional stations can have on a topographic map. If Figure 8-18 represents the property, then any design or planning using the map in Figure 8-16 would have serious flaws.

DRAWING TOPOGRAPHIC MAPS

Drawing topographic maps by hand is a multistep process. A computer program should be used if maps will be drawn frequently or if large maps are being used. The advantages of computer drawn maps include better quality and more accurate interpretation of the data. The disadvantage is the program's restraints on input data format and map output.

Hand Drawing Maps

Drawing maps by hand is a useful skill for small areas or for quick low-precision maps that can be used for planning and estimating. Drawing a topographic map by hand is a two-step process that requires two layers of paper. The first layer is used to record all of the information. This includes the boundaries, the locations of the stations, and the elevations of the stations. The second layer is used to draw the map.

Mapping Supplies

Before drawing a topographic map, insure the necessary materials are available. Both topographical methods require pens or pencils and a scale or ruler. A scale or ruler that is graduated in tenths is easier to use than one graduated in fractions. The angle and distance method will also require a protractor and compass. Printer paper can be used for the first layer, but 10 × 10 graph paper is very useful, because the sheets can be taped together because it can be difficult to accurately determine the size of the map. The top layer, final map, should have paper large enough to contain the map using one piece. Thinner drawing or tracing paper makes it easier to transfer the map from the first layer. Typing and drawing paper can be used if a good light source is available like a light table or a window with strong light behind it.

Drawing Topographic Maps by Grid

The first step is to draw the boundaries to scale. It is very important to coordinate the scale of the map and the size of paper that will be required. Before selecting the paper size the mapmaker must analyze the date and determine the dimensions of the parcel of land being mapped. The grid method expedites this process because it is easy to tell from the data the number of rows and columns that were used. This information and the grid spacing provide the information necessary for determining the size of the map. Data collected by angle and distance must be carefully studied to determine the outlying points. This may require drawing a sketch of the area before the size of the paper can be determined. In addition, the drawing must be orientated on the paper. Remember the convention for maps is north to the top of the page.

The second step is to lay out the grid and write in the elevations for every point. The third step is to draw the contours on the first paper. Next, the tracing paper or second sheet is taped over the first. The second paper should only be taped along one edge so it can be lifted out of the way when it becomes necessary to refer to the data on the first sheet. The first step in drawing the map is transferring the boundary lines from the first sheet to the second sheet. The last step is drawing the contour lines and any other features that are important for the map.

Drawing Contour Lines

Contour lines can start at any one of the desired intervals, but it is important to study the elevations to insure that some are not missed. Review the elevations for the minimum and maximum numbers and pick one of the desired contour elevations that fall within this range. Start on a boundary or at an interior point and trace the line across the map. Remember that contour lines must close inside the map or extend to a boundary line. A contour line *cannot* stop at any interior point. Interpolating between stations extends the line.

A confusing part of drawing contour lines is determining their route between stations. Remember a contour line is a series of points with equal elevation. Study Figure 8-19. The 100-foot contour starts between A1 and B1. The difficulty is determining the next point. Remember that for a contour to cross between two stations, one station must have an elevation greater than the elevation of the contour, and the other station elevation must be less than the elevation of the contour.

The contour is extended by comparing the elevations of the next pair of stations that have higher and lower elevations. The pairs of stations that should be considered are the four sides of the box and the diagonals. In Figure 8-19 the elevation of stations B1 and B2 are greater than 100 feet. The contour line doesn't cross between these two stations. Also, note that the elevations of stations A1 and A2 are less than 100 feet. The 100-foot contour will not cross between these two. That leaves stations A2 and B2. The

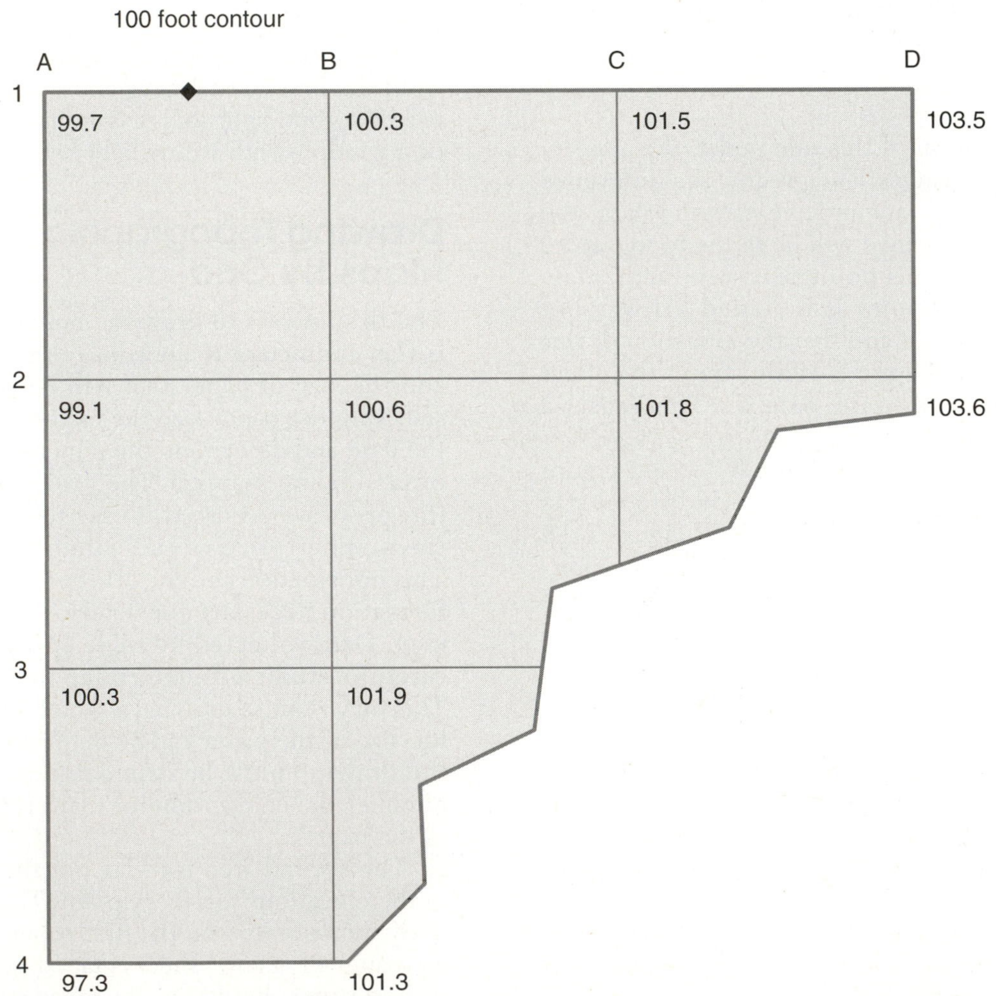

Figure 8-19 First point for 100 foot contour line.

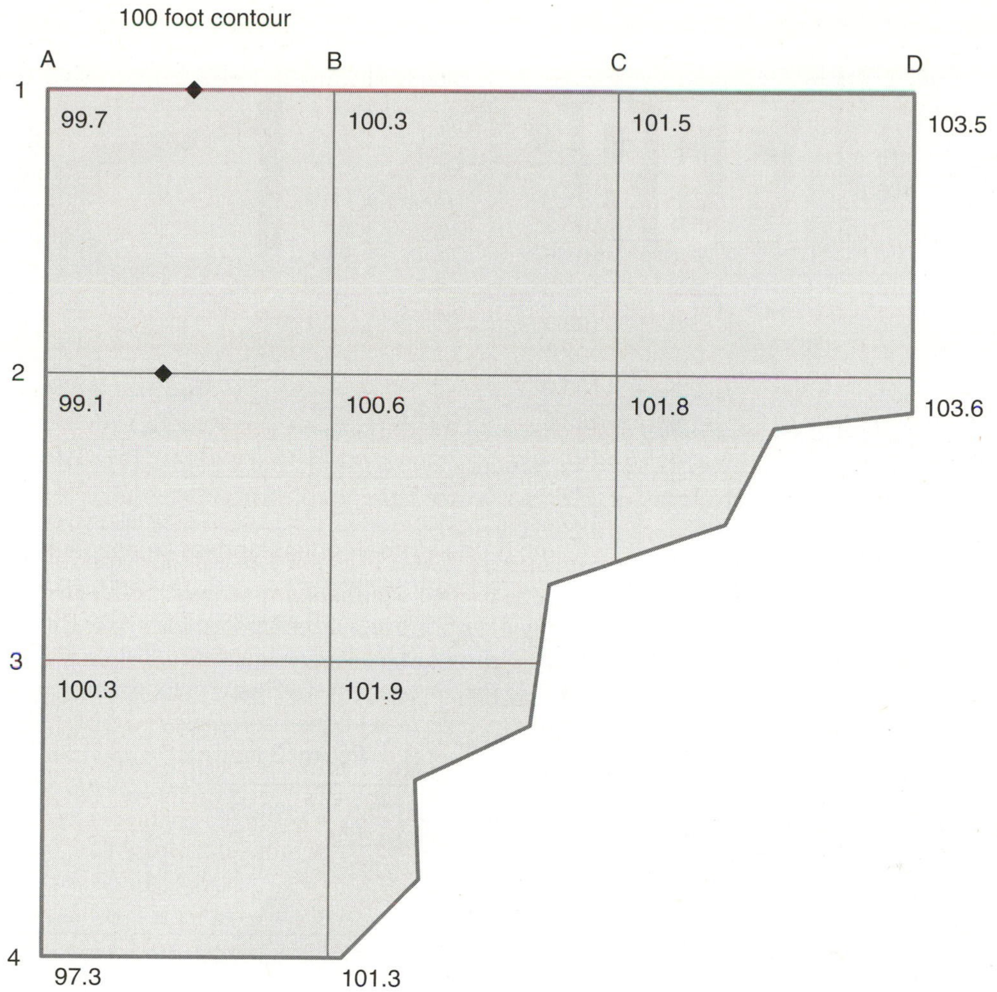

Figure 8-20 Second point for 100 foot contour line.

elevation of station A2 is less than 100 feet and the elevation of station B2 is greater than 100 feet. The 100 foot contour crosses the imaginary line between these two stations. Before extending the contour to this point it is important to check the diagonal corners also. In this example A1 is less than 100 and B2 is greater than 100 feet, this diagonal is ok. B1 is greater than 100 and A2 is less than 100, this diagonal is also ok. It is not unusual for a situation to exist that requires extending a contour line between two stations that are both either higher or lower elevations than the contour line. When this occurs, it may be necessary to draw another one or two contours to determine the best fit. Remember contour lines tend to be parallel. The intersection is located using interpolation, Figure 8-20.

The 100 foot contour line is continued one grid at a time until in intersects with the border or closes on itself. Continue drawing the contour lines until the map is complete. Remember to look for small hills and depressions that are complete within the map and for contour lines that may enter and exit the map a short distance apart. The process of drawing contour lines will be illustrated with a sample problem.

SAMPLE PROBLEM—TOPOGRAPHIC BY GRID

Study the page from the field notes illustrated in Figure 8-21. In many situations, the person drawing the map was not part of the surveying crew and the only information they have is the notes and sketch from the field book. This is why it is so critical that the notes are complete.

Conceptualizing Map

The first step in drawing a map is to picture in your mind what the area looks like. The two primary clues are the data table and the sketch. The sketch

TOPO BY GRID

LOCATION:

STAT.	BS	HI	FS	IFS	ELEV.
BS	5.6	105.6			100.0
A1				5.4	100.2
A2				4.1	101.5
A3				3.2	102.3
A4				3.0	102.6
A5				2.7	102.9
A6				2.4	103.2
A7				1.8	103.8
B1				3.8	101.8
B2				7.3	98.3
B3				5.1	100.5
B4				5.7	99.9
B5				5.9	99.7
B6				6.0	99.6
B7				6.1	98.5
C1				5.3	100.5
C2				7.3	98.3
C3				4.5	101.1
C4	6.4	111.6	1.0.		104.6
C5				5.8	105.2
C6				11.3	99.7
C7				7.2	103.8
D1				8.7	102.3
D2				8.8	102.2
D3				9.8	101.9
D4				2.4	104.3
D5				6.5	104.5
D6				6.9	104.1
D7				6.8	104.2
E1				8.4	102.6
E2				8.4	102.6
E3				8.4	102.6
E4				5.5	104.5
E5				4.5	106.5
E6				6.8	104.2
E7	4.8	108.8	7.0		104.0
BM			8.8		100.0
Σ	16.8		16.8		*Notes OK*

IBS - FSI = IBM - BMI

I16.8 - 16.8I = I100.0 - 100.0I

0.0 - 0.0

Cloudy 75°

Transet # 34591

Rod 62

Chain 01

Flags

Stakes

Jim	Notes
Bill	Rod
Mary	Chain
Henry	Flags

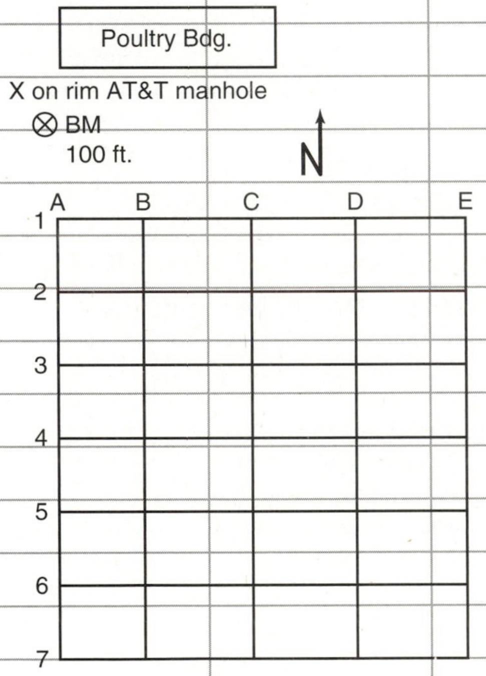

Grid spacing = 100 feet.

A - E and 1 - 7 fence

SIGNATURE:

Figure 8-21 Field notes for sample problem.

shows that the field is rectangular and that it is orientated north to south. A note with the sketch also states that a fence defined two of the boundaries. These are important clues when you start to draw the map. The data table shows that the stations range from A1 to E7. The data includes a benchmark that was not part of the grid. The data also shows that station C4 was used as a turning point. What is not clear is whether the structure used for the turning point represented the surface of the earth. This information should be included with the sketch. In this example, we will assume the turning point can be included in the map.

Also note the rows were identified by numbers and columns were identified by letters. This is important to remember when the elevations are written on sheet one of the map. An additional factor that must be remembered when drawing this map is that no man-made or natural features were identified and measured during the survey other than the AT&T utility entrance used for the benchmark.

The purpose of this sample problem is to show the process of drawing a topographic map; therefore, a simple rectangular area was used. The process is the same for larger sets of data and shapes that are more complex; it just takes more time and care to layout the boundaries and the station locations on the drawing.

Paper Size and Scale

The next step is to determine the size of paper that will be needed to draw the map to the desired scale. The note below the sketch, Figure 8-21, indicates that a 100-foot grid was used. Five columns and seven rows equate to an area of 400 ft. × 600 ft. A map scale of 1:100 would produce a map that measured 4 in. × 6 in. This would be a small map and it would not be very useful for construction or planning purposes. An alternative is to use a different scale. A scale of 1:10 would result in a map that was 40 inches by 60 inches. This is not a practical size. The options are to use a small map or a scale that is different than a multiple of ten. For example, a scale of 1:50 would result in a map that is eight inches by twelve inches. For this example, the scale of 1:100 will be used because the small map is large enough to demonstrate the principles of drawing contours and it will fit on standard 8-½ × 11 paper.

Sheet One

The first task is to locate station A1 on the paper so that the map will be centered on the paper when it is complete. For this example, the width of the map is 4 inches. Subtracting the width of the map from the width of the paper, 8.5 inches, leaves 4.5 inches. Dividing 4.5 inches by 2 produces a left and right margin of 2.25 inches. Completing the same calculations for the top and bottom margin results in 2.5 inches ((11 − 6)/2). This results in a location of station A1 2.25 inches from the left margin and 2.5 inches from the top of the paper.

The next task is to draw the boundaries. The top boundary, A to E, is drawn parallel, starting with the location of A1, with the top edge of the paper. The left boundary, rows 1 to 7, is drawn starting at the location of A1 and parallel with the left side of the paper. Station E7 is located by measuring 4 inches from station A7 and 6 inches from station E1. The point where these two dimensions meet is the location of station E7. Drawing a line connecting station E7 to E1 and A7 completes the boundary. Next the grid is drawn. A scale of 1:100 and a grid distance of 100 feet result in a grid distance on the paper of one inch.

After the grid is drawn, the station elevations are added to the grid. The elevations must be plain and legible because they must be read through the second layer when the contour lines are drawn.

Once the grid and elevations have been entered on the first layer, the intersections of the contour lines and the grid lines can be interpolated. This is accomplished by placing a mark at the appropriate spot where the contour line intersects a horizontal, vertical, or diagonal grid line. The location of the contour line should also be marked for where the contour intersects an imaginary diagonal between either two pairs of stations.

Figure 8-22 and Figure 8-23 show the process of locating the first two intersections. For intersection one, the equation produces a proportion of 0.72.

$$1.\ \text{Proportion} = \frac{103.8 - 100}{103.8 - 98.5} = 0.72$$

The scale of the map is one inch equals 100 feet; therefore, the location of intersection one from the station with the highest elevation is 0.72 × 1 inch or 0.72 inches, Figure 8-22.

The procedure for locating intersection two is slightly different because it is a diagonal. The interpolation equation produces a proportional distance of 0.68.

$$2.\ \text{Proportion} = \frac{103.2 - 100.0}{103.2 - 98.5} = 0.68$$

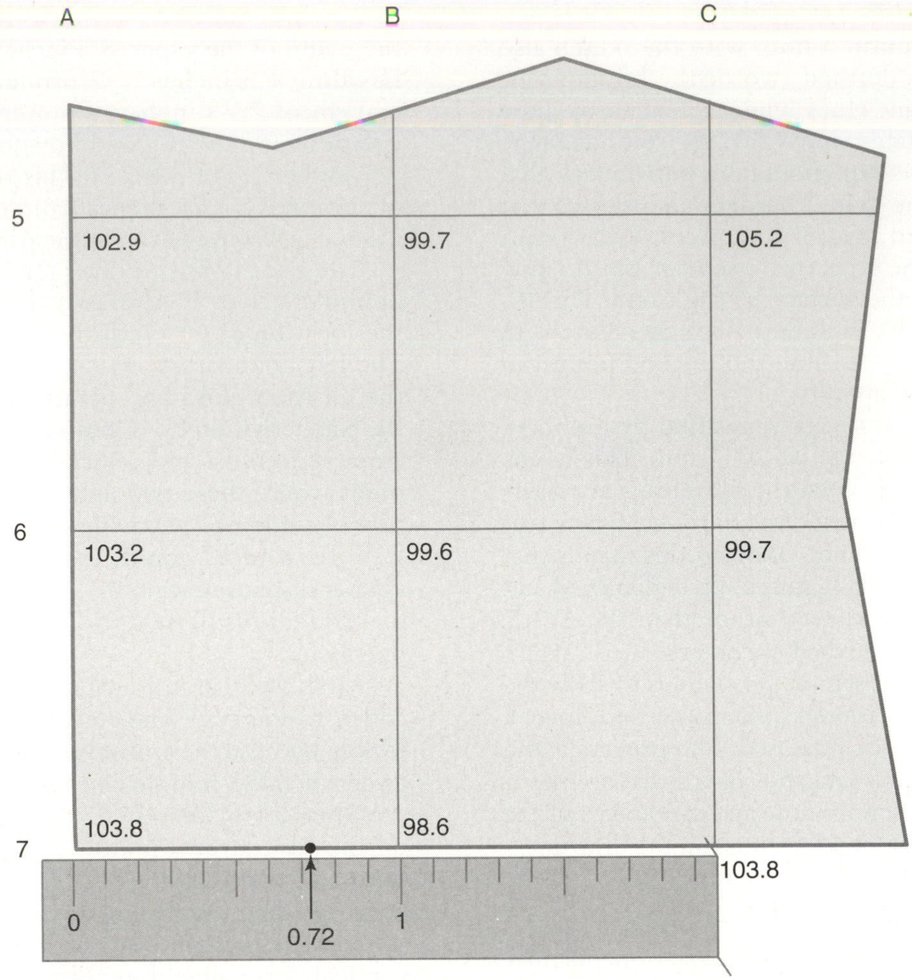

Figure 8-22 Locating the first intersect point.

Intersect two is on the diagonal, therefore the distance between A6 and B7 is greater than one inch. Using the Pythagorean Theorem:

$$\begin{aligned}\text{Diagonal distance} &= \sqrt{a^2 + b^2}\\ &= \sqrt{1^2 + 1^2}\\ &= \sqrt{2}\\ &= 1.4\end{aligned}$$

The distance between stations A6 and B7 is 1.4 inches. Therefore, the distance between station A6 and the intersection with the contour line is 0.68 × 1.4 inches or 0.95 inches, Figure 8-23.

The remaining intersections are located and marked using the same method. The math required to locate the intersections is not difficult—just tedious. The calculations for a 100-foot contour are included to aid in the understanding of the example problem. The intersections were numbered starting with the location between A7 and B7 as 1 and then moving counterclockwise around the contour until the contour leaves the map between stations B7 and C7.

3. $\dfrac{103.2 - 100.0}{103.2 - 99.6} = 0.89$

4. $\dfrac{102.9 - 100.0}{102.9 - 99.6} = 0.88$

$0.88 \times 1.4 = 1.2$

5. $\dfrac{102.9 - 100.0}{102.9 - 99.7} = 0.91$

6. $\dfrac{102.6 - 100.0}{102.6 - 99.7} = 0.90$

$0.90 \times 1.4 = 1.2$

7. $\dfrac{102.6 - 100.0}{102.6 - 99.9} = 0.96$

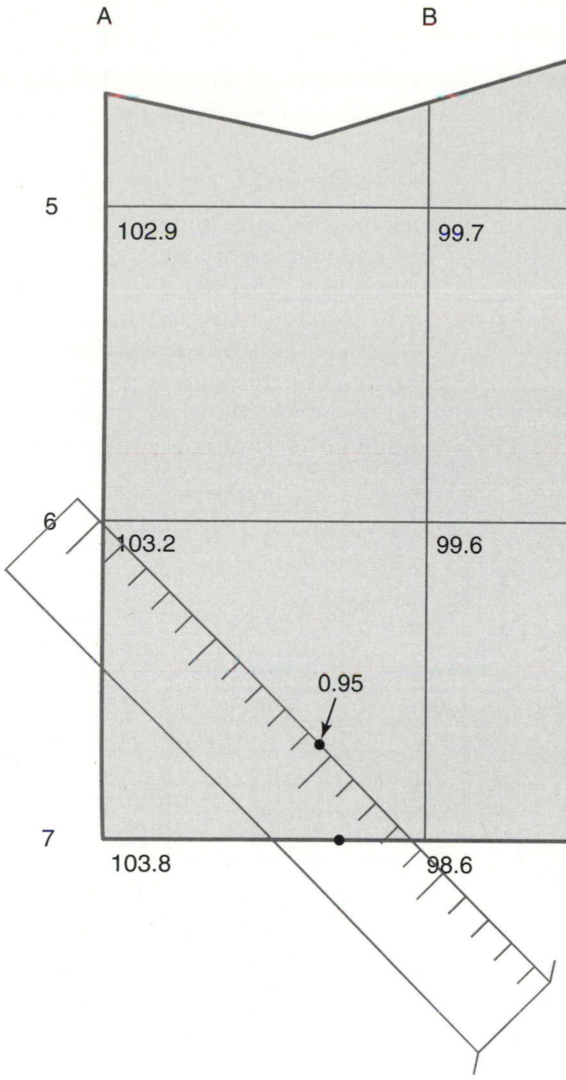

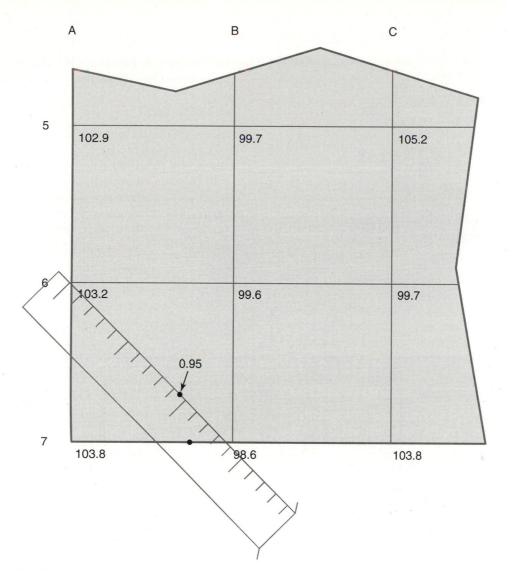

Figure 8-23 Locating the second point.

8. $\dfrac{102.3 - 100.0}{102.3 - 99.9} = 0.96$

$0.96 \times 1.4 = 1.3$

9. $\dfrac{100.5 - 100.0}{100.5 - 99.9} = 0.83$

10. $\dfrac{101.1 - 100.0}{101.1 - 99.9} = 0.92$

$0.92 \times 1.4 = 1.3$

11. $\dfrac{104.6 - 100.0}{104.6 - 99.9} = 0.97$

12. $\dfrac{104.6 - 100.0}{104.6 - 99.7} = 0.94$

$0.94 \times 1.4 = 1.3$

13. $\dfrac{105.2 - 100.0}{105.2 - 99.7} = 0.94$

14. $\dfrac{105.2 - 100.0}{105.2 - 99.6} = 0.93$

$0.93 \times 1.4 = 1.3$

15. $\dfrac{105.2 - 100.0}{105.2 - 99.7} = 0.94$

16. $\dfrac{104.5 - 100.0}{104.5 - 99.7} = 0.94$

$0.94 \times 1.4 = 1.3$

17. $\dfrac{104.1 - 100.0}{104.1 - 99.7} = 0.93$

18. $\frac{104.2 - 100.0}{104.2 - 99.7} = 0.93$

$0.93 \times 1.4 = 1.3$

19. $\frac{103.8 - 100.0}{103.8 - 99.7} = 0.93$

20. $\frac{103.8 - 100.0}{103.8 - 98.5} = 0.90$

$0.90 \times 1.4 = 1.3$

21. $\frac{103.8 - 100.0}{103.8 - 98.5} = 0.72$

A smooth curve line connects the intersection points. The map with the completed 100 foot contour can be seen in Figure 8-24. The remaining contour lines should be completed following the same steps.

Sheet Two

The next step is to tape one edge of the second piece of paper over the first sheet. Both sheets of paper should be taped to a light table or to a surface that will provide enough backlight to see the boundary lines and contour lines through the second sheet. Make sure the second sheet can be lifted up, without removing the tape, to expose the first sheet. The boundaries should be traced onto the second sheet

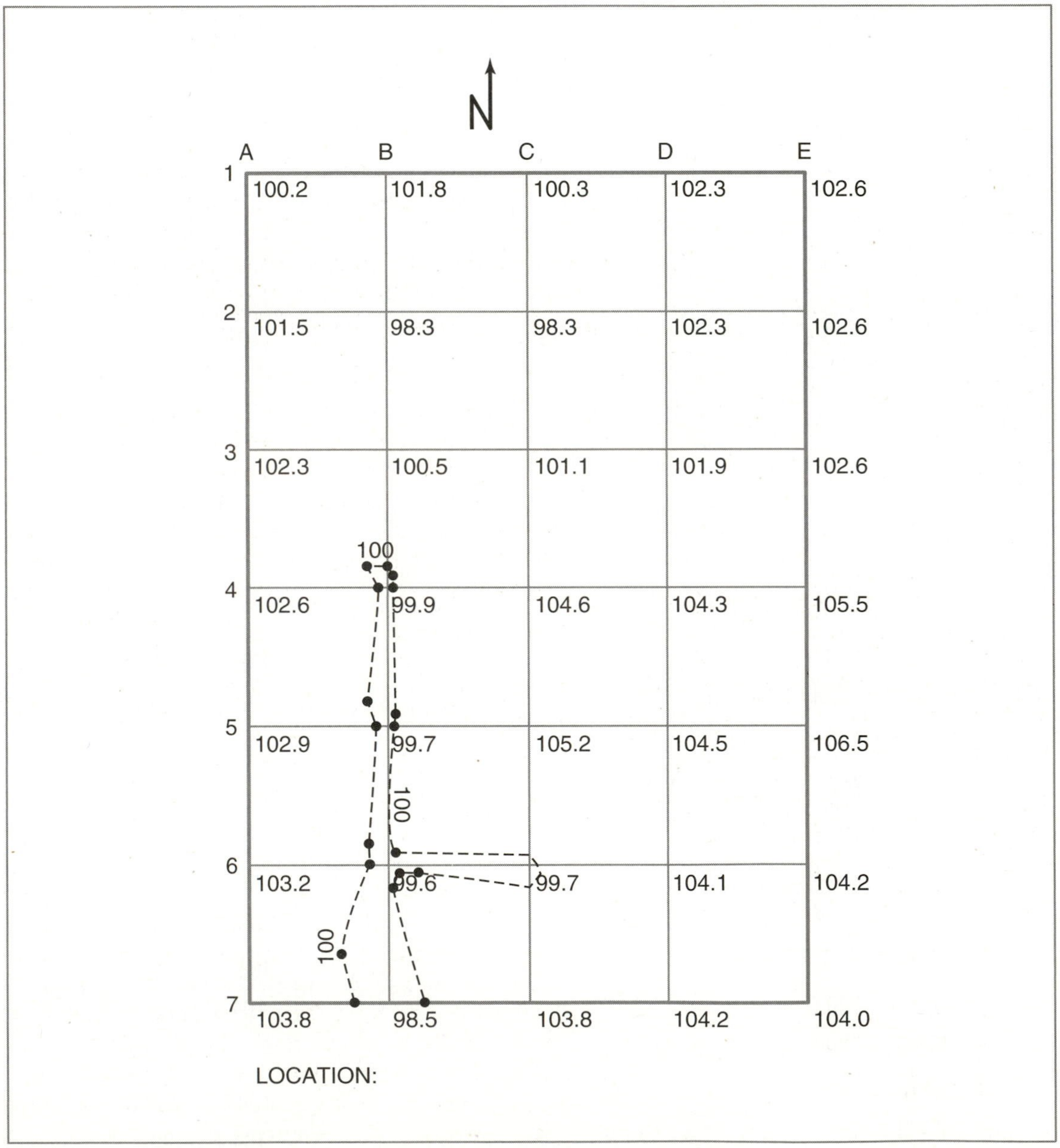

Figure 8-24 First layer of sample problem with 100 foot contour.

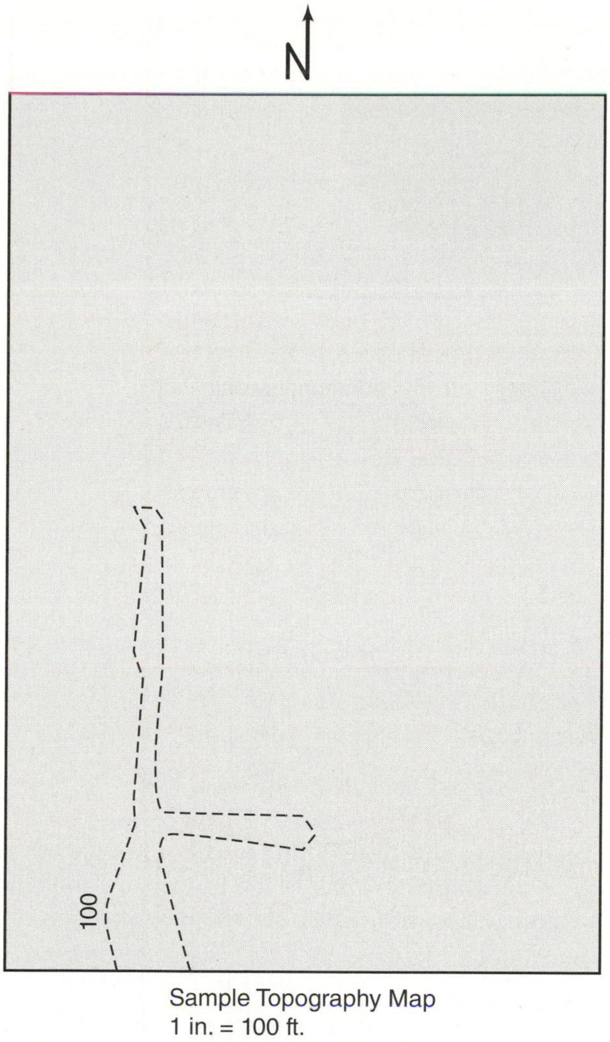

Figure 8-25 Second layer with 100 foot contour.

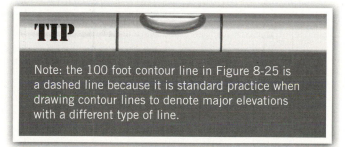

TIP

Note: the 100 foot contour line in Figure 8-25 is a dashed line because it is standard practice when drawing contour lines to denote major elevations with a different type of line.

and then the contour lines can be traced. The finished map with the 100 foot contour line can be seen in Figure 8-25.

PROFILE GRAPH FROM TOPOGRAPHIC MAP

Topographic and profile surveys have one thing in common. They both are used to define the topography of the earth's surface. The difference is topographic surveys collect the information to define the surface of an area and profile surveys only define the topography for a route. Because of their similarity, a profile graph can be drawn from a topographic map. There are two common methods of producing the profile of a route from a topographic map.

- Determining the values for distance and elevation from the map.
- Transferring the information from the topographic map to the profile.

It must be pointed out that neither of these two methods is as precise as a profile survey. Both methods are useful for selecting a route and initial planning and design work. Many situations would require the completion of a profile survey along the desired route before final designs would be completed.

Drawing a Profile by Determining Values

In Chapter 6 the requirements for drawing a profile were determined to be a value for distance and a value for elevation. Numbers for these two values can be determined from a topographic map. The task is to determine the appropriate stations on the topographic map, and a distance and elevation for each one.

Determining Stations

Stations must be points where the distance from the initial point and the elevation can be determined. Distances can be determined at any point along the route by measuring with a rule and using the map's scale to convert the measurement to the distance, but the only elevations are the contour lines. Therefore, the stations are the junctions of the route and a contour line. The stations can be identified in the same way, by their distance from the initial point, but any system is acceptable as long as it is consistent and clear. Refer to Appendix I for more information on drawing graphs by hand.

EXAMPLE PROFILE BY DETERMINING DISTANCE AND ELEVATION

Figure 8-26 illustrates a topographic map with the route of the desired sidewalk drawn on the map. This illustration contains all of the information needed to draw a profile of the proposed sidewalk.

The first step is to determine the distance for each station from the initial point of the survey. This is accomplished by placing a ruler along the route and determining the distance for each junction with a contour line, Figure 8-27.

For this example a table for the data will help organize it and reduce the chance of errors.

Knowing the distance from the initial point and the elevation of the station, the information can be plotted on a graph, Figure 8-28.

This is another example of where the use of a computer would save time in surveying. Most spreadsheets will draw a line graph. One big advantage is the ability to change the position of the route and almost instantaneously see the new profile.

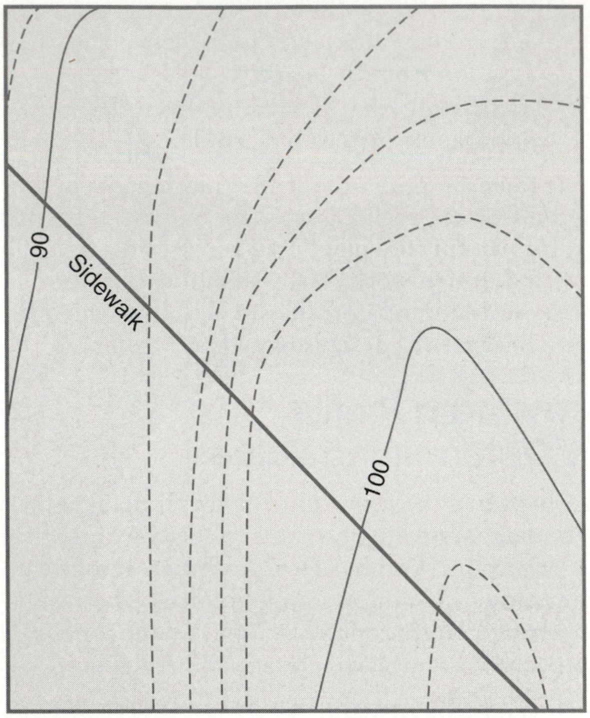

Figure 8-26 Profile from topographic map.

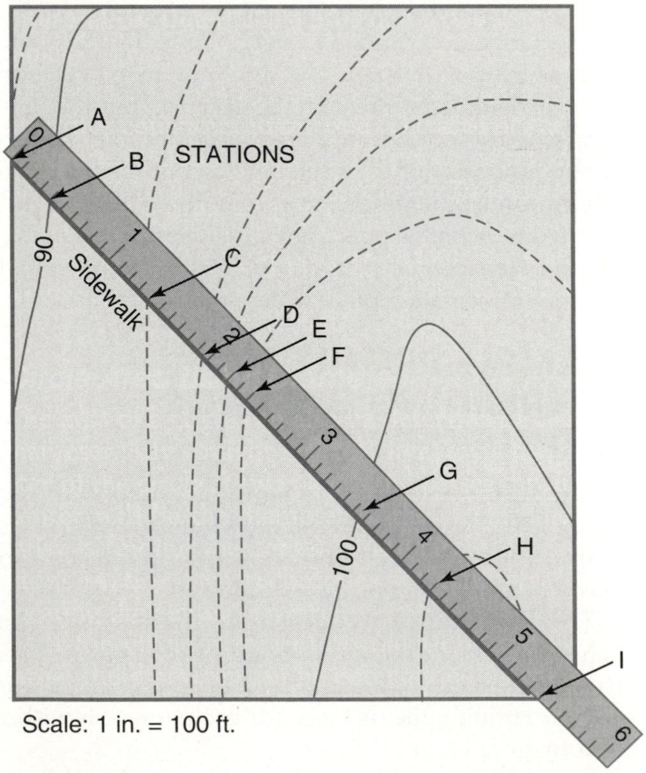

Figure 8-27 Determining stations and distances for sidewalk.

TIP

Note: The start of the sidewalk is not at a contour line. The elevation of this point must be interpolated. In this example, the elevation of station A is 89.4 feet. The information for all of the stations is included in Table 8-2.

Table 8-2 Profile from topographic data

Station	Inches	Distance (ft)	Elevation
A	0.0	0	89.4
B	0.4	40	90
C	1.4	140	92
D	1.95	195	94
E	2.2	220	96
F	2.4	240	98
G	3.55	355	100
H	4.3	430	102
I	5.4	540	102

DRAWING A PROFILE BY TRANSFERRING THE INFORMATION

This method of producing a profile graph from a topographic map is easier to produce, but it is not as useful because the distance for each station is unknown. It also requires the profile graph to be the same width as the length of the route on the topographic map. It is still useful for visualizing what the profile will look like.

In this method, the X-axis and Y-axis are drawn on the paper and the paper is folded along the X-axis. The folded edge is aligned with route of the sidewalk on the topographic map, Figure 8-29. Insure that the corner of the X-Y axis is aligned with the initial point of the sidewalk. Move along the sidewalk and mark the location of each contour line on the X-axis of the graph.

Remove the graph paper and determine the scale for the Y-axis. In this example, the Y-axis scale is the same as the scale for the map. This eliminates the vertical exaggeration that is normally used with profile graphs, but it does provide a visual image of the profile. Complete the graph by connecting the elevations with a straight line and attaching the appropriate labels and titles, Figure 8-30.

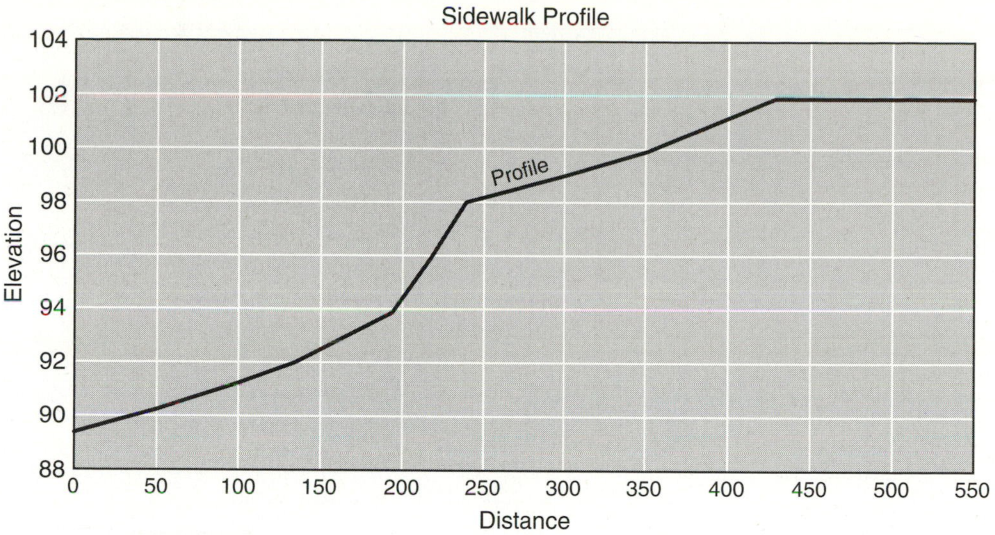

Figure 8-28 Graph of profile from topographic map.

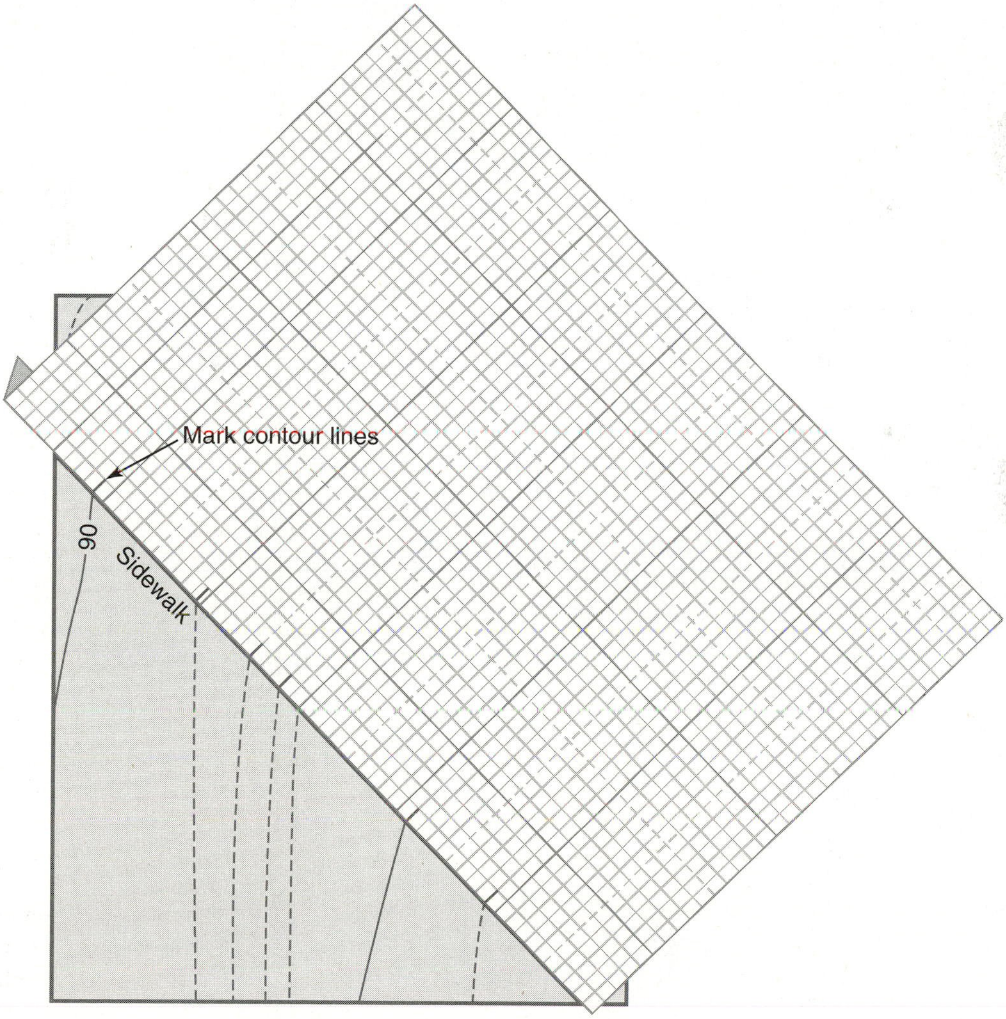

Figure 8-29 Drawing profile from topographic map by transferring stations.

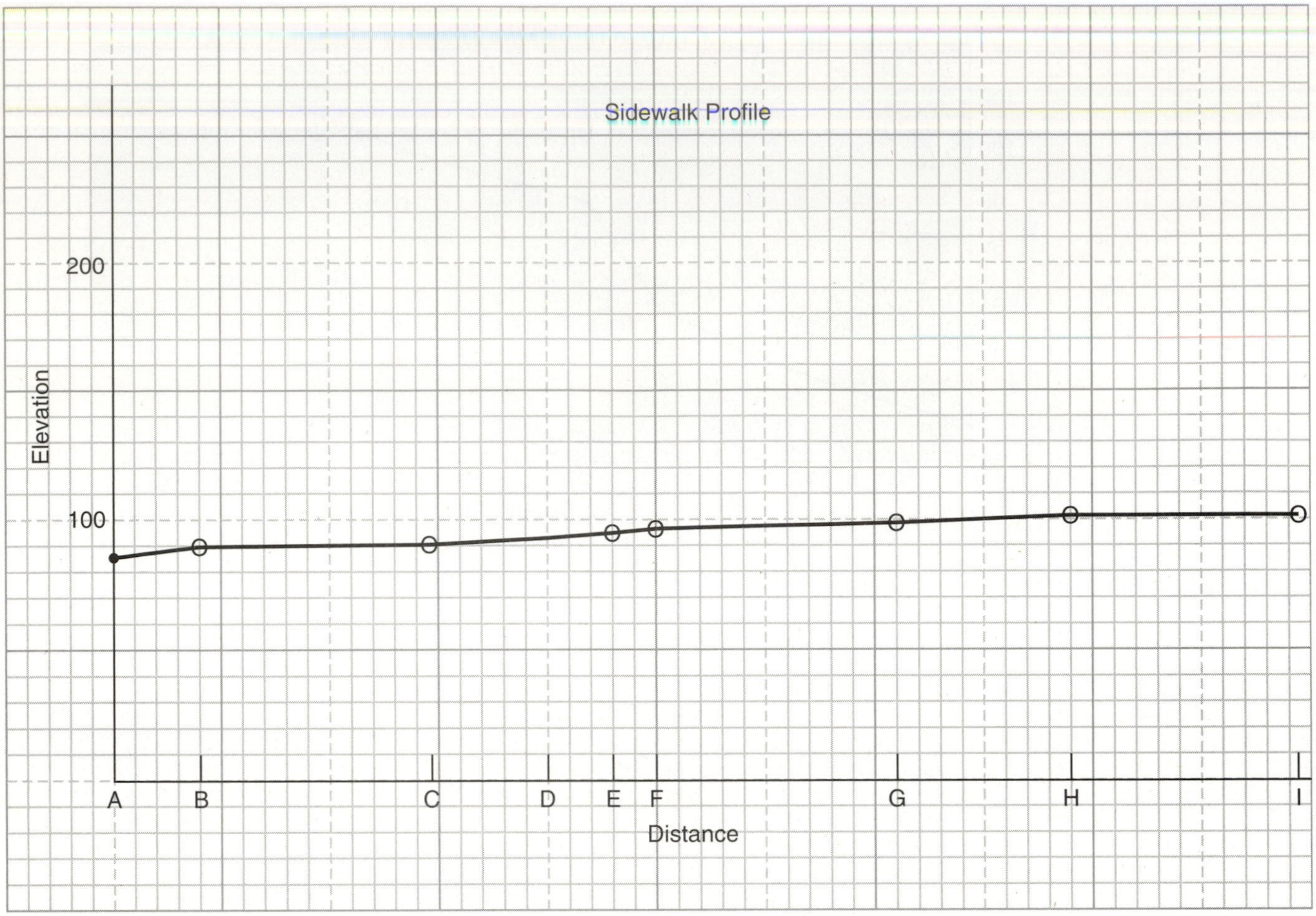

Figure 8-30 Graph of profile completed by transferring stations.

Summary

The object of this chapter was to acquaint the reader to the principles and methods of topographical surveys and topographic maps. These are the preferred maps for construction, landscape architects, and for land use planning. Producing a topographic map requires a large amount of resources. Careful planning must be completed before any equipment is set up and used. As with most surveying activities many of the aspects of topographical surveying is as much of an art as a science. There is no substitute for experience, but if a beginning surveyor takes care to adhere to recommended principles and practices they will be able to produce useable topographic maps.

Student Activity

1. Draw a topographic map for the data in the following table. Start with even elevations and use a contour interval of 2 feet. Grid spacing was 50 feet.

STA	BS	HI	FS	ELEV
BM	2.3	52.3		50.0
A1			14.3	38.0
A2			12.3	40.0
A3			1.1	51.2
A4			11.1	41.2
A5			12.7	39.6
A6			14.8	37.5
B1			13.8	38.5
B2			12.2	40.1
B3			10.0	42.3
B4			6.5	45.8
B5			6.4	45.9
B6			4.1	48.2
C1			16.3	36.0
C2			10.8	41.5
C3			10.3	42.0
C4			7.9	44.4
C5			3.8	48.5
C6			2.8	49.5
D1			18.3	34.0
D2			9.09	42.4
D3			10.3	42.0
D4			7.7	44.6
D5			3.5	48.8
D6			2.3	50.0
E1			13.8	38.5
E2			9.5	42.8
E3			8.8	43.5
E4			6.8	45.5
E5			4.3	48.0
E6			2.3	50.0
F1			11.1	41.2
F2			10.0	42.3
F3			8.5	43.8
F4			7.1	45.2
F5			3.8	48.5
F6	3.3	53.5	2.1	50.2
BM			3.5	50.0

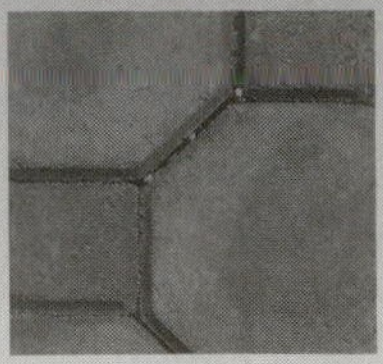

CHAPTER 9

Traverse Survey

Objectives

After reading this chapter the reader should be able to:

- Understand the need for balancing a traverse.
- Balance the interior angles of any geometric shape.
- Understand how to use latitudes and departures.
- Balance a traverse using the compass rule method.

Terms To Know

Traverse
Closed traverses
Open traverses
Balancing a traverse
Compass rule
Departures
Law of parallel lines
Angle of declination

INTRODUCTION

A **traverse** is a sequence of known distances connected to each other by known angles. The primary uses of traverses are to establish a series of control points, define the boundaries of property, determine the area of a parcel of land, establish control points for additional surveys or to locate/establish a route. A traverse may or may not close to form a complete shape, and may or may not include elevations.

Traverse surveys are more complex than differential, profile, and topographical surveys and when they are used to establish controls or boundaries error control is more critical. This chapter will explain the principles of traverses, explore their uses, and present the steps in balancing a traverse using the compass rule method.

SELECTING PRECISION

One of the decisions that must be made before completing a traverse is the appropriate precision for the measurements. The appropriate method depends on the use of the information. Legal property surveys have stringent requirements. They may require distances measured to thousandths of a foot and angles to the nearest second. A traverse for a sidewalk or a road could be measured in tenths or hundreds of a foot and angles measured to the nearest minute. A traverse used for planning a housing site or landscaping project could be measured in tenths or hundredths of a foot and angles measured to the nearest minute. State and local laws or client preferences should be considered when determining the precision for the survey. It is the responsibility of the survey crew to insure that the data is collected with the appropriate precision.

OPEN AND CLOSED TRAVERSES

Two types of traverses are used, closed traverses and open traverses. A **closed traverse** is a sequence of angles and distances that form a polygon. An example is the boundary lines of a parcel of land. An **open traverse** is a sequence of angles and distances that define a line or route, but it does not form a polygon. When elevations are recorded for the stations, the traverse survey should be closed back to the starting point or benchmark. This provides data for error of closure calculations. The correct type of traverse must be selected for the use of the data.

Open Traverse

An open traverse is a route where the angles and distances are carefully measured, but the finished route does not form a polygon. Open traverses do not have a means for checking for error or for balancing. Therefore greater care must be taken to control measuring errors. Distances and angles should be measured multiple times. In open traverses, the angles are usually measured as deflection angles (see Figure 9-1.)

The open traverse originates at a point with a known or assumed horizontal position, and terminates at a point other than the starting point. The use of the data may require that elevations also be recorded. In this case the elevation of the starting point must be known or assumed. Because an open traverse does not have the checks for error that are part of a

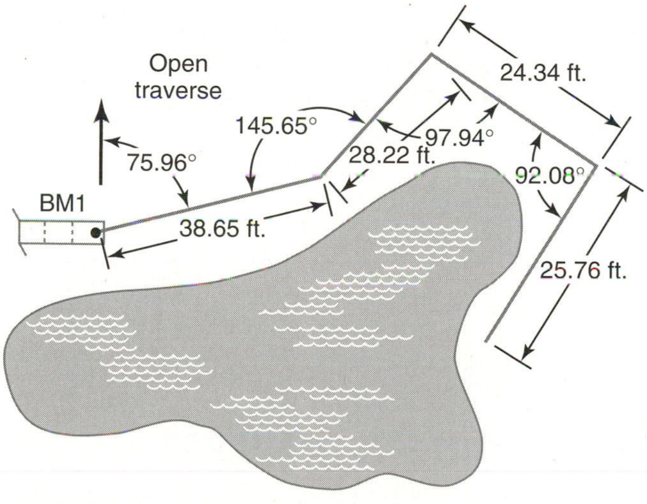

Figure 9-1 Open traverse.

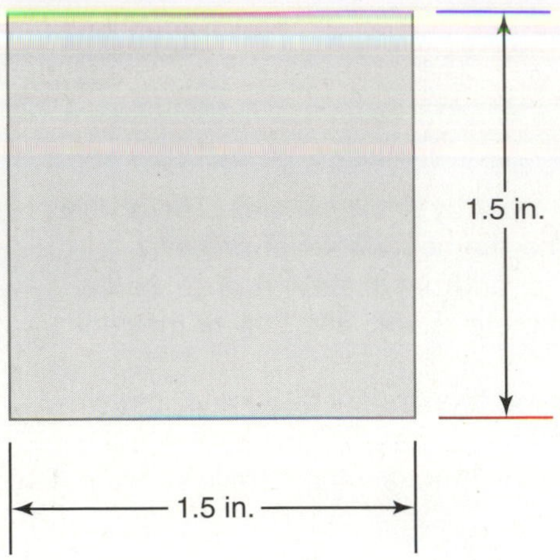

Figure 9-2 A closed polygon.

closed traverse, they are primarily used for exploratory or estimating routes. One technique used to reduce error is to survey back to the starting point using the same stations.

Closed Traverse

A closed traverse is a series of distances and angles that form a polygon. For any geometric shape and size of shape, there is only one combination of distances and angles that will form a closed figure, Figure 9-2. The angles and distances of a closed traverse can be checked for error and adjusted because the end product is a closes shape and any error in measuring distances and angles will prevent it from closing. The amount of closure error can be determined and the angles and distances adjusted accordingly.

Figure 9-2 will not form a square unless all four sides are the same length, 1.5 inches, and all four angles are 90 degrees. If any one of these factors is not true, the shape is not a square. This principle is also true of polygons with more angles and lengths of sides.

It is not possible to complete a traverse survey without some error in the measurement of the distances and angles. The process of determining the amount of error that occurred and correcting the angles and distance is called **balancing a traverse**. When a traverse is balanced, the angles and distances are adjusted so that the combination of angles and distances form a closed geometric shape.

BALANCING A TRAVERSE

More than one method can be used to balance a traverse. The "best" method is related to use of the data and the relative precision of the angle and distance measurements. Three possible precision situations can exist.

1. The angle measurements are more precise than the distance measurements.
2. The angle measurements are less precise than the distance measurements.
3. The angle measurements have the same precision as the distance measurements.

An example of number one is a traverse that was completed using an instrument with a least count of 20 seconds to measure angles and using an odometer wheel to measure the distances. The distance measurements would have much more uncertainty than the angle measurements. An example of number two is measuring angles with an instrument with a least count of one degree and the distances are measured with an accurate EDM. The third situation is preferred. This situation exists when the accuracy of the angle and distance measurements is similar. For example, the angles are measured with a least count of one minute and distances are determined by horizontal chaining. The preferred conditions also exist when a total station is used because they are designed with similar precision for both angle and distance measurements.

Different methods of balancing the traverse are recommended for each one of these situations. One method is to arbitrarily adjust the angles and the distances until the traverse balances. This method would be appropriate for the analysis of an irregular shaped lot to determine its dimensions and shape for selecting a site for a house. The arbitrary method would be appropriate as long as the surveyor uses reasonable care recording the measurements. The arbitrary method is also appropriate for low-precision surveys when the surveyor knows that one or more measurements, either distance or angle, are suspect. In situations when conditions at one or more stations, such as terrain, mud, trees, are much worse than the other stations, the survey would adjust these stations until the traverse is balanced. In other situations, such as determining the legal boundary, mathematical balancing will be required. Mathematical balancing methods are based on probability theory, for example, least squares, proportional adjustments, or compass rule method.

For a traverse that meets situation one or two, an engineering surveying text should be consulted to determine the best method to use. Situation three is typical for general surveys that follow good standard practices. It is also important to consider the surveying conditions. A traverse survey that met the criteria for situation three and was completed under extreme conditions would also require additional investigation to determine the best method to use. For surveys that fall into situation three and are not completed under extreme conditions, the compass rule method is commonly used.

Balancing a Closed Traverse Using the Compass Rule Method

This method is called **compass rule** because it uses a compass to orient the traverse to North. It uses the interior angles, and latitudes and departures to balance the traverse. Latitudes are movements to the North and South; movements to the East and West are called departures. The process of balancing a traverse is not difficult to understand, but the balancing process can be challenging because of the number of steps that must be completed in sequence. The complexity of the process is directly related to the number of angles in the figure. The greater the number of angles the greater the complexity and the more time it will take to balance the traverse. The number of calculations that must be completed provides many opportunities for making mistakes. For these reasons they are usually completed using a spreadsheet or by using a computer program developed for this purpose.

Balancing a traverse using the compass rule method requires nine steps. In situations where the traverse will be staked out, the coordinates for each station must be determined after the traverse is balanced.

1. Collect the data
2. Sum the angles
3. Balance the angles
4. Convert the interior angles to bearings
5. Determine the sine and cosine values for the bearings
6. Determine the latitudes and the departures
7. Determine the differences in the sum of the latitudes and departures
8. Correct the latitudes and the departures
9. Calculate the corrected distances

These steps will be discussed in the following sections and demonstrated in an example problem.

Collecting Data

The compass rule method requires a North orientation for the traverse. It is important to indicate in the notes whether magnetic or geographic North was used. The decision should be made based on the needs of the client, the capability of the instrument and the amount of resources that will be required to locate North.

Table 9-1 Traverse example data

STA	Distance (ft.)
AB	193.00
BC	176.50
CD	248.49
DE	122.38
EA	120.00

Angle	Degrees
AB:N	126.57
ABC	−115.13
BCD	−79.63
CDE	−100.48
DEA	−138.17
EAB	−106.55

The orientation to North is accomplished by setting the instrument over one of the corners of the boundary being surveyed. The horizontal angle scale is zero set on North and the angle is turned to either one of the adjacent corners. The angle can be turned to either the right or left. The only recommendation is that the traverse is completed in the same direction. When a right angle is turned from North to the first corner, the traverse would be completed clockwise and counterclockwise when a left angle is used. The traverse is completed by moving the instrument to consecutive corners, zero setting the instrument on the previous corner and turning the interior angle. Setting up the instrument on each corner makes it easy to measure distance between corners by stadia or EDM.

Table 9-1 illustrates that two types of data must be collected, the length of the side and the interior angle.

Summing the Angles

The sum of the interior angles for a closed polygon is a constant. Any difference between the sum of the measured interior angles and the constant for that shape was caused by errors when measuring the angles. Care must be taken to insure that only interior angles are used. It is also important to remember that the angle turned from North is not included in the sum of the angles. The angles can be summed using DMS or DD. Standard practice has been to balance the angles in the units they were measured, but because of the use of calculators and spreadsheets the standard practice has switched to using decimal degrees.

Balancing the Angles

Balancing angles adjusts for any error that occurred during the measurement of the angles. The balancing procedure used in the compass rule method makes the assumption that the error was equal for each angle measured. This may or may not be a safe assumption, but it is the procedure used by the compass rule method. The first step in balancing the angles is summing the interior angles and comparing the result to the theoretical sum of the angles. The theoretical sum of the angles for any closed shape is determined by the equation:

$$\text{Sum of interior angles} = (\text{number of angles} - 2) \times 180°$$

The theoretical sum of the angles is subtracted from the sum of the interior angles. The difference is the amount of error that occurred when the angles were measured. Experience has shown that because of limitations of the measuring equipment, the environmental conditions and human error, it is normal for a small amount of error to occur. Therefore, standards have been adopted to determine if the error is acceptable.

Sum of the Angles Standard

The error in the sum of the angles is used as a checkpoint for the accuracy of the survey. The balancing process will produce a closed geometric shape regardless of the magnitude of the sum of the angle error, but if the angle error is large, the balanced shape will not represent the boundaries of the area. The standard for allowable angle error is determined by the required precision of the survey and these can be found in any engineering surveying textbook. Two recommended standards are:

1. The sum of the angle error must be *equal to* or *less than* the least count of the instrument.
2. The sum of the angle error must be *equal to* or *less than* the least count of the instrument times the square root of the number of angles.

Either standard is acceptable. The second standard is more liberal and will permit a larger acceptable error. An excessive sum of the angle error means the data is suspect and the survey should be repeated. Great care and strict adherence to angle measuring procedures are required to prevent excess sum of the angle error.

Balancing Angles by Averaging The averaging method balances the angles by determining a correction for each angle. This correction is determined by subtracting the theoretical sum of the interior angles from the sum of the measured angles and dividing this by the number of interior angles. Expressed as an equation:

$$\text{Correction} = \frac{\text{Measured Sum of Angles} - \text{Theoretical Sum of Angles}}{\text{Number of Angles}}$$

The angles are adjusted by subtracting the correction from each angle. It is important to keep track of the sign (+ or −) for the correction. As a check, remember that if the sum of the interior angles is less than the theoretical sum of the angles, the corrected angle should be larger than the measured angle. If the sum of the interior angles is greater than the theoretical sum of the angles, the corrected angles should be smaller than the measured angle. Expressed as an equation:

$$\text{Corrected angle} = \text{Measured angle} - \text{Correction}$$

A potential problem when using the averaging method is ending up with an artificial level of precision. To illustrate this potential problem, assume that four angles are measured with a least count of one minute and the error for the four angles is plus one minute. The correction would be one minute divided by four or 15 seconds. Fifteen seconds subtracted from each angle will change the angle from degrees and minutes to degrees, minutes, and seconds. The reader could conclude that the instrument least count was 15 seconds instead of one minute. In this situation, an averaging method that maintains the level of precision should be used or the surveyor should arbitrarily adjust the largest angle to correct the sum of the angles.

Converting Angles to Bearings Using Compass Rule

Another characteristic of the compass rule method is the use of bearings to indicate the direction of the movement from one corner to the next. A bearing is a way of expressing a direction relative to another point. Bearings are used because all bearing angles are measured from a north-south line. This sets up a system where the bearing line between two corners and the latitude and departure traversing from one corner to another forms a right triangle. Once the boundary of the property is defined as a series of right triangles, the lengths of the sides can be balanced.

Bearings for a traverse balanced using the compass rule method are determined by starting with the bearing of the first side from North and then calculating the bearings of the remaining sides in sequential order. A review of the section on bearings in Chapter 7 would be helpful if the reader is not confident in converting angles to bearings.

The first step is to determine the quadrant for the bearing. This determines the N-S and E-W designation. Next, the angle of the bearing from the North-South compass line is determined.

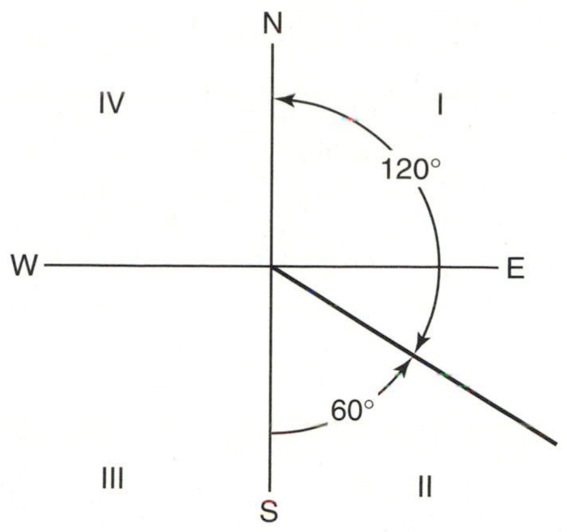

Figure 9-3 Determining the bearing from azimuth angle.

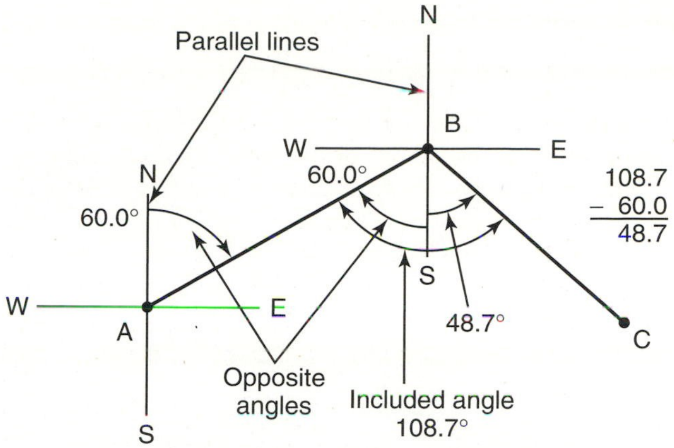

Figure 9-5 Application of the law of parallel lines.

Example: Determine the bearing for angle AB:N with a value of 120 degrees.

Each bearing quadrant encloses 90 degrees. Therefore, an angle of 120 degrees from North will fall in quadrant II and will have a designation of S and E. In the SE quadrant the bearing angle is measured from South. Half of a circle is 180 degrees. Subtracting the angle, 120 degrees, from 180 degrees leaves an angle from South of 60 degrees, resulting in a bearing of S60E, Figure 9-3.

The bearings of the remaining sides are determined in sequence. The sequence is usually determined by the direction of the first angle from North. The law of parallel lines is used to determine the bearings of the remaining sides.

Law of Parallel Lines The **law of parallel lines** states: when a line connects two parallel lines, the opposite angles are equal. The angles A_1 and A_2 in Figure 9-4 are equal angles.

The law of parallel lines can be used to determine bearings because each bearing is determined from a north-south reference line at each station. The north-south reference lines for two adjoining stations form parallel lines. The side of the traverse becomes the line that connects the two parallel lines.

In the example in Figure 9-5, the bearing of line A:B at station A is N60.0E, which means it is 60 degrees from North. Following the law of parallel lines,

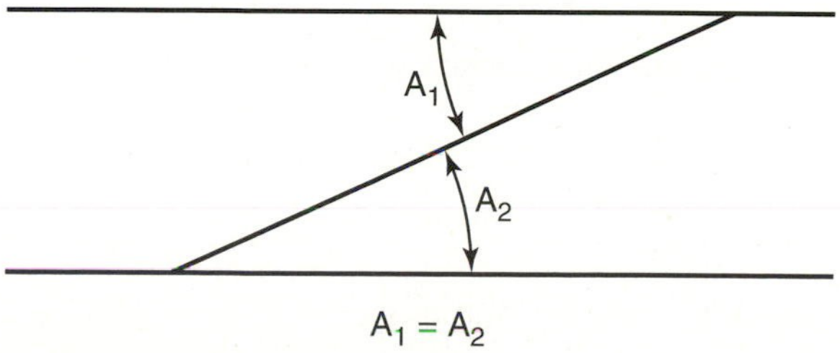

Figure 9-4 Law of parallel lines.

the angle of line B:A from South, at station B, is also 60 degrees. The interior angle of ABC at station B is 108.7 degrees. From this, we can determine that the bearing of line B:C from South, at station B, is 48.7 degrees, 108.7 − 60 = 48.7. The bearing of line B:C is S48.7E. The bearings for the remaining sides are determined in sequence from the bearing of the AB:N angle. Two items of information are required to determine the bearings, the bearing of the previous side and the included angle.

Example: Assume the bearing for AB is S60.0E and the interior angle for ABC, at station B, is 63.0 degrees. What is the bearing for line BC?

The bearing for side BC is S57W.

On occasion, it may be difficult to determine the quadrant for the new line. Drawing a sketch as in the Figures 9-6 and 9-7 will help reduce the chance of a math error.

In some situations, there are alternative mathematical paths that can be taken to arrive at the correct solution. The path is not important. Just remember to write down what you know, identify what you need to solve for, and complete the required math.

Determining the Sine and Cosine Values

The next step is to determine the sine and cosine values for the bearing angles. These values are determined by entering the bearing angle into a calculator using the appropriate buttons to produce the sine and cosine ratio for each angle. The values are recorded in the appropriate column and row of the table.

Determining Latitudes and Departures

Using latitudes and departures to define the movement from one station to another converts the angle and distance information for each station to grid distances. Movement to the north or south is called latitude and east and west movement is called departure. Movement to the east is a plus (+) departure and movement

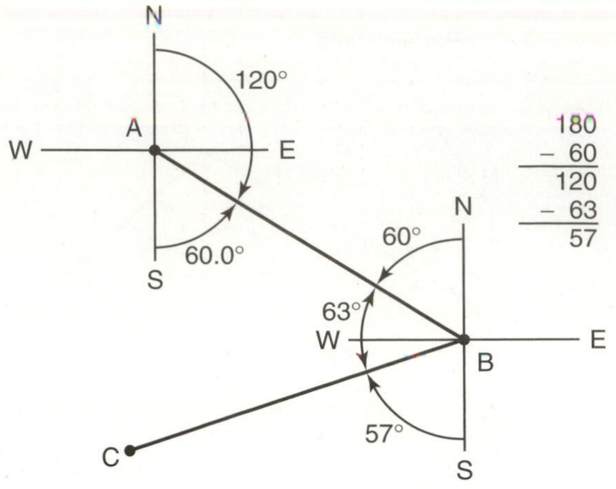

Figure 9-6 Determining the second bearing.

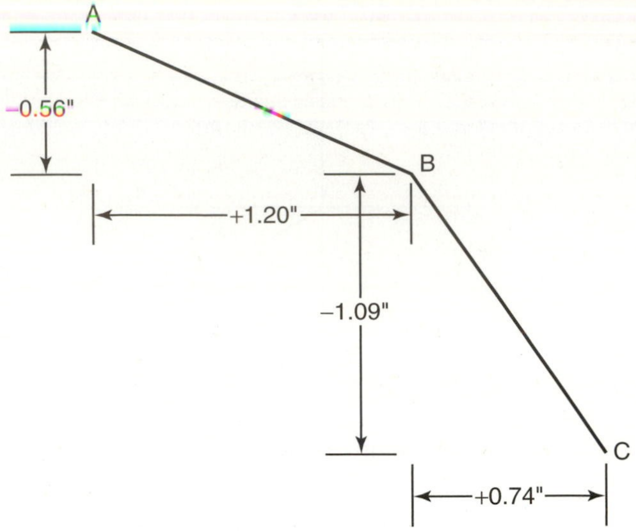

Figure 9-8 Determining movement by latitudes and departures.

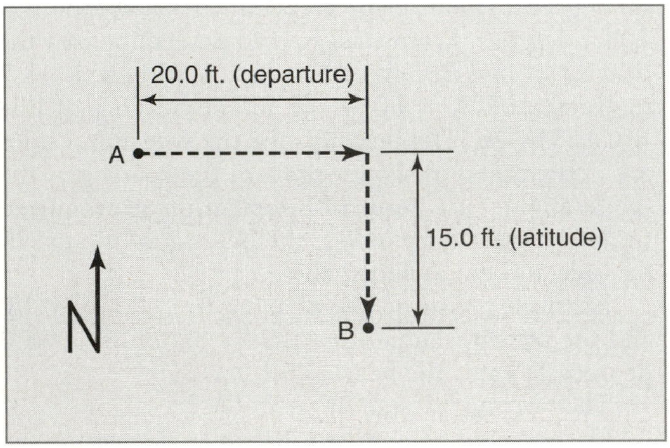

Figure 9-7 Latitude and departure.

Table 9-2 Sum of latitudes and departures

A-B	+Lat	0.63
	+Dep	1.37
B-C	+Dep	0.50
	−Lat	−0.75
C-D	−Lat	−0.75
	−Dep	−0.88
D-A	−Dep	−1.00
	+Lat	0.88
	Sum	0.0

to the west is a minus (–) departure. Movement to the north is a plus (+) latitude and movement to the south is a minus (–) latitude.

Using this information we can say that in Figure 9-7 point B is +20 feet departure and a –15 feet latitude from point A.

Figure 9-8 illustrates clockwise sequential movement from station to station for a traverse measured using latitudes and departures. The latitude and departure method is used to correct the distances because when all of the latitudes and departures of a closed geometric shape are added together, the sum is zero.

Table 9-2 shows the results of summing all of the latitudes and departures for the balanced polygon in Figure 9-9.

The sum of the latitudes and departures will not equal zero if any errors occurred during measurement of the lengths of the sides of the traverse. If the

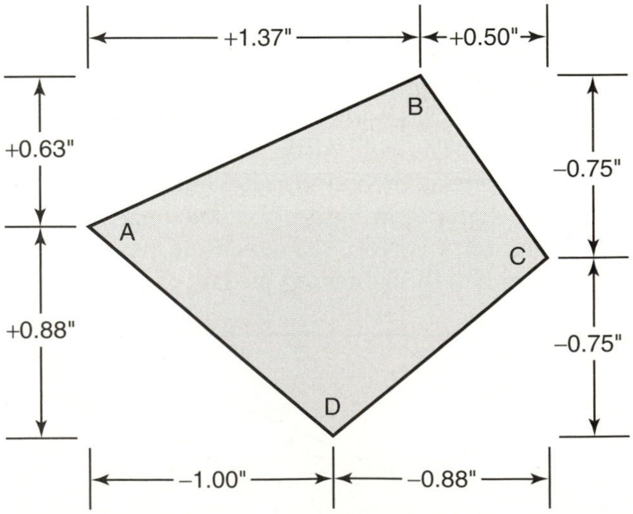

Figure 9-9 Latitudes and departures for a balanced shape.

Table 9-3 Column headings for traverse balancing table

Line	Bearing			Length	Cosine	Sine	Latitudes		Departures		Lat. Cor.	Dep. Cor.	Corrected Latitudes		Corrected Departures		Corrected Distance
	N/S	Degrees	E/W				+	−	+	−			+	−	+	−	
A																	
B																	

latitudes and departures are adjusted so that their sum is equal to zero, any error that occurred when measuring the lengths of the sides is corrected.

Balancing Latitudes and Departures

Balancing the latitudes and departures for complex polygons is not complicated, but it requires several steps and careful attention to detail. The chances for making errors in the process are greatly reduced if the data is organized in a table, Table 9-3.

The values for latitude and departure between two points are determined by the angle of the line connecting the two points. In Figure 9-10 the distance from A to B and B to C is the same, but the latitudes and departures are different because the angles of the lines are different.

The latitude and departure method uses the bearing of the side as the angle for the hypotenuse. If the angle (bearing) and the length of the hypotenuse are known, the latitude and departure can be calculated using the sine and cosine trigonometric functions.

$$\text{Sin } \Phi = \frac{\text{opp}}{\text{hyp}} \qquad \text{opp (latitude)} = \text{Sin } \Phi \times \text{hyp}$$

$$\text{Cos } \Phi = \frac{\text{adj}}{\text{hyp}} \qquad \text{adj (departure)} = \text{Cos } \Phi \times \text{hyp}$$

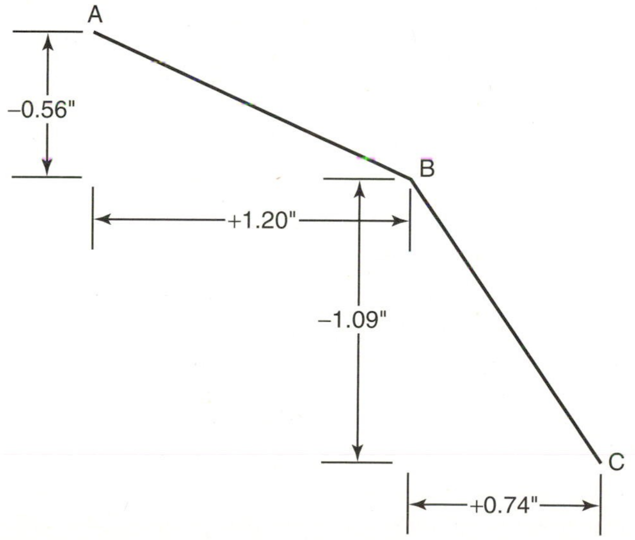

Figure 9-10 Latitude and departure illustration.

Sorting the Latitudes and Departures

During the process of calculating and recording the latitudes and departures in the table, they must be sorted into plus and minus columns. The beginning and ending letter of the bearing determine the correct column. The beginning letter indicates the direction of the latitude. Therefore, a "N" latitude is positive and would be placed in the positive latitude column of the data table. An "S" latitude is negative and would be placed in the negative latitude column of the date table. The same principle is true of the departures. An "E" departure is positive, it would be placed in the plus column. A "W" departure is negative, it would be placed in the negative departure column.

Determining Latitudes and Departure Error After the latitudes and departures are calculated and sorted into the appropriate columns, each column is summed. The amount of latitude and departure error is determined by subtracting the sum of the minus column from the sum of the plus column for both the latitudes and the departures.

Correcting the Latitudes and Departures

Correcting the latitudes and departures is based on a different assumption than the one used for balancing angles. When balancing angles, the error was distributed equally for each angle. When balancing latitudes and departures, it is assumed that the error is proportional to the length of the side. The error correction for distance is based on the ratio of the length of the side compared to the perimeter of the traverse.

$$\text{Lat. Cor.} = \frac{\text{Sum of Latitude Error} \times \text{Distance}}{\text{Perimeter}}$$

This calculation must be completed for each side of the traverse. It is important to remember that the mathematical sign of the latitude correction must be opposite of the sign of the latitude error. A positive latitude error will indicate a negative correction. A negative latitude error means the latitude correction must be positive. The same procedure is used to correct the departures.

$$\text{Dep. Cor.} = \frac{\text{Sum of Departure Error} \times \text{Distance}}{\text{Perimeter}}$$

The latitude correction column in the data table should be summed and the results compared to the latitude error. Occasionally, due to rounding error, the sum of the corrected latitude column does not equal the latitude error by a small amount. When this occurs, the largest latitude correction should be arbitrarily adjusted until the sums are equal. A large difference indicates a problem with the math or a number has been placed in the wrong column.

Calculating Corrected Distances

The correct distances (lengths of the sides) are determined by using the corrected latitudes and departures and the Pythagorean Theorem. The corrected latitudes are the vertical side of a triangle and the corrected departure is the horizontal side of a right triangle. The hypotenuse of the triangle is the corrected distance. Modifying the Pythagorean Theorem for determining corrected distances gives:

$$\text{Corrected Distance} = \sqrt{\left(\text{Corrected Latitude}\right)^2 + \left(\text{Corrected Departure}\right)^2}$$

After the corrected distances are calculated they should be summed and this total compared to the measured perimeter of the traverse. A small difference is acceptable, but a large difference indicates a procedure problem or a math error. When a large difference occurs between the sum of the corrected distance and the measured perimeter of the traverse, start by checking the math and the sorting of latitudes and departures.

Once the corrected distances are calculated and checked, the traverse is balanced and the balanced angles and corrected distances can be used to draw the boundary of the area.

TRAVERSE SAMPLE PROBLEM

Table 9-4 contains the data for a closed traverse. This traverse will be balanced using the compass rule method.

In this example, a transit compass was used to measure the angles from north. When a compass is used, it is important to record if magnetic north was used and if it was corrected for the angle of declination. The **angle of declination** is the angular difference between true north and magnetic north. This number is available on USGS maps and other sources that use maps.

Figure 9-11 is a drawing of the raw data. It shows that errors in collecting the data resulted in a geometric shape that does not close at corner A.

Table 9-4 Data for sample traverse

STA	Distance (ft.)
A-B	702.51
B-C	669.90
C-D	593.97
D-E	469.19
E-A	365.05

STA	Angle (DMS)
AB:N	59° 5′
ABC	−121° 44′
BCD	−30° 15′
CDE	−216° 25′
DEA	−91° 55′
EAB	−79° 35′

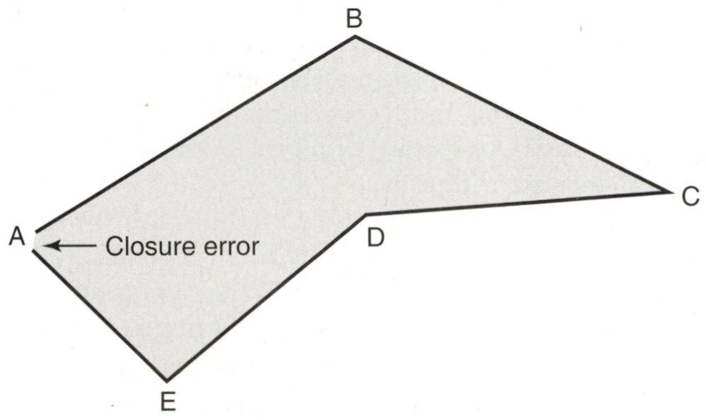

Figure 9-11 Plot of Table 9-4 data.

It is not a common practice to draw the raw data to determine the presence and magnitude of the errors. The uncorrected data for this traverse was plotted to show the effect of errors. This error is called closure error or error of closure. All closed traverses for precise work should be balanced before the data is used. The first step in balancing a traverse is balancing the angles.

Balancing Angles

In this example the angles were measured in DMS. The first step is to convert the DMS angles to DD to facilitate the use of a calculator. Table 9-5 shows the result of this process.

Significant Figures When Converting Angles

The conversion of angles from DMS to DD requires a decision on the number of decimal places that should be retained in the DD. One recommended method is

Table 9-5 Determining DD angles

Angle	Measurement (DM)	Math	DD
AB:N	59° 5′	59 + 5/60 =	59.08
ABC	−121° 45′	121 + 45/60 =	121.75
BCD	−30° 15′	30 + 15/60 =	30.25
CDE	−216° 25′	216 + 25/60 =	216.42
DEA	−91° 55′	91 + 55/60 =	91.92
EAB	−79° 35′	79 + 35/60 =	79.58

dividing the number one by the number of possible decimal numbers and the number of significant figures is the first column that has a non zero number. In this example the angles were measured in degrees and minutes, but not seconds. Therefore, the number of significant decimal places is $1/60 = 0.0166$. The first digit other than zero is in the hundredths column, therefore two decimal places should be used. When angles are measured in degrees, minutes and seconds, then the number of significant decimal places would be $1/3600 = 0.00027$ or four significant decimal places.

The next step is to balance the angles. The angles are balanced using the equation:

$$\text{Sum of Angles} = (N - 2) \times 180$$
$$= (5 - 2) \times 180$$
$$= 540$$

This equation shows that the sum of the interior angles should be 540 degrees. The correct sum is subtracted from the sum of the angles to determine the error. The error is averaged and subtracted from each angle to determine the corrected value for each angle, Table 9-6.

For this example, Table 9-6, the theoretical sum of the angles is 540 degrees. Therefore, 539.92 degrees −540 degrees = −0.08°. The sum of the angles

Table 9-6 Balancing angles for sample traverse

Angle	Angle (DD)	Correction	Balanced Angle
AB:N			
ABC	121.75	−0.016	121.77
BCD	30.25	−0.016	30.27
CDE	216.42	−0.016	216.44
DEA	91.92	−0.016	91.94
EAB	79.58	−0.016	79.60
Σ	539.92 −540.00		540.02
	−0.08		

TIP

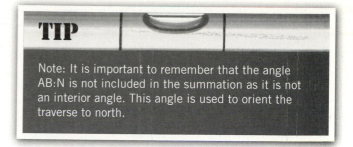

Note: It is important to remember that the angle AB:N is not included in the summation as it is not an interior angle. This angle is used to orient the traverse to north.

error is −0.08°. The compass rule method assumes the same amount of error occurred for each angle. The correction is determined by dividing the sum of the angle error by the number of angles. For this example, the correction is:

$$\text{Correction} = \frac{-0.08}{5} = -0.016$$

After −0.016 is subtracted from each angle, the corrected angle column was summed. In this example, because of rounding, the sum of the interior angles still has a difference of 0.02 degrees. When this occurs, the largest angle is arbitrarily adjusted until the sum of the balanced angles column equals the theoretical sum of the angles. In this example, the 216.44 degree angle is reduced by 0.02 to balance the angles. This adjustment is illustrated in Table 9-7.

The interior angles are balanced.

Data collected by someone else should be carefully scrutinized. It is important to look at the data to determine the process that was used, the least count of the instrument, the direction the angles were turned and any other feature of the data that might indicate the process used to collect the data. In addition, it must be determined if there are any unique aspects that need to be included in the calculations.

Table 9-7 Balancing angles for sample traverse

Angle	Angle (DD)	Correction	Balanced Angle
AB:N			
ABC	121.75	−0.016	121.77
BCD	30.25	−0.016	30.27
CDE	216.42	−0.016	(216.44) 216.42
DEA	91.92	−0.016	91.94
EAB	79.58	−0.016	79.60
Σ	539.90 −540.00		540.00
	−0.10		

Bearings for Sample Problem

It is important to remember that when determining bearings the corrected interior angles are used, not the measured angles. The first bearing is the bearing for line A:B from North. In the data table this angle is recorded as 59°5'. This angle must be converted to DD. Manually: 59 + 5/60 = 59.80°. This is the DD angle of line A:B from north. The angle, 59.08°, is listed as a positive angle, turned to the right, and it is less than 90 degrees. Therefore, the bearing is in the NE quadrant. The bearing for the line A:B is N59.08E, Figure 9-12.

An individual well versed in trigonometric functions and angles can determine the bearings from the angle data table. This method is not recommended. The number of mistakes that are made will be reduced if a sketch of each angle is produced before completing the calculations.

Study Figure 9-13. The balanced interior angle is 121.77°. The bearing for line A:B is N59.08E, which means that the angle for line B:A at station B from south is also 59.08 degrees. The bearing for B:C is determined by subtracting the angle of the bearing for line A:B from the interior angle, 121.77 – 59.08 = 62.69. The bearing angle for line B:C, 62.69 degrees, is less than 90 degrees, therefore the bearing for line B:C is in the SE quadrant. The bearing for B:C is S62.69E.

The bearing for each distance is determined using the same techniques. Different math steps may be required depending on the size of the included angle, the bearing of the previous line and the quadrant of the line in question.

The bearing for side C:D is S87.04W, Figure 9-14.

This is a situation where more than one set of mathematical steps can be used to determine the bearing. The different paths are not right or wrong; one will usually be more efficient. For example, the bearing for line C:D could also be determined by subtracting the opposite angle, 62.69 degrees, from 180 degrees, which leaves 117.31 degrees, and then subtracting the interior angle, 30.27 degrees, from 117.31 degrees, 117.31°−30.27° = 87.04°.

Side D:E bearing is S50.62W, Figure 9-15.

Determining the bearing for line D:E requires two steps. It is helpful to remember that a circle can

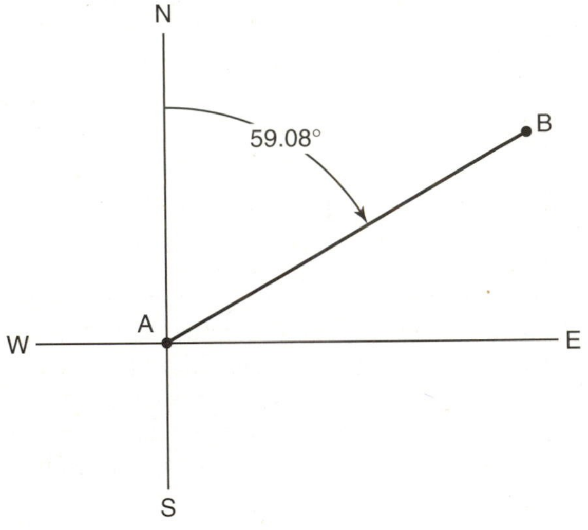

Figure 9-12 Bearing for line A:B.

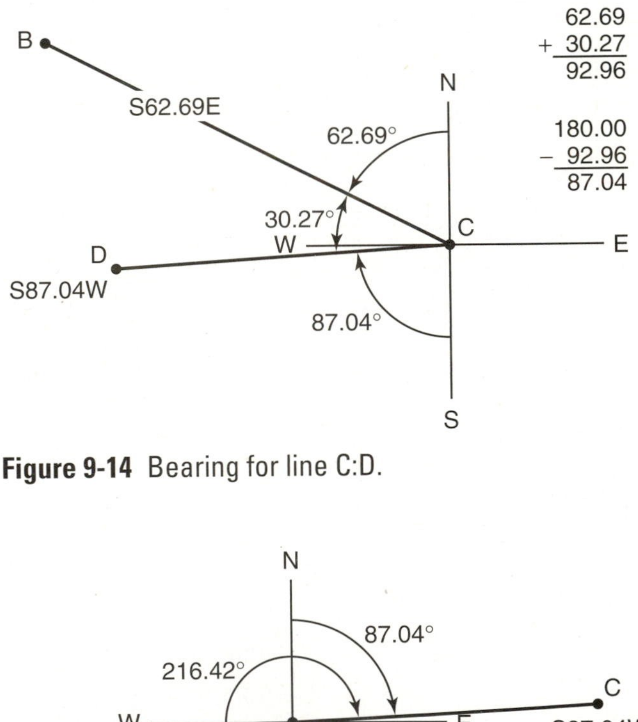

Figure 9-14 Bearing for line C:D.

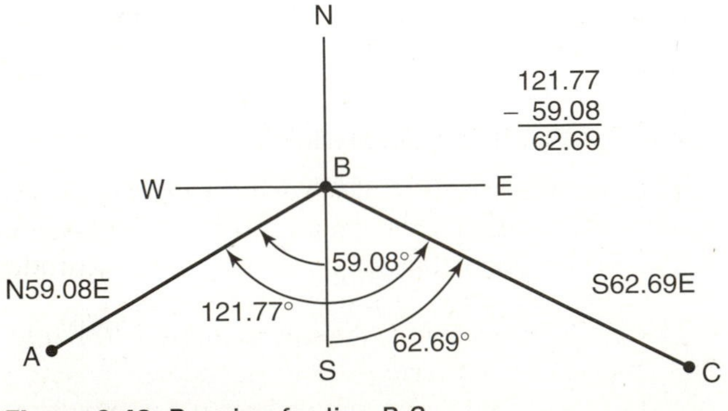

Figure 9-13 Bearing for line B:C.

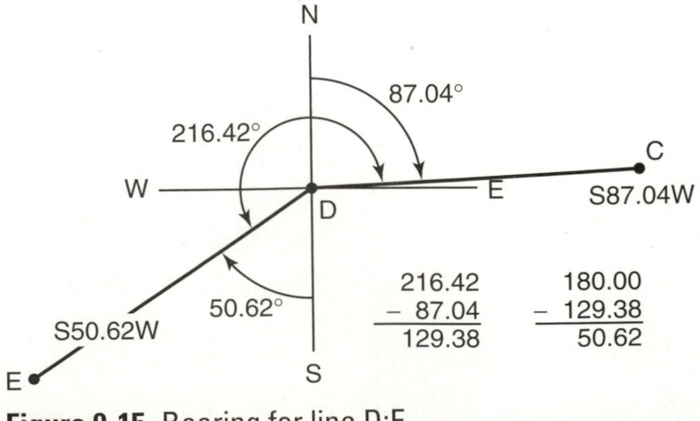

Figure 9-15 Bearing for line D:E.

be divided up into four quadrants of 90 degrees. In the first step, 216.42° − 87.04° was used to determine the quadrant. In this situation, subtracting the previous bearing angle from the interior angle reveals the angle of the side D:E from North. A minus angle from North of 129.38° places the bearing in the SW quadrant. The second step, 180.00°−129.38°, determines how many degrees separated the D:E line from South. This angle, 50.62 degrees, is the bearing for the side D:E.

Side E:A bearing is: N41.32W, Figure 9-16.

Finding the bearing for line E:A completes the traverse. This bearing is determined by subtracting the angle of the previous bearing from the included angle. The result is the angle from North, which is the bearing, 91.94°−41.32° = 50.62°. See Figure 9-17.

After the bearing for each side is determined, it is entered into the table. Determining the bearings is the first step in the compass rule method for correcting the lengths of the sides. The lengths of the sides are corrected by balancing the latitudes and departures and then calculating the correct lengths of the sides. The first step in balancing the latitudes and departures is determining the cosine and sine values for the angle of the bearings and entering them in the traverse table, Table 9-8.

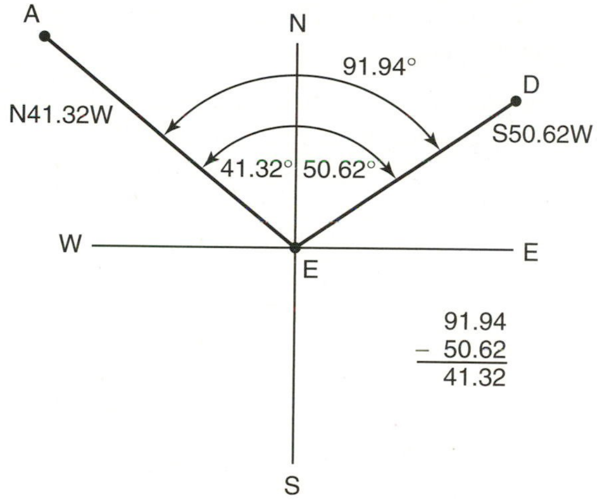

Figure 9-16 Bearing for line E:A.

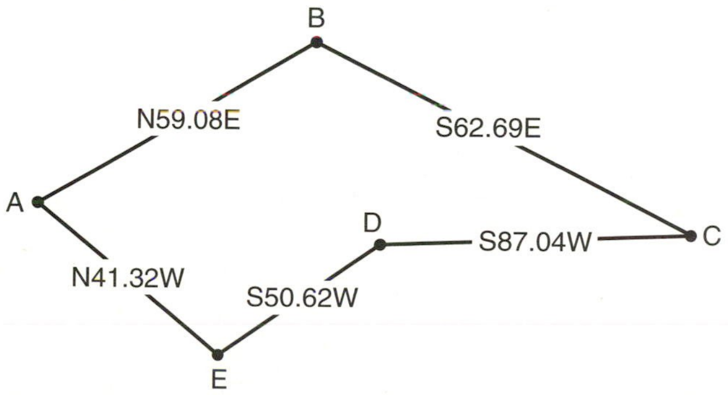

Figure 9-17 Bearings and complete traverse.

Table 9-8 Cosine and Sine values

Line	N/S	Bearing Degrees	E/W	Length	Cosine	Sine
A						
	N	59.08	E	702.51	0.5143	0.8576
B						
	S	62.69	E	669.60	0.4593	0.8883
C						
	S	87.04	W	593.97	0.0524	0.9986
D						
	S	50.62	W	469.19	0.6348	0.7727
E						
	N	41.32	W	365.05	0.7513	0.6600
A						
			$\sum$	2800.32		

Note: The sine and cosine values will differ slightly depending on the brand of calculator that is used. Calculations completed on a computer spreadsheet will be different than those from calculators. All of the calculations for this sample problem were completed using a spreadsheet.

Balancing Latitudes and Departures

Balancing the latitudes and departures is explained together because the process is the same for each. Latitudes and departures are balanced by calculating the lengths of latitudes and departures for each side, calculating the amount of closure error and using the closure error to determine the correction. The correction is used to calculate the corrected latitudes and departures. The sum of the + and − corrected latitudes and corrected departures are compared to determine if they are balanced. The distances for the latitudes and departures are calculated using the cosine and sine trig functions.

Calculating Latitudes

The latitudes are determined by multiplying the length of the side by the cosine value of the bearing angle. At this point it is very critical that the latitudes are sorted into plus (+) and minus (−) columns.

Study Figure 9-18. The bearing angle is the interior angle of a right triangle. The latitude is the length of the opposite side. The hypotenuse is the measured

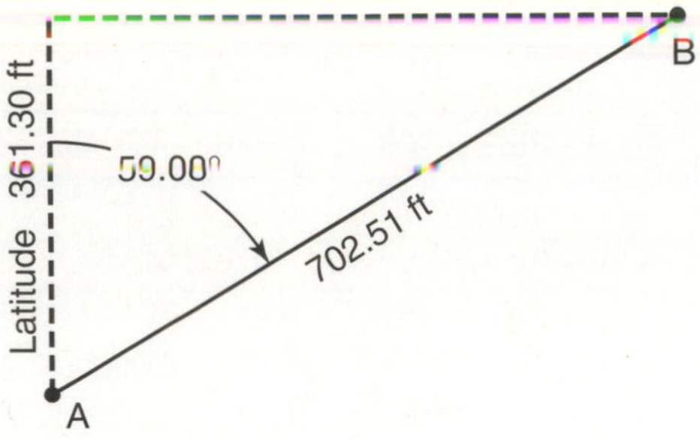

Figure 9-18 Latitude calculation.

distance between stations A and B. Using the cosine trig function the latitude is 361.3 feet.

Line A:B has a "N" latitude, therefore the value of 361.30 is recorded in the + latitude column. A "S" latitude would be recorded in the – latitude column. The latitudes are calculated for the remaining sides.

These values are recorded in the appropriate column of the traverse table, Table 9-9.

Calculate the Departures

Departures are calculated by multiplying the length of the side by the sine value of the bearing angle. They must also be sorted into the appropriate + or – columns. The appropriate column is determined by the direction of the bearing. If the bearing is "E," the departure is placed in the + departures column and if the bearing is "W," the departure is placed in the – departures column.

$$AB: 702.51 \text{ ft.} \times 0.8576 = 602.47 \text{ ft.}$$

$$BC: 669.60 \text{ ft.} \times 0.8883 = 594.81 \text{ ft.}$$

$$CD: 593.67 \text{ ft.} \times 0.9986 = 593.14 \text{ ft.}$$

$$DE: 469.19 \text{ ft.} \times 0.7727 = 362.54 \text{ ft.}$$

$$EA: 365.05 \text{ ft.} \times 0.6600 = 240.93 \text{ ft.}$$

The departures are recorded in the appropriate columns, Table 9-9.

Closure Error

The next step is determining the closure error for the latitudes and the departures. The closure error for latitudes is determined by subtracting the sum of the – latitudes column from the sum of the + latitudes column. The closure error for the departures is determined by subtracting the sum of the – departures from the sum of the + departures, Table 9-9.

Table 9-9 Latitudes and departures with their corrections

	Latitudes		Departures			
Line	+	–	+	–	Lat. Cor.	Dep. Cor.
A						
	361.30		602.47		0.24	–0.17
B						
		307.55	594.81		0.23	–0.16
C						
		31.12		593.14	0.20	–0.14
D						
		297.84		362.54	0.16	–0.11
E						
	174.26			240.93	0.12	–0.09
A						
$\sum$	635.56	636.51	1197.28	1196.61	0.95	–0.67
	636.51		1196.61			
–	–0.95		0.67			

Balancing Latitudes and Departures

The latitudes and departures are balanced by determining the proportion of the closure error for each one. The balancing of the latitudes and departures assumes the amount of error is proportional to the length of the side. This is accomplished by the following equations:

$$\text{Correction}_{\text{lat.}} = \frac{\text{Error}_{\text{lat.}} \times \text{Length}}{\text{Perimeter}}$$

$$\text{Correction}_{\text{dep.}} = \frac{\text{Error}_{\text{dep.}} \times \text{Length}}{\text{Perimeter}}$$

These equations are used to calculate a correction for each latitude and departure.

Latitude corrections:

$$AB: \frac{0.95 \times 702.51 \text{ ft.}}{2800.32} = 0.24 \text{ ft.}$$

$$BC: \frac{0.95 \times 669.60 \text{ ft.}}{2800.32} = 0.23 \text{ ft.}$$

$$CD: \frac{0.95 \times 593.97 \text{ ft.}}{2800.32} = 0.20 \text{ ft.}$$

$$DE: \frac{0.95 \times 469.19 \text{ ft.}}{2800.32} = 0.16 \text{ ft.}$$

$$EA: \frac{0.95 \times 365.05 \text{ ft.}}{2800.32} = 0.12 \text{ ft.}$$

Departure corrections

$$AB: \frac{0.67 \times 702.51 \text{ ft.}}{2800.32} = 0.17 \text{ ft.}$$

$$BC: \frac{0.67 \times 669.60 \text{ ft.}}{2800.32} = 0.16 \text{ ft.}$$

$$CD: \frac{0.67 \times 593.97 \text{ ft.}}{2800.32} = 0.14 \text{ ft.}$$

$$DE: \frac{0.67 \times 469.19 \text{ ft.}}{2800.32} = 0.11 \text{ ft.}$$

$$EA: \frac{0.67 \times 365.05 \text{ ft.}}{2800.32} = 0.09 \text{ ft.}$$

The corrections for the latitudes and departures are shown in the appropriate columns in Table 9-9.

Corrected Latitudes and Departures

The next step is to use the latitude and departure corrections to calculate the corrected latitudes and departures. These corrections are subtracted from each + and – latitude and departure and the results are recorded in the appropriate columns in Table 9-10.

Table 9-10 shows the calculated latitudes and departures, the closure error, correction, and the balanced latitudes and departures. The latitudes and departures are balanced because adding the sum of the minus (–) latitudes to the sum of the plus (+) latitudes results in zero. The same is true for the departures. Occasionally, mathematical rounding will cause a slight difference between the sum of the plus and minus columns. When this occurs, the largest latitude and departure should be adjusted until the sums equal zero.

Determining Corrected Distances

The last step is to calculate the corrected distances. This is accomplished by using the Pythagorean Theorem and the balanced latitudes and departures. The math for this step is:

$$AB: \sqrt{(361.54 \text{ ft.})^2 + (602.30 \text{ ft.})^2} = 702.48 \text{ ft.}$$

$$BC: \sqrt{(307.32 \text{ ft.})^2 + (594.65 \text{ ft.})^2} = 669.37 \text{ ft.}$$

Table 9-10 Corrected and balanced latitude and departures

	Latitudes		Departures				Corrected Latitudes		Corrected Departures	
Line	+	–	+	–	**Lat. Cor**	**Dep. Cor.**	+	–	+	–
A										
	361.30		602.47		0.24	0.17	361.54		602.30	
B										
		307.55	594.81		0.23	0.16		307.32	594.65	
C										
		31.12		593.14	0.20	0.14		30.92		593.28
D										
		297.84		362.54	0.16	0.11		297.68		362.65
E										
	174.26			240.93	0.12	0.09	274.38			241.02
A										
Σ	635.56	636.51	1197.28	1196.61	0.95	0.67	635.92	635.92	1196.95	1196.95
–	635.51		1196.61				635.92		1196.95	
	–0.95		0.67				0.0		0.0	

$$CD: \sqrt{(30.92\text{ ft.})^2 + (593.28\text{ ft.})^2} = 594.09\text{ ft.}$$

$$DE: \sqrt{(297.68\text{ ft.})^2 + (362.65\text{ ft.})^2} = 469.18\text{ ft.}$$

$$EA: \sqrt{(274.38\text{ ft.})^2 + (241.02\text{ ft.})^2} = 365.21\text{ ft.}$$

Recording these values in the traverse table and summing the corrected distance column balance the traverse, Table 9-11.

A final check for the accuracy of the procedure is to compare the sum of the balanced distance column with the sum of the length column. A small difference is acceptable. If there is a large difference, it is an indication that a math error was included or a procedure was completed wrong. For this example, the sum of the balanced distances is 2800.33 ft. and the perimeter of the polygon that was measured was 2800.32 ft. This is an acceptable difference.

Table 9-11 Corrected distances

	Corrected Latitudes		Corrected Departures		Corrected
Line	+	−	+	−	**Distance (ft.)**
A					
	361.54		602.30		702.48
B					
		307.32	594.65		669.37
C					
		30.92		593.28	594.09
D					
		297.68		362.65	469.18
E					
	274.38			241.02	365.21
A					
Σ	635.92	635.92	1196.95	1196.95	2800.33
–	635.92		1196.95		
	0.0		0.0		

Student Activity

1. Balance the traverse for the data in the following table.

Angle	Degrees	Minutes	Seconds	Distance (ft.)
AB:N	74	32	20	
ABC	103	25	31	598.361
BCD	55	27	11	414.012
CDE	256	14	35	382.860
DEA	54	42	8	271.498
EAB	70	15	37	517.215

CHAPTER 10

Global Positioning

Objectives

After reading this chapter the student should be able to:

- Know the history of the global positioning system.
- Understand the difference between triangulation and trilateration.
- Know how the global positioning system works.
- Understand the limitations and applications of GPS for surveying.
- Determine the distance between two points using UTM.

Terms To Know

Radio Direction Finding
LORAN
Navigation Signal Timing and Ranging Global Positioning System
Global Positioning System
Universal Transverse Mercator
Triangulation
Trilateration
Wide Area Augmentation System
GLONASS

INTRODUCTION

The previous chapters presented the principles of surveying and some of the equipment that has been traditionally used. Professional surveyors still use these principles and methods of surveying, but the equipment has changed. As late as the mid 1900s, surveying used a chain to measure distance, a compass to determine direction, theodolite or transit to determine angles and elevation. For the professional surveyor it is no longer cost effective to conduct surveys of large areas or long distances using traditional mechanical equipment. Mechanical equipment is still appropriate for small projects in construction, landscaping, and agriculture.

The movement away from mechanical instruments started with electronic distance measuring (EDM) devices and electronic theodolites. These two technologies were combined to produce an instrument called the total station. Total stations almost eliminated the four- to five-person crew required for chaining because they can measure distance, elevation, and angles. All that is required is a line of sight. One person with the newest total station can do the same job as a crew of four to five people using transits, levels, and chains. Total stations have also been replaced as the surveyors preferred instrument. The newest technology is using the global positioning systems for surveying.

GPS HISTORY

There are many examples in which history would have occurred differently if a military commander had up-to-date information on the location of his or her, or the opposition's, troops. Accurate compasses and maps were a big improvement, but they did not provide the military with the locational and navigational ability they needed. The first step in the development of global positioning systems was called **radio direction finding** (RDF). RDF systems use a directional antenna to determine the bearing of two or more radio stations. Using a directional antenna and a book that listed the location and frequencies of stations, a person could calculate his or her position. For many situations this was an improvement to a map and a compass but it was time-consuming, it required at least two stations within range, enemy forces could use the same technology to locate the broadcast stations and eliminate them, it required the book of stations, and the operator had to complete the calculations correctly. The search continued for a better system.

Radio direction finding evolved into long range navigation systems. The currently used land base radio navigation systems were first tried during the 1940s. One of the early systems was called **LORAN** (Long Range Navigation). The LORAN system provided valuable service during WWII, but the military was not satisfied with the range and accuracy limitations. The LORAN system was improved in the 1950s and called LORAN-C. It has been a very useful system in marine applications for locating the position of sea vessels, but was not widely available, and it did not include elevation information.

Note: The US government has discontinued supporting LORAN-C.

In the search for a better system, researchers looked up.

GLOBAL POSITIONING SYSTEMS

Scientists early recognized the advantages of a space-based system. The **global positioning system** is a group of satellites dispersed in known orbits around the earth. Each satellite broadcasts a unique signal and by trilaterating multiple signals, the receiver can calculate its location. Several satellites would provide a wide area of coverage, and signals traveling back and forth in space have a lower incident of being blocked by terrain, jamming, or structures. Several systems were developed and tried before the **Navigation Signal Timing and Ranging Global Positioning System** (NAVSTAR GPS) was put into place. NAVSTAR GPS became operational in 1995 and has rapidly become the GPS of choice. It didn't take long after the DOD activated NAVSTAR GPS for civilian scientist and engineers to realize that they could develop receivers to use the same signals for surveying and mapping. Because of the capabilities of the NAVSTAR GPS as a tool for surveying and mapping it rapidly became the system of choice for almost all surveying and other navigational, mapping and locating needs. The NAVSTAR system in the U.S. has become so popular that for every day use, the NAVSTAR acronym has been dropped and the system is called the global positioning system (GPS). This usage will be used for the remainder of this text.

The initial success of NAVSTAR GPS navigational/surveying system as prompted several other countries and the Economic Union to establish their own systems. The three world wide systems are:

- **GLONASS** (GLObal NAvigation Satellite System)—Russia
- Galileo—European Economic Community
- COMPASS—China

Several regional systems are also in place. One of these is the Wide Area Augmented System (WASS) being developed by the U.S. Federal Aviation Administration (FAA).

WIDE AREA AUGMENTATION SYSTEM

Wide area augmentation system (WAAS) compensates for GPS errors caused by ionosphere disturbances, satellite orbital errors, and timing errors. The WAAS uses several ground reference stations positioned across the United States to constantly monitor the GPS satellite data. These reference stations communicate with two master stations located on the coasts. The reference stations determine the amount of error in the GPS signals and broadcast this information to two communication satellites in geostationary orbits at the equator. These satellites broadcast the corrections to any GPS receiver capable of receiving the signal. The current WAAS provides an accuracy of plus or minus one meter. The combination of GPS and WASS will produce precisions of less than one foot with survey or mapping quality GPS units.

One advantage of the WAAS is that the signal is compatible with the NAVSTAR GPS signal structure. This means it does not require a different type of receiver, as differential GPS does. WAAS receivers have the same features and capabilities as NAVSTAR GPS receivers. The trade off is cost. The newer systems replace manpower with technology. Technology can be expensive.

All global positioning systems have three major segments: space, control, and user. The remainder of this chapter will explain the global positioning system used in the United States.

Space Segment

The global positioning system is a space based system of satellites that provides all-weather, 24-hour, three-dimensional, one-way, proprietary, world-wide signals. The number of satellites varies as old ones retire and new ones are added, but it is usually 24 or more satellites. The satellite orbits are managed to insure at least four satellites are available at all times because that is the minimum number required for a receiver to function, but usually more than four are available. The actual number of satellites that a receiver can read depends on the number of channels built into the receiver, the terrain, and local obstacles that block the signals. The space segment consists of satellites in six orbital planes, with four satellites in each plane, dispersed around the globe. Each satellite is constantly told its location and the precise time. It continuously broadcasts this information and all other information required by the receiver.

Control Segment

The accuracy of the receiver's calculation of its location and elevation is dependent upon the satellite knowing its position. One of the functions of the control segment is to insure the satellite position information is accurate. The control segment consists of a master control center at Schriever Air Force Base in Colorado and several monitoring stations dispersed around the world. The monitor stations constantly track the orbits and elevation of the satellites and relay this data to master control. At master control, the position and elevation of each satellite are constantly monitored. Corrections are determined and the appropriate information is relayed to each satellite. The master control center also constantly monitors the clocks on each satellite and insures they are synchronized with the master clock.

User Segment

The user segment is a receiver that gathers and processes the satellite information. The receiver is the heart of the user segment. They are available in many different sizes, level of precision, and capabilities. The cost of GPS receivers can range from less than one hundred dollars to several thousand dollars depending on the precision required and the processing speed, Figure 10-1.

The standard GPS receivers are passive and must get all of the information they need from the satellites. Survey quality GPS receivers can be passive but the majority use a process called **Real Time Kinematic** (RTK). A RTK system is in constant communication with a base station. This can be another unit stationed at a benchmark or a **Continuously Operating Reference Station** (CORS). A CORS is a stationary highly precise GPS receiver that continuously compares its know location to the current calculated location from the satellite information. It broadcasts

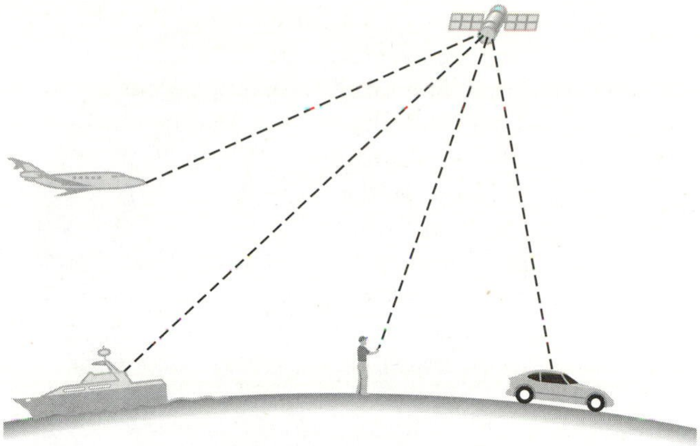

Figure 10-1 User segment.

this information so that individual GPS receivers can adjust their calculated position for the current errors. The data can also be made available over the internet. The heart of all receivers is a microprocessor and this allows manufacturers to design receivers with many different capabilities. Some of these include:

- Determining locations
- Determining elevations
- Determining distances between points
- Storing maps
- Determining velocity
- Recording routes and how to return
- Storing coordinates
- Selection of geodetic model
- Determining direction

Discussing all of the features and capabilities of GPS receivers is beyond the scope of this text because each model has different features and uses different keys and windows of information. Before using a GPS receiver the operator should study the instructions for the receiver they will be using. One feature that is important to understand is the coordinate system used for locating points. GPS receivers can provide the user with location information in more than one coordinate system. Latitude and longitude, and Universal Transverse Mercator (UTM) are two that are commonly used. They are explained in the following sections.

LATITUDE AND LONGITUDE

Latitude and longitude is a coordinate system that was developed to locate points on the surface of the earth. A coordinate system has a reference point or line and a means to locate points away from the reference. All locating systems must use one of or a combination of three methods, three dimensions, an angle and a distance, or two angles. The system of latitude and longitude uses two angles.

Latitude

Latitude is a measurement of a position north or south of the equator. Visualize a line extending from the equator to the center of the earth. This is the reference line for the angle and the center of the earth is the vertex. A line drawn from the center of the earth to the unknown position on the earth's surface forms the second line of the angle. The latitude is the angle formed by these two lines, Figure 10-2. Common practice has been to express latitude in units of degrees, minutes, and seconds. The use of calculators and computers has increased the use of expressing angles in decimal degrees.

Latitudes form a series of circles that gradually decrease in size as you move from the equator to the poles, Figure 10-3.

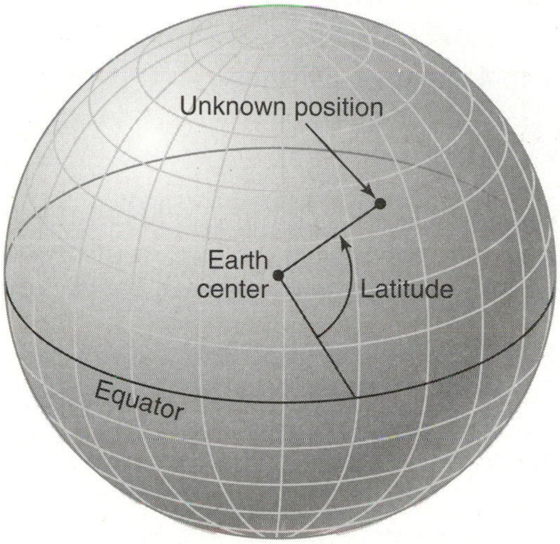

Figure 10-2 Latitude.

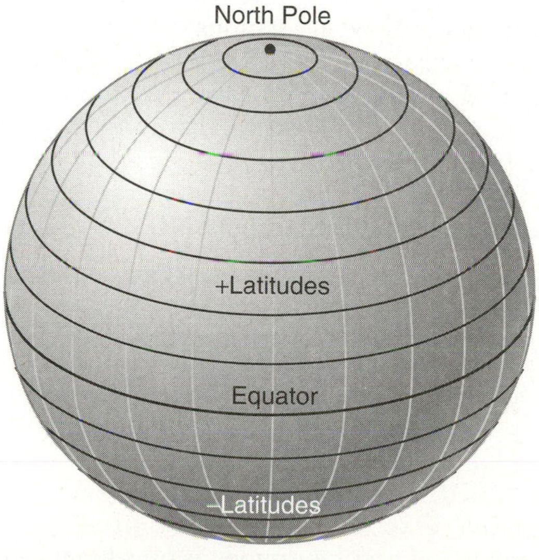

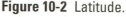

Figure 10-3 Latitude lines.

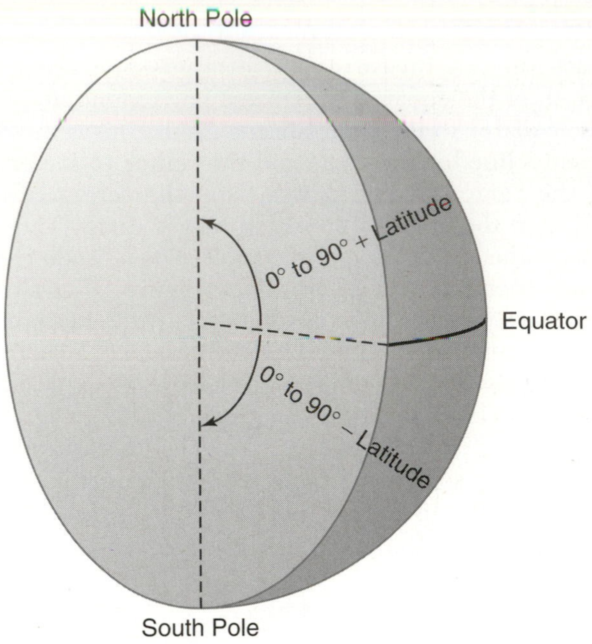

Figure 10-4 Range of latitudes.

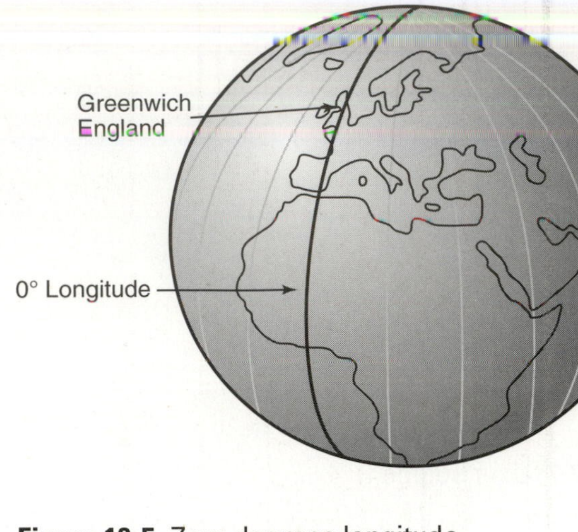

Figure 10-5 Zero degrees longitude.

In Figure 10-3 it can be seen how the latitudes north of the equator are considered plus angles and latitudes south of the equator are called minus angles.

Zero latitude is at the equator and maximum latitude is the North Pole, therefore latitude angles can range from 0 to 90 degrees north (+) and 0 to 90 degrees south (−), Figure 10-4.

Longitude

Longitude is an angular measurement of how far a position is east or west. Longitude is an angle measured along the equator. The longitude of a position is determined by where its meridian line crosses the equator, Figure 10-6. The use of the equator to measure angle of longitude means all points along the longitude or meridian line will have the same degrees of longitude. Historically, zero meridian angle, and therefore zero longitude, has been the observatory in Greenwich, England, Figure 10-5. At this site is a brass disc with a line scribed on it. The line is zero longitude.

The longitude angles are numbered eastward and westward from this point, Figure 10-6. The range of longitudes is from zero to 180 degrees west (−) and from zero to 180 degrees east (+). It is important to record longitudes correctly so the correct hemisphere is identified. West and east are commonly used. An alternative is to identify west longitudes as minus longitudes and east longitudes as plus longitudes.

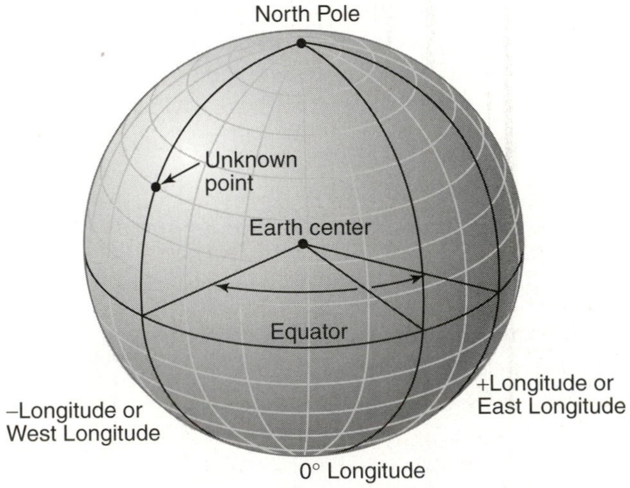

Figure 10-6 Longitude.

The result is a grid system that can be used to locate any point on the earth. Figure 10-7 illustrates this grid for the northern hemisphere.

Identifying positions by latitude and longitude is ideal for shipping and locating points on a map but it is not as useful for surveying because determining distance between two points using latitude and longitude angles is a cumbersome mathematical process. If we assume the earth is a sphere, there are 69.167 miles per degree, 24,900 miles/360°, at the equator. Determining horizontal distances between points along the equator can be accomplished by subtracting the two longitudes and multiplying by 69.167 miles per 0° of longitude. Geodetic distance (surface distance) could be determined using one of the geodetic models of the earth. Unfortunately, most unknown distances do not lie on the equator. Because

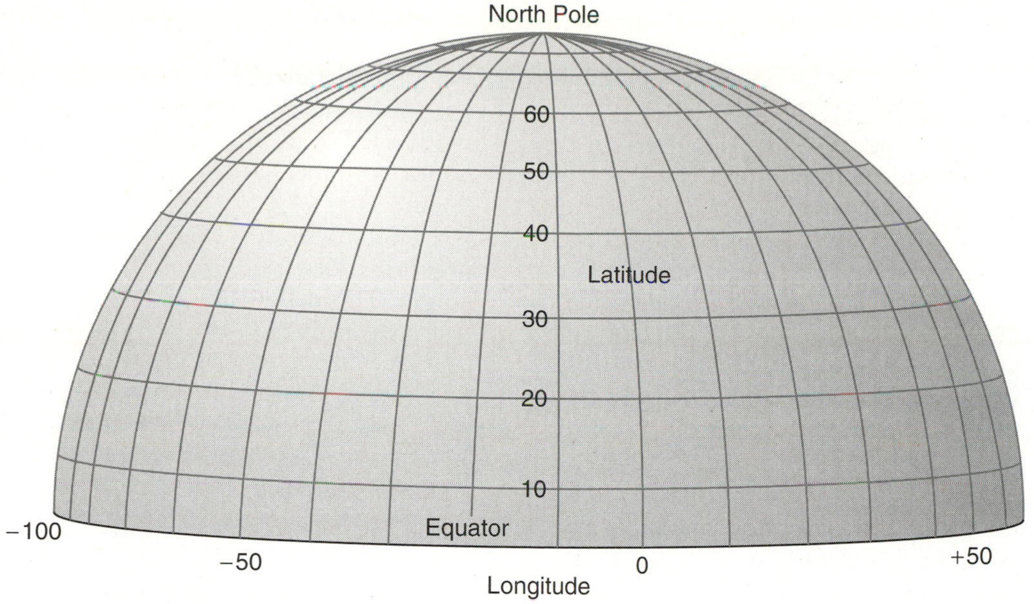

Figure 10-7 Latitude and longitude grid.

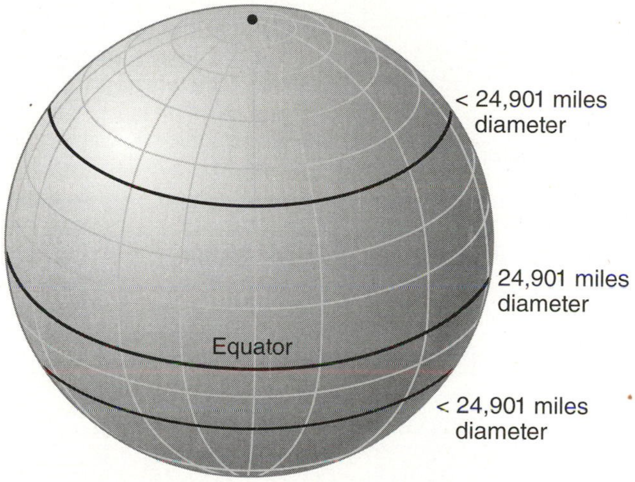

Figure 10-8 Change in miles per degree of longitude.

of convergence, the diameter of the longitude circles decreases as latitudes move away from the equator. This means the number of miles per degree of longitude also decreases as latitudes move away from the equator, Figure 10-8.

The miles per degree of longitude can be determined by mathematical equation, but this is beyond the scope of this text.

Manually determining distances between points by latitude and longitude requires rigorous use of math that includes using an equation to determine the diameter of the longitude at the latitude of the unknown points. If the unknown points are at different latitudes, additional equations must be used.

The latitude and longitude system is not the best system to use when trying to determine distances or areas, but it is the preferred system when the desire is to locate individual points.

GPS receivers have the capability of outputting the data in either system. It must be remembered that the results produced by the GPS receiver are based on one of several mathematical models of the earth's shape. The operator must insure the GPS receiver is set for the desirable model. An alternative method is Universal Transverse Mercator (UTM).

UNIVERSAL TRANSVERSE MERCATOR

Universal Transverse Mercator (UTM) is a coordinate system used to project the surface of the earth onto a map. Mapmakers have always been challenged by the necessity of representing a curved earth's surface on a flat map surface. The different methods of projecting either distort the size of continents, the shape of the continents, or both. Gerardus Mercator produced a map in the 1500s based on the projection that all latitude and longitude lines form straight lines. He produced his map by visualizing a cylinder around the earth with the same diameter as the earth and with the centerline of the cylinder parallel with the earth's poles. He then projected points on the earth to the inside of the cylinder and then unrolled the cylinder to form a map. The Mercator projection has been the standard mapping projection for several centuries.

Although the Mercator projection is not universally accepted by map makers as the best projection, it is the basis for the system developed by the U.S. Army in the 1940s. This system is called the transverse Mercator because the Army system is a projection of the latitude and longitude lines onto a cylinder that is mounted transverse to the centerline of the earth instead of parallel. Because the center meridian of the map can be selected for each map, very detailed maps, with a narrow east west distance, can be produced. This projection is called the Universal Transverse Mercator System (UTM), Figure 10-9.

In the UTM system the 360 degrees along the equator is divided into 60 sections starting at 180 degrees west from the prime meridian. The slices are numbered from 1 to 60 eastwards and extend from the equator to the polar regions. The polar regions are covered by a different system called Polar Stereographic. Three hundred and sixty degrees divided by 60 results in 6 degrees for each section. Using 40,000 kilometers for the diameter of the earth results in 667 kilometers per section at the equator, Figure 10-10.

Within each section, the north distance is the distance from the equator and the east distance is determined from an imaginary point 500,000 meters west of the junction of the centerline of the section and the equator, Figure 10-10. Standard form for writing UTM locations is to list the section number first, then the east distance and then the north distance. In the UTM system an east distance less than 500,000 meters will be west of the centerline and an east distance greater than 500,000 meters will be east of the centerline of the UTM section.

The UTM system has several advantages over latitudes and longitudes. It is a grid system with metric values and all of the values are positive. The math is less confusing when working with positive values. It is the preferred system to use when determining distances and areas.

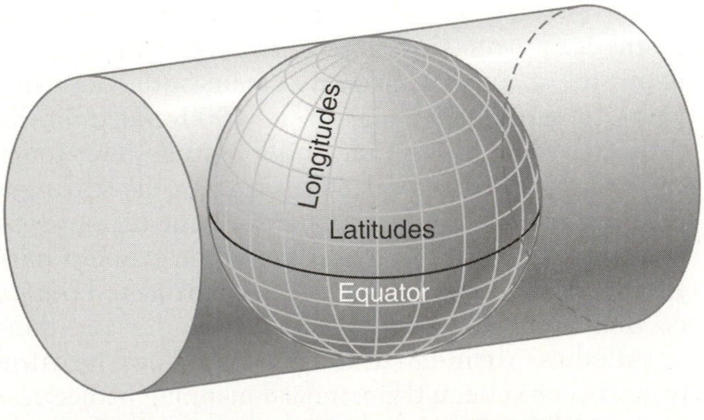

Figure 10-9 Transverse Mercator projection.

The UTM system combined with a GPS receiver is very useful for surveying because it is easy to determine distances between two or more points. For example, determine the distance between a point with a UTM of 14, E556,000, and N780,000 and a point with a UTM of 14, E348,000, and N243,000, Figure 10-11.

The solution is based on a right triangle. The unknown distance forms the hypotenuse of the triangle. The difference between the east values for the two points is the length of one side and the

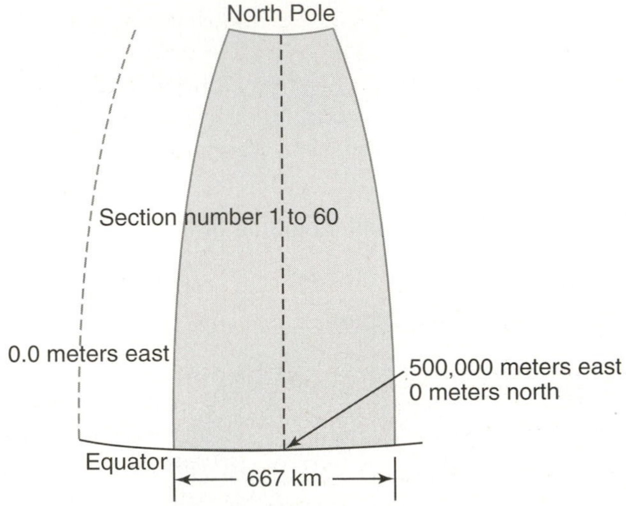

Figure 10-10 North half of UTM section.

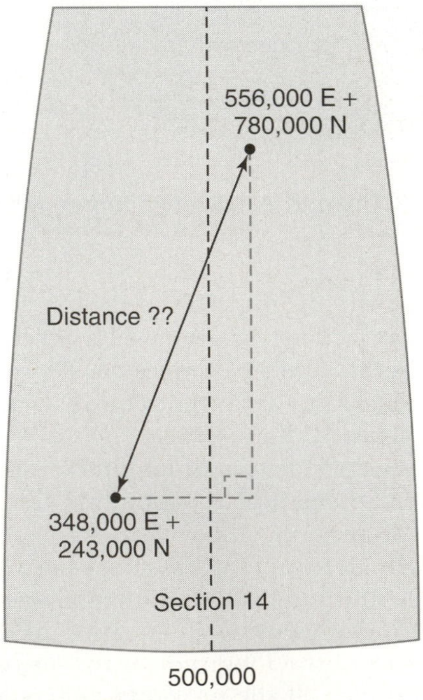

Figure 10-11 UTM distance example.

difference between the north values for the two points is the length of the second side. The Pythagorean Theorem can be used to solve for the unknown distance.

$$\begin{array}{r} 556{,}000 \\ -348{,}000 \\ \hline 208{,}000 \end{array} \qquad \begin{array}{r} 780{,}000 \\ -243{,}000 \\ \hline 537{,}000 \end{array}$$

$$\text{Distance} = \sqrt{208{,}000^2 + 537{,}000^2}$$
$$= 575{,}875.85\ldots \text{ or } 575{,}876 \text{ M}$$

In this example the horizontal distance between the two points is 575,876 m or 576 kM. Remember this is the plane distance. The geodetic distance could be determined by using an equation for one of the geodetic models of the earth. This example illustrates that UTM is easy to use to determine the distance between points.

POSITIONING

To understand the operation of a GPS receiver it is important to remember two characteristics of the system. 1) It was developed by the Department of Defense (DoD) and one of their requirements was a passive receiver. If the receiver was required to send out a signal for the system to work, that signal could be used to locate the receiver. This would allow enemy forces to locate any receiver being used. 2) The task of the receiver is to use the signals from several satellites to locate itself. The theory of operation is the same that is used to draw a map of an area. The location of a tree or some other structure is determined by using other points as a reference. Two different methods are used, triangulation and trilateration.

Triangulation

Triangulation is a process of locating a position by using at least one angle and distance from known points, Figure 10-12.

In Figure 10-12, the point 'C' is located by turning an angle of 49.2 degrees from the line A:B and measuring the distance of 177.39 feet along this line. In this example the positions of stations A and B must be known and used as the baseline to turn the angle. This is a useful method when locating points on the earth's surface. It does not work for GPS because of the problems associating with maintaining a constant baseline that can be used as the reference for the angle. The second method is trilateration.

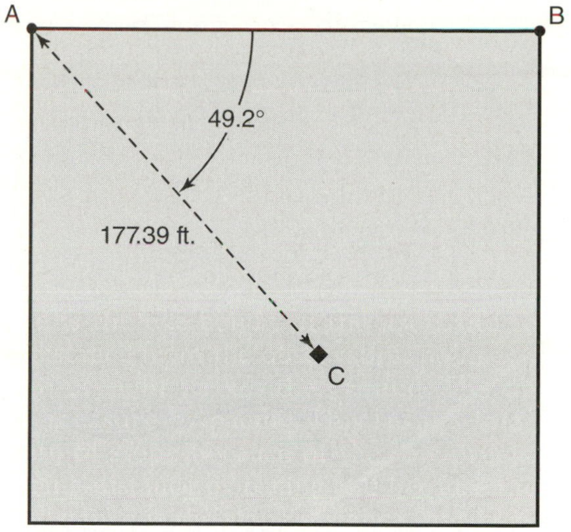

Figure 10-12 Locating a point by angle and distance.

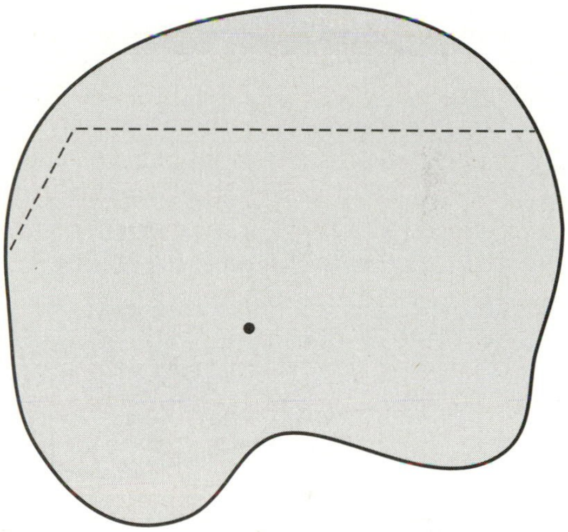

Figure 10-13 Locating a point using trilateration.

Trilateration

Trilateration locates a point by knowing the position of at least two reference points and three distances. Study Figure 10-13. In this example an unknown point is positioned to the side of a dashed line.

The dashed line could represent a fence, sidewalk, property line, or any type of reference line. It is important that the reference line is a permanent feature that will not be damaged or removed during construction. The first step in using trilateration is to establish two stations along the reference line. In Figure 10-14 stations A and B are located along a reference line.

Stations A and B will not be useful for locating station C unless their locations are known and can be repeated. The next step is to measure the location of

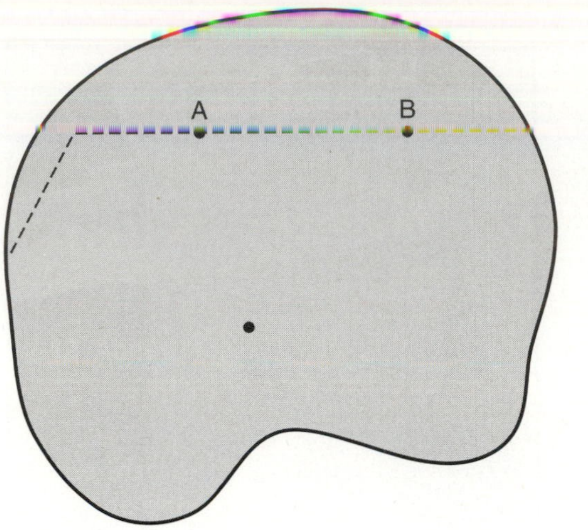

Figure 10-14 Two reference points for trilateration example.

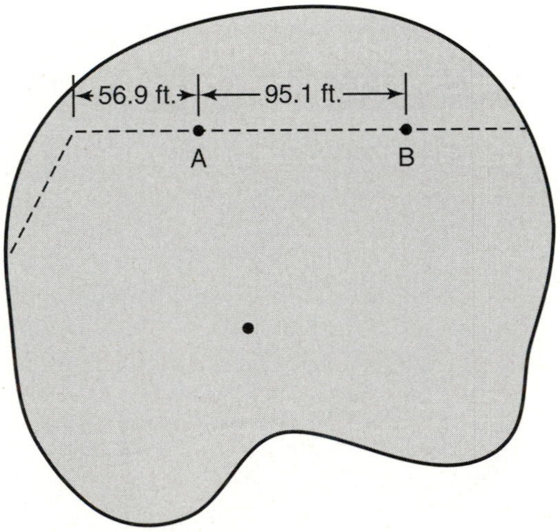

Figure 10-15 Locating stations A and B along reference line.

stations A and B from a reference point. In this case they are measured from a corner in the reference line, Figure 10-15.

The last step is measuring the distances from stations A and B to the unknown point, station C.

In this example station C is located 95.7 feet from station A and 117.2 feet from station B. As long as stations A and B can be located, station C can be located.

GPS receivers use the same process. They are at station C in Figure 10-16. They determine their location by knowing the distance from their position to several satellites, reference points. In the example, Figure 10-16, all three points were located on the same plane. This is known as a two-dimensional

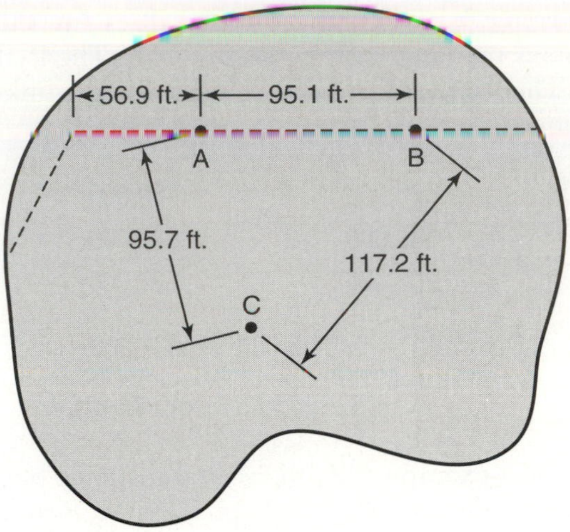

Figure 10-16 Locating station C.

system. GPS is a three-dimensional system, Figure 10-17. Each satellite knows its distance from the center of the earth and each receiver calculates its distance from the satellite based on the time it takes for a signal to travel from the satellite to the receiver.

Study Figure 10-17. The circle on the earth's surface represents all of the possible positions that are an equal distance from the earth's surface to one satellite. Calculating a position from one satellite reduces the possible locations to the circle, but it would not be very useful.

Figure 10-18 shows that by knowing the distance to two satellites, the possible number of positions is narrowed down to just two on the surface of the earth, a and b.

In Figure 10-19 it can be seen that with the addition of a third satellite the receiver can determine its position because the three distances from the three satellites only converge on the earth's surface at one point, b. In theory three satellites are all that are needed for a GPS receiver to determine its location, but most require at least four to insure the clock times and other information is correct.

Figure 10-20 shows the effect of the number of satellites in three dimensions. The surface of the sphere around each satellite represents the distance from the satellite to the receiver. As the distances to more and more satellites are included in the calculations, the possible number of locations is reduced and the precision of the receiver is improved. Survey quality GPS receivers are capable of receiving signals from many satellites and several different systems.

The precision of mapping and survey quality GPS receivers is enhanced because they are capable of receiving signals from more than one global positioning system.

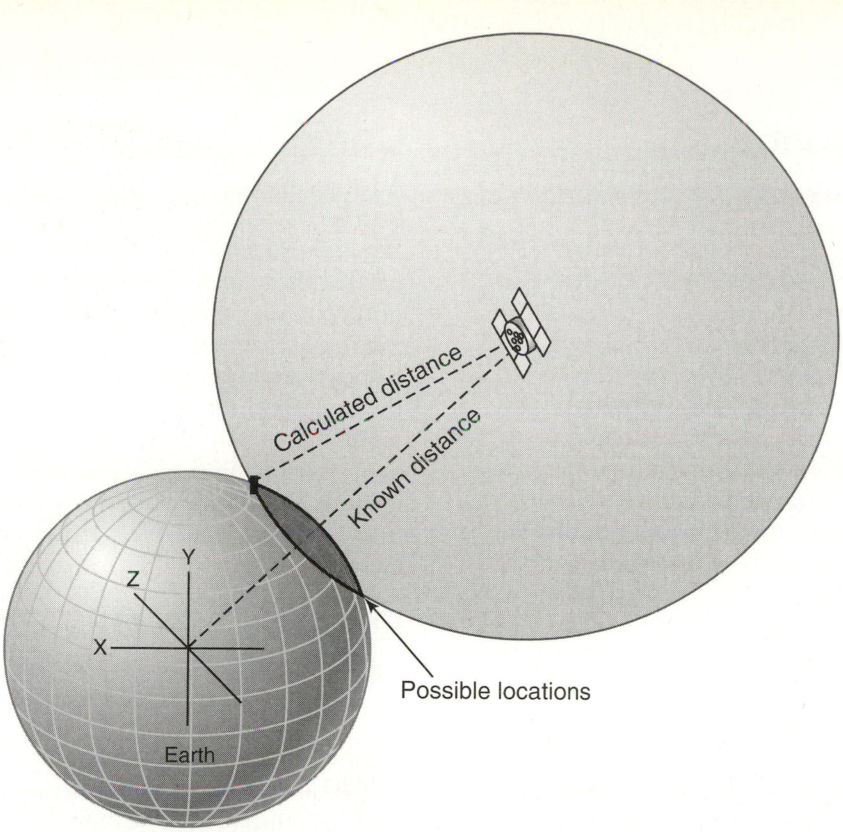

Figure 10-17 GPS system.

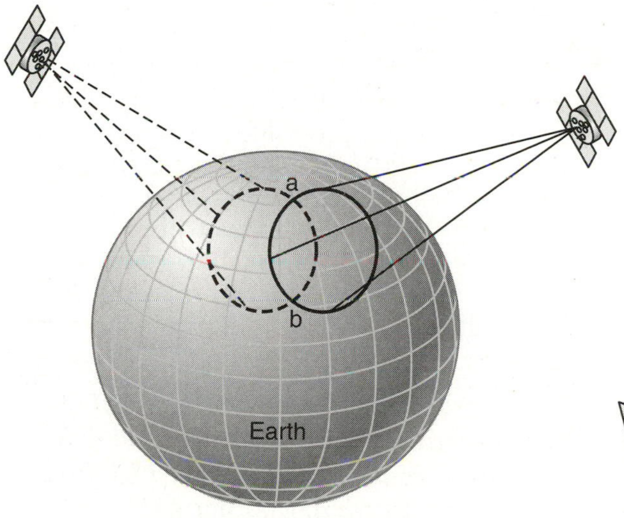

Figure 10-18 Possible positions with information from two satellites.

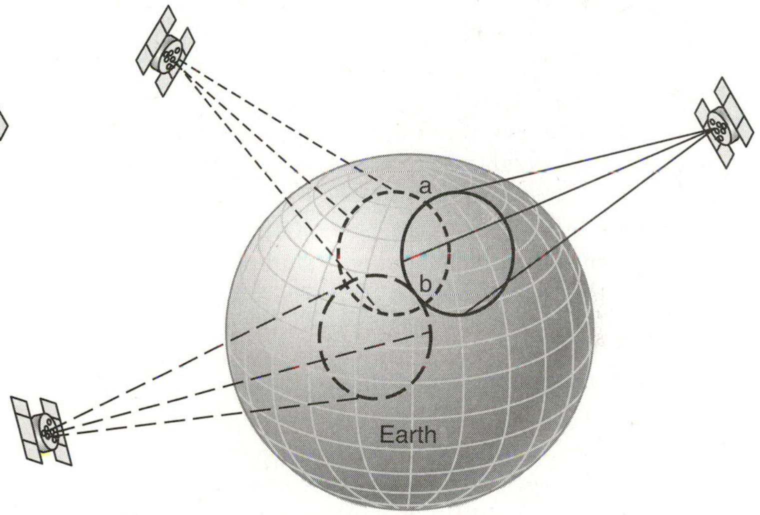

Figure 10-19 Possible positions with information from three satellites.

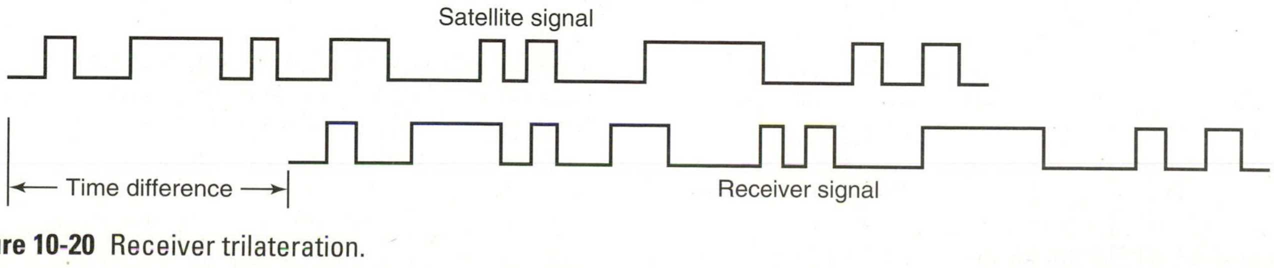

Figure 10-20 Receiver trilateration.

Determining Distance

Distance is determined by rearranging the velocity equation:

$$\text{Velocity} = \text{Distance} \times \text{Time}$$

$$\text{Distance} = \frac{\text{Velocity}}{\text{Time}}$$

In theory, the receiver knows the velocity of the signals, the speed of light, and the travel time from each satellite to the receiver. With this information it can calculate the distance to each satellite. The time required for the signal to travel from the satellite to the receiver is determined by using a pseudorandom noise (PRN) signal.

Pseudorandom Noise

The PRN code is a combination of signals that use a series of ones and zeros, and it appears to be a random signal, but it is designed to repeat at a predetermined rate. Each satellite has its own code and each receiver has a copy of each satellite code in its memory. The receiver determines the time it takes for the signal to pass from the satellite to the receiver, called arrival time, by calculating the shift in the code that occurred from when the satellite told the receiver it was sending the signal and when the receiver received the signal. Any error in measuring the travel time will cause an error in the measurement. In addition, errors can occur in the transmission of the signal. A distance computed using pseudorandom noise is referred to as a pseudo-range, Figure 10-21.

A key component of this process is time. The clock in the receiver is not as accurate as the clock in the satellite. To achieve the best accuracy with this process the receiver must be able to acquire signals from at least four satellites and it must continuously recompute the distances until the affect of the clock errors are reduced to an acceptable level. Experience with pseudo-ranging has shown that some receivers must be stationary for at least 20 to 30 minutes to achieve reasonable accuracy. Precision of pseudo-ranging with modern non-surveying quality receivers is 3 to 10 meters.

Carrier Phase GPS

Another method for improving the precision of GPS is called carrier phase GPS. The carrier is transmitted at a higher frequency. The receiver can determine the partial phase of the carrier wavelength, but to determine the distance the receiver must also know how many complete cycles have occurred from the time the carrier signal left the satellite until it reached the receiver. The number of complete cycles is determined by constantly tracking several satellites and post-processing the data. Combining the principle of using a differential station and the carrier signal produces an accuracy of less than one inch.

SOURCES OF GPS ERROR

The global positioning system is a very complex system with many opportunities for errors. The common errors can be categorized as:

- Poor satellite geometry
- Satellite orbits
- Multipath effect
- Atmospheric effect
- Clock inaccuracies and rounding errors

Satellite Geometry

GPS receiver accuracy and precision is linked to both the number of satellites available and the arrangement of the satellites. The ideal condition is when the satellites are well distributed across the sky. A receiver that has signals from six satellites will produce poor quality data if the satellites are all overhead or along N-S or E-W planes. Mapping and survey grade GPS receivers will express the geometry error in dilution of position values (DOP). Common DOPs are:

- Geometric Dilution Of Precision (GDOP)—overall-accuracy; 3D-coordinates and time
- Positional Dilution Of Precision (PDOP)—position accuracy; 3D-coordinates
- Horizontal Dilution Of Precision (HDOP)—horizontal accuracy; 2D-coordinates

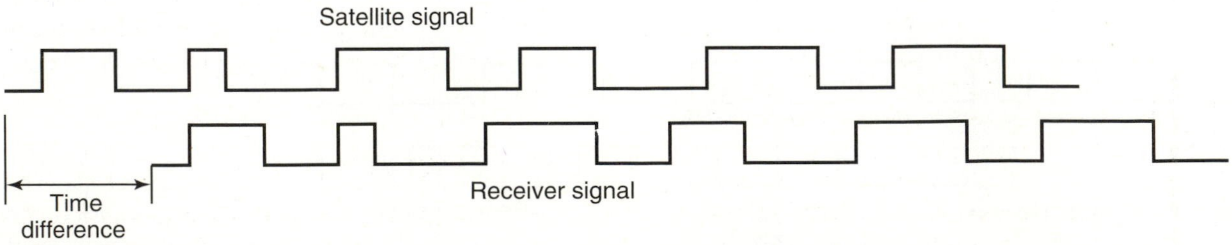

Figure 10-21 PRN signals.

- Vertical Dilution Of Precision (VDOP)—vertical accuracy; height
- Time Dilution Of Precision (TDOP)—time accuracy; time

The appropriate values for all of the dilution of positions depend on the desired level of precision for the survey. The operator of the GPS unit should determine the desired level of precision and appropriate values before starting the survey.

Satellite Orbits

Even though the satellite orbits are carefully monitored, small errors can occur. The GPS receiver downloads the latest orbital information every time it is turned on. This information is called the satellite almanac or ephemeris data. If the receiver has been shut off for a long period of time and especially if it has been moved a large distance since it was last used, it can take several minutes for the receiver to update the satellite almanac and connect with the satellites.

Multipath Effect

The signal from the satellite can be reflected off of buildings and other structures. A reflected signal will take a longer time to reach the receiver. Because the receiver calculates is position based on the travel time of the signal, a false travel time will cause an error in the calculated position.

Atmospheric Effect

The satellites are in space where the signals travel at the speed of light. To reach the receiver, the signals must pass through ionosphere and troposphere. The layers of the ionosphere tend to refract the signals from the satellites, which extends the travel time. Low frequency signals are slowed down more by the ionosphere than high frequency signals. Military and survey quality GPS receivers account for this error by comparing the travel time of a low frequency and a high frequency signals and calculating a correction. Consumer quality GPS receivers do not usually have this ability.

The troposphere can also cause error because the signal speed is affected by the density of the air and air density changes with temperature and humidity. GPS receivers cannot determine a correction for this error. Some high quality receivers use a mathematical model to estimate a correction.

The atmosphere can also cause error in the data because the signal from satellites closest to the horizon travel through more miles of atmosphere than the signals from satellites overhead. This is why most receivers allow the operator to eliminate the signal from satellites close to the horizon.

Clock Inaccuracies and Rounding Errors

The global positioning system uses an atomic clock that is very accurate, but some timing errors still occur. The receiver central processing unit (CPU) does many calculations per second. These calculations can result in a small rounding error.

SURVEYING WITH GPS

The purpose of GPS surveying is to provide the position and elevation of the receiver. Position and elevation data can be absolute or relative. The method used depends on the equipment that is available and the required accuracy and precision.

Absolute Data

When a receiver is determining absolute data, the reference system must be uniquely defined and inaccessible to the user. Total reliance is placed on the integrity of the reference system. As discussed earlier, a receiver determines its location by determining the distance to multiple satellites. The receiver <u>cannot</u> determine if the satellite position or time information is accurate. The accuracy of the data produced by the receiver is only as good as the information the receiver receives from the satellites.

A single receiver, operating independent of any other source of information, is producing absolute positions. This is the situation for most civilian uses of GPS. A backpacker or fisherman with a single consumer-grade GPS receiver is getting absolute positions. Depending on the quality of the receiver, the number of channels it uses and the number of positioning systems it can utilize, the accuracy of the data is in the range of 1 to 10 meters. The accuracy and precision of absolute positioning data improves with static time. The longer the receiver stays stationary on one point, the better the accuracy and precision. Figure 10-22 is an example of the static time versus positional error curve for a GPS receiver. All GPS receivers have a similar shaped curve; the value of the error per unit of time will be different for each instrument.

Relative Data

Relative data is produced when the GPS receiver uses information from an external source to adjust its calculations. Base stations continuously calculate their

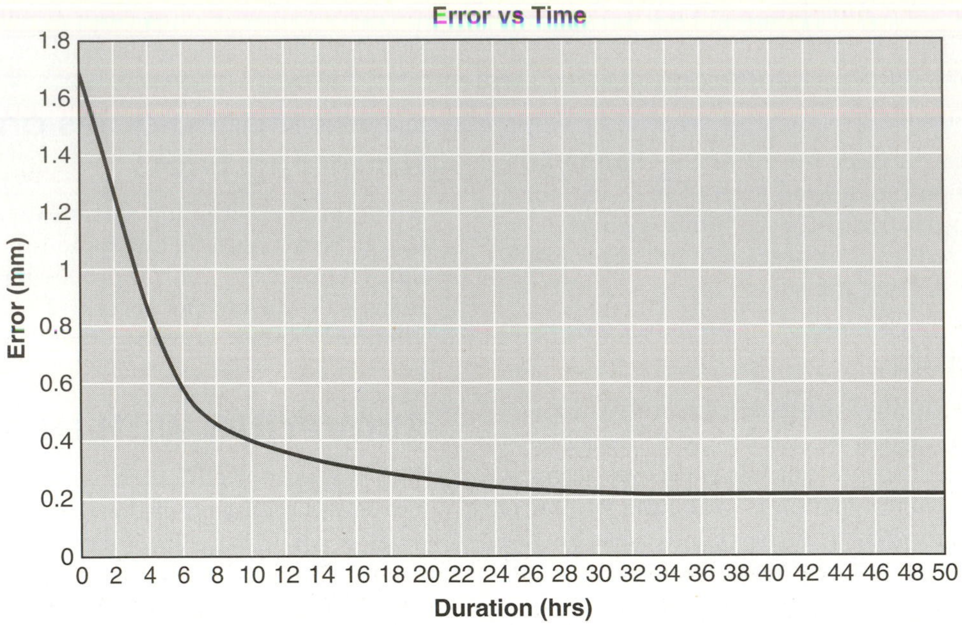

Figure 10-22 GPS positional error compared to static time.

GPS position and compare the results with their static position. The difference between their instantaneous calculated position and their known static position is used to correct the absolute data collected by a rover. When RTK is used, the rover position and elevation calculations are adjusted as they are determined. The relative data from another source can also be used to post process the rover data.

The position data for a base station may be based on absolute or relative position. As discussed in the previous section, any high quality GPS receiver will produce accurate and precise position information if left in a static position for a long period of time. When a new GPS control station is established, using absolute positioning, it may be operated continuously for several weeks, or longer, before it is added to the network or made available to roving units.

Government agencies, private organizations, and commercial engineering and surveying firms started establishing base stations several years ago. It was a natural progression to combine these base stations into a network. Almost all areas with high population density have some type of GPS base station network. Access to these networks is usually available through an annual fee. The National Geodetic Survey (NGS), an office of NOAA's National Ocean Service, coordinates a system called CORS (Continuously Operating Reference Station). The CORS network is available to any receiver or computer that has Internet access. One advantage of relative positioning is that an individual with a single rover and access to a base station or network of stations can produce very precise and accurate position data. GPS receivers can use networks in two ways, real time kinematic (RTK) and post processing (PP).

Real Time Kinematic Surveying

Real Time Kinematic (RTK) surveying is the method of choice for professional surveyors. With this system the mobile receiver is in continuous communication with at least one base station or network of stations. As the rover evaluates its position, it adjusts its calculated location by the correction determined by a base station. The base station may be another GPS receiver set up on a station that has been carefully located, or the base station may be part of a network. Several different means of communication have been tried and are available for communication between GPS receivers and between the receivers and base stations. The key is that for RTK surveying the rover must be able to communicate with the base in real time while collecting data. One advantage of RTK is that it can be used for layout work because the position data is continuously being corrected. Depending on the desired level of accuracy and precision, some RTK receivers still require static time on each station.

Post Processing

Post processing corrects the receiver's data after it has been collected. Every time a receiver saves data the time is recorded with it. When post processing, it is common to download the receiver data into a networked computer and then use CORS data or base station data to correct the receiver's data. Post processing has the capability of producing very accurate and precise results, but because the corrected data is not available until after the data is transferred from the rover, it is usually limited to measuring existing structures, not for real time layout work.

Differential GPS

Differential GPS (DGPS) is a type of relative data system. A stationary differential DGPS station is used with a roving receiver equipped for DGPS. The communication between the base and rover may be open access or by cell phone, Wi-Fi, Bluetooth, etc. or proprietary. The DGPS base station is fixed at a location that has been carefully surveyed using conventional methods or multiple GPS measurements. The differential base GPS station receives the same satellite signals as the receivers and computes its position. It compares the computed position with its known position and determines the amount of error. The error information is broadcast and any roving receivers that are connected and within range of the differential GPS station use the information to correct their data. The first differential DGPS systems produced an accuracy of 1.5 to 2.5 meters. This number has continued to decline as improvements have been made in equipment and techniques. DGPS has two major disadvantages. It requires a differential base station and differential receiver. Standard GPS receivers are not capable of using the differential signal. The second disadvantage is that the differential signal is limited to about 50 miles. DGPS is not available in many parts of the country. Because of these limitations and the preference for RTK, the use of differential GPS has declined.

Another DGPS use is as mobile receivers. One is designated as the base and the other the rover. The base unit is set up over a benchmark with a known location. The rover is moved from station to station collecting the necessary data. The base unit continuously compares its calculated position with the known position and determines the error. Both units record the time when data is recorded. The data from both units is post processed and the error calculated by the base unit is used to correct the data from the rover.

Summary

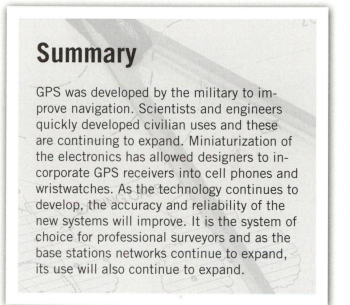

GPS was developed by the military to improve navigation. Scientists and engineers quickly developed civilian uses and these are continuing to expand. Miniaturization of the electronics has allowed designers to incorporate GPS receivers into cell phones and wristwatches. As the technology continues to develop, the accuracy and reliability of the new systems will improve. It is the system of choice for professional surveyors and as the base stations networks continue to expand, its use will also continue to expand.

Student Activity

1. Determine the distance between two points with the UTM coordinates of 13, 782453, 822943, and 13, 783576, 823482.
2. Determine the area of a triangle with the UTM coordinates of 14, 489570, 847343; 14, 490743, 847343; and 14, 489982, 848481.

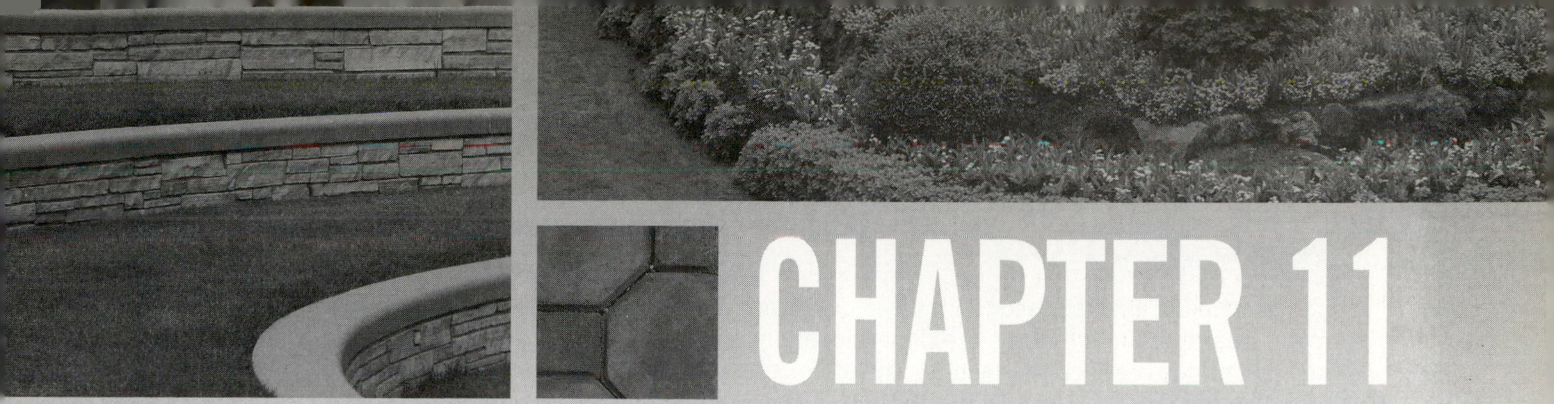

CHAPTER 11

Determining Area

Objectives

After reading this chapter the student should be able to:

- Determine the area of standard geometric figures.
- Determine the area of complex shapes.
- Determine the area of land parcels with irregular boundaries.

Terms To Know

Triangle
Square
Rectangle
Parallelogram
Trapezoid
Circle
Sector
Segment
Irregular shape
Coordinate squares
Coordinates

INTRODUCTION

The term area is used to describe the two-dimensional size of a surface. Because area is two-dimensional, the units for area are dimensional units squared. In land measurement the common dimension units are feet and meters. Therefore, square feet and square meters are common units for area. When measuring large areas, it is a common practice to convert square feet to acres and square meters to hectares. An acre is equivalent to 43,560 ft.^{2} and a hectare is equivalent to 1000 m^2.

The ability to determine area is an important skill for many occupations. The area may be a standard geometric shape, Figure 11-1.

It may be a complex shape or have an irregular boundary caused by a stream, Figure 11-2. Common methods for determining area are:

- Area of standard geometric shape
- Division into standard shapes
- Coordinate squares
- Offsets from a line
- Coordinates

This chapter will explain these methods of determining area.

STANDARD GEOMETRIC SHAPES

In some situations, such as determining the area of a driveway or roof, the unknown area can be in the shape of a standard geometric figure, Figure 11-1. If the unknown shape is a standard figure, or close enough to a standard figure that it is a good approximation of the land parcel, a standard area equation can be used.

This section will explain the equations and their use for the following eight standard figures.

- Triangles
- Squares

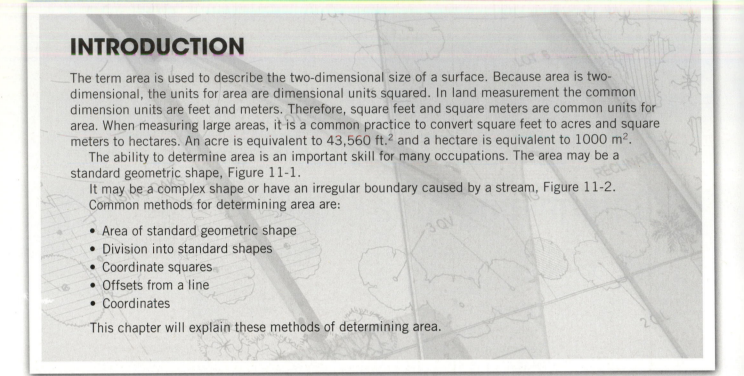

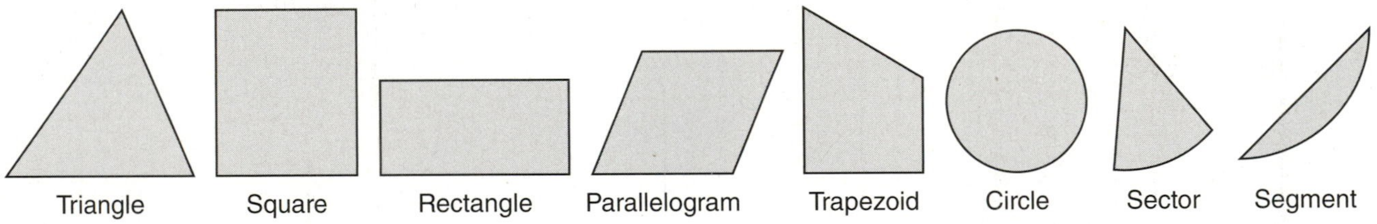

Figure 11-1 Standard geometric shapes.

Figure 11-2 Complex and irregular shapes.

- Rectangles
- Parallelograms
- Trapezoids
- Circles
- Sectors
- Segments

Triangle

A triangle is a three-sided polygon whose sum of interior angles is 180 degrees. The next section will explain the many types of triangles. Three equations can be used to determine the area of a triangle. Any one of the three equations can be used for any one of the types of triangles, but some equations and triangles are a better fit depending on the type of triangle and the measurements that have been made. The three equations are called the common equation, Heron's equation, and the trig function equation.

The common equation is:

$$\text{Area} = \frac{\text{base} \times \text{height}}{2}$$

Heron's equation is:

$$\text{Area} = \sqrt{s(s - a)(s - b)(s - c)}$$

But the constant s must be determined first

$$s = \frac{a + b + c}{2}$$

The trig function equation is:

$$\text{Area} = \frac{a \times b \times \text{sine } \theta}{2}$$

The following sections will explain the use of these equations and the best fit between equations and the many types of triangles.

Types of Triangles

Triangles are categorized by the relative lengths of the three sides and the interior angles. Three different types of triangles are categorized by the lengths of their sides, and three different types of triangles are categorized by their interior angles. This results in a large number of possible triangles and all of these are beyond the scope of this text. The following section will explain how the three area equations are used with a sample of these triangles.

Categorized by Length of Side The three common types of triangles classified by length of sides are:

- Equilateral
- Isosceles
- Scalene

An equilateral triangle is unique because all three sides are the same length and all three angles are 60 degrees, Figure 11-3.

An isosceles triangle is a triangle with two sides that are the same length, Figure 11-4. If two sides of the triangle are the same length, then two of the interior angles must also be the same.

The distinguishing characteristic of a scalene triangle is that no side lengths or interior angles are the same, Figure 11-5.

Triangles can also be classified by the type of interior angles. The three common types of triangles classified by the interior angles are:

- Right
- Acute
- Obtuse

Right triangles are unique because one angle is always 90 degrees, Figure 11-6. The lengths of the sides adjoining the right angle can vary, but they are always shorter than the side opposite the right angle. A right triangle can be part of either a scalene or an isosceles triangle.

An acute triangle has at least one angle less than 90 degrees. Isosceles, scalene, and right triangles can have an acute angle, Figure 11-7.

If a triangle has one angle greater than 90 degrees, it is an obtuse triangle. Only isosceles and scalene

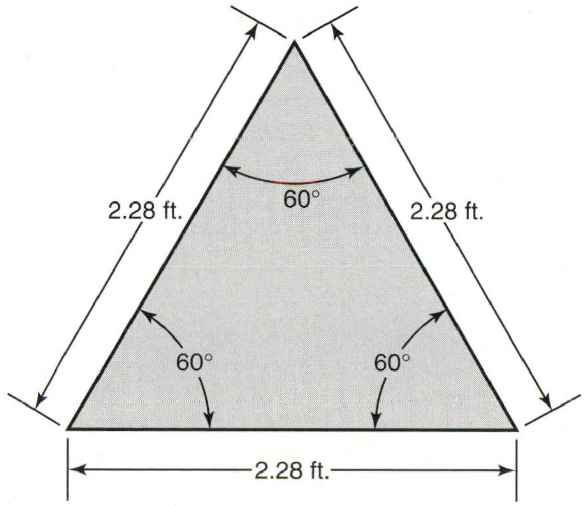

Figure 11-3 Equilateral triangle.

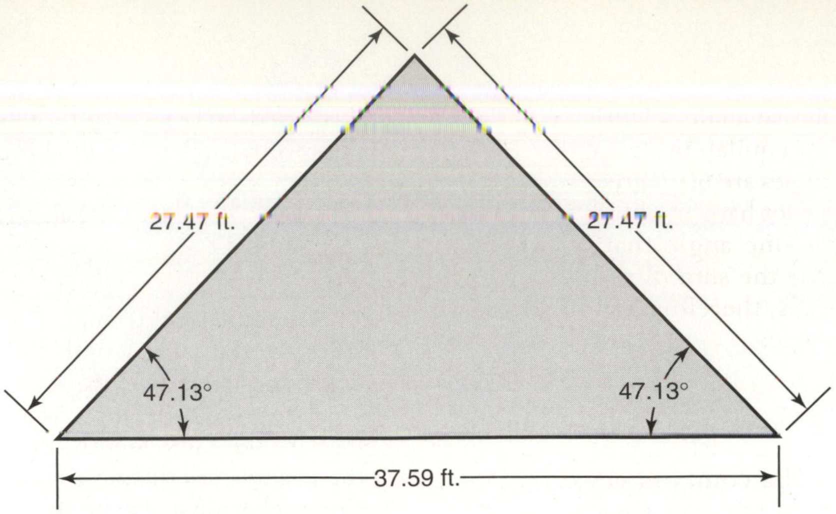

Figure 11-4 Isosceles triangle.

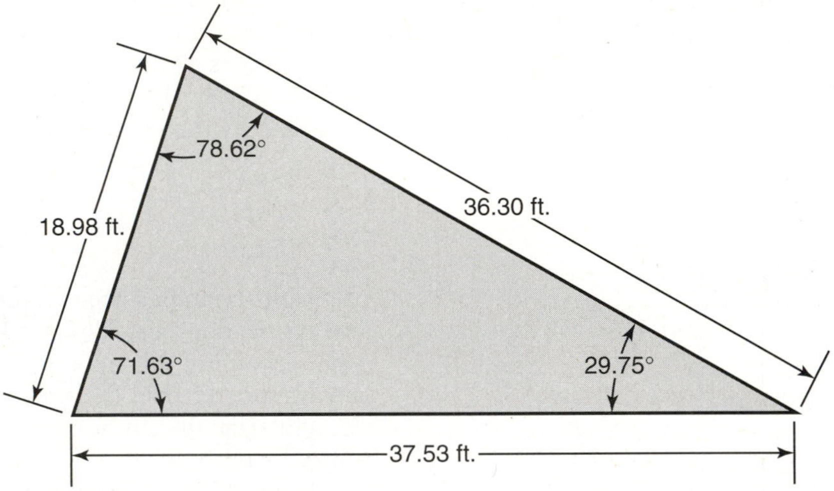

Figure 11-5 Scalene triangle.

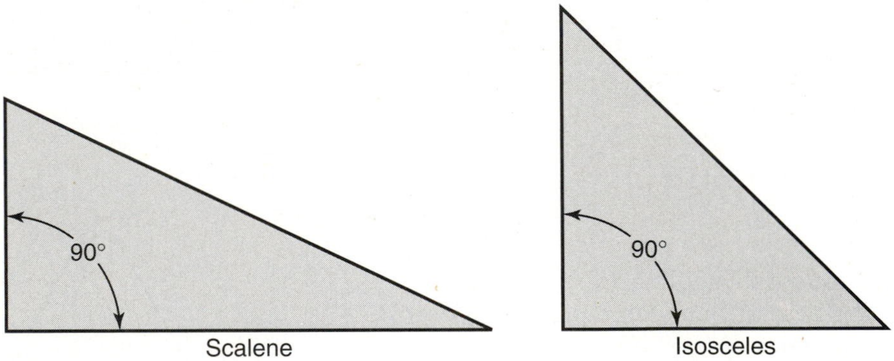

Figure 11-6 Right triangles.

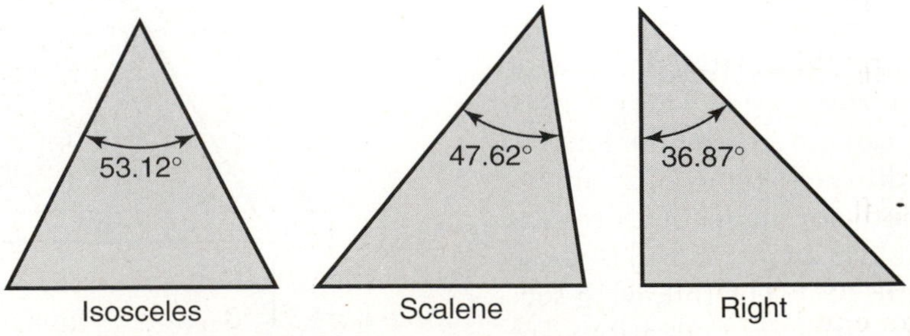

Figure 11-7 Isosceles, scalene, and right triangles with an acute angle.

triangles can have an obtuse angle, Figure 11-8. The lengths of the sides of an equilateral triangle are the same, all three interior angles are 60 degrees, therefore an equilateral triangle cannot have an obtuse angle.

Right triangles have one angle that is fixed at 90 degrees, which means the sum of the other two angles must be 90 degrees, therefore right triangles cannot have an obtuse angle.

Areas of Triangles

Equilateral Triangle The common equation can be used with an equilateral triangle when the height is known. To measure the height the baseline is measured and marked at the midpoint. The distance between the midpoint and the apex is the height, Figure 11-9.

For the equilateral triangle in Figure 11-9 the area is:

$$\text{Area} = \frac{\text{base} \times \text{height}}{2}$$
$$= \frac{27.50 \text{ ft.} \times 23.76 \text{ ft.}}{2}$$
$$= 326.70 \text{ ft.}^2$$

Often in the field a tree, pond, building, or some other structure will obstruct measuring the height of an equilateral triangle. When the height of the triangle cannot be measured, one of the other two equations can be used.

When the lengths of the sides can be measured, Heron's equation is used, Figure 11-10.

The area of the triangle in Figure 11-10 using Heron's equation is:

$$s = \frac{a + b + c}{2}$$
$$= \frac{27.50 \text{ ft.} \times 27.50 \text{ ft.} \times 27.50 \text{ ft.}}{2}$$
$$= 41.25 \text{ ft.}$$

$$\text{Area} = \sqrt{s(s-a)(s-b)(s-c)}$$

$$= \sqrt{41.25 \text{ ft.} (41.25 \text{ ft.} - 27.50 \text{ ft.})(41.25 \text{ ft.} - 27.50 \text{ ft.})(41.25 \text{ ft.} - 27.50 \text{ ft.})}$$

$$= \sqrt{107233.8867 \text{ ft.}^4}$$

$$= 327.465\ldots \text{ or } 327.46 \text{ ft.}^2$$

This example shows that Heron's equation can be used to find the area of an equilateral triangle, but when doing the calculations by hand on a calculator it is usually the last method selected because of the complexity of the math. When set up on a spreadsheet, it is a popular method to use because it requires only three boundary measurements.

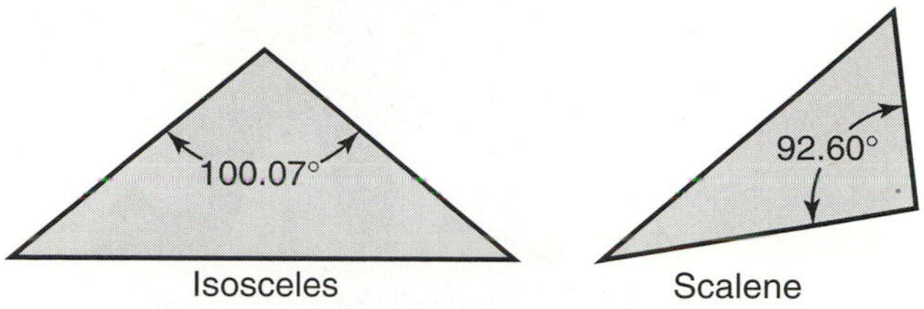

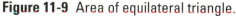

Figure 11-8 Obtuse triangles.

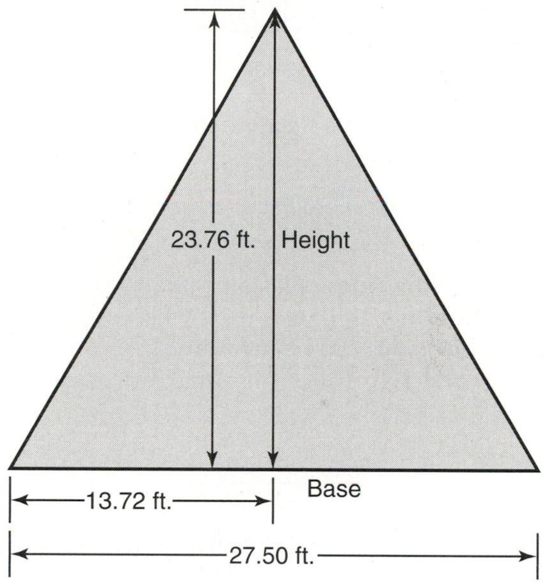

Figure 11-9 Area of equilateral triangle.

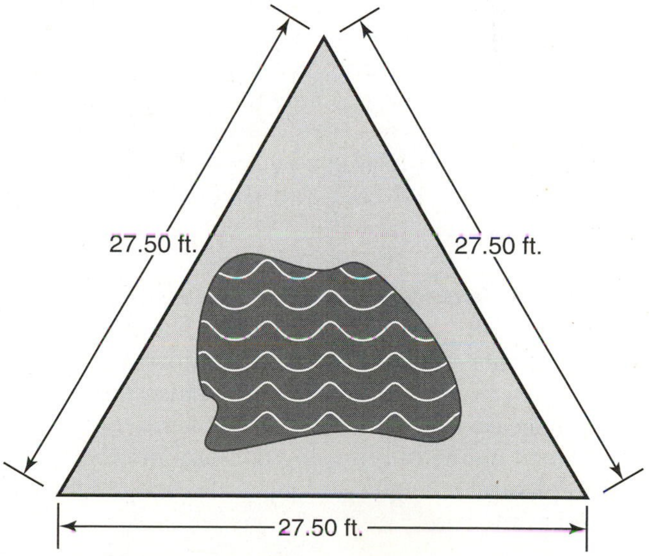

Figure 11-10 Area of equilateral triangle using Heron's equation.

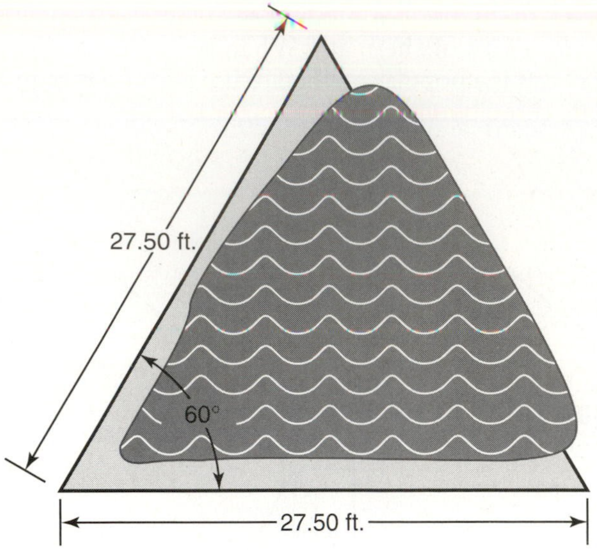

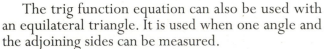

Figure 11-11 Area of equilateral triangle using trig function equation.

The trig function equation can also be used with an equilateral triangle. It is used when one angle and the adjoining sides can be measured.

Using the trig function equation, the area for Figure 11-11 is:

$$\text{Area} = \frac{\text{a} \times \text{b} \times \text{sine}\ \theta}{2}$$

$$= \frac{27.5\ \text{ft.} \times 27.5\ \text{ft.} \times 0.8660}{2}$$

$$= 327.4658\ldots \text{ or } 327.46\ \text{ft.}^2$$

When only two sides and one angle can be measured, the trig function equation will determine the area of the triangle. This method has two disadvantages: it requires a calculator with trig functions and the angle must be measured. Refer to Chapter 7 for indirect and direct methods for measuring angles.

Isosceles Triangle Isosceles triangles will have either an acute or an obtuse angle. The process for using the three equations is the same for both acute and obtuse isosceles triangles. Because two of the sides of an isosceles triangle are the same length, the height can be measured using the same method used for equilateral triangles, as long as the odd length side is used as the base.

When the odd length side is used as the base, a line from the midpoint to the apex of the triangle is

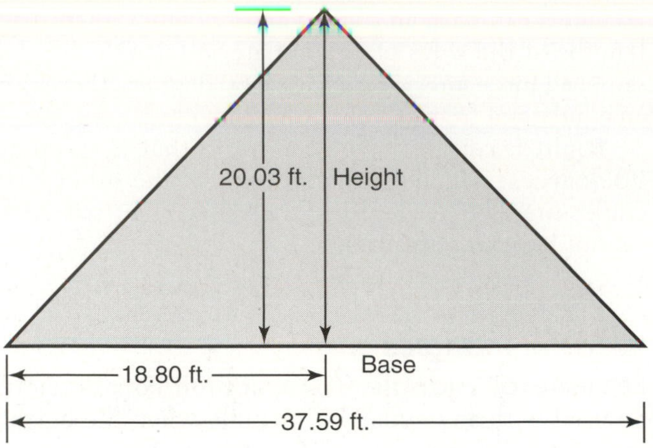

Figure 11-12 Area of isosceles triangle using common equation.

the height of the triangle. The area for the triangle in Figure 11-12 using the common equation is:

$$\text{Area} = \frac{\text{base} \times \text{height}}{2}$$

$$= \frac{37.59\ \text{ft.} \times 20.03\ \text{ft.}}{2}$$

$$= 376.463\ldots \text{ or } 376.46\ \text{ft.}^2$$

When the height cannot be measured or the odd length side cannot be used as the base, Heron's equation can be used to determine the area of an isosceles triangle, Figure 11-13.

The area of the triangle in Figure 11-13 using Heron's equation is:

$$\text{s} = \frac{\text{a} + \text{b} + \text{c}}{2}$$

$$= \frac{27.42\ \text{ft.} \times 27.42\ \text{ft.} \times 37.59\ \text{ft.}}{2}$$

$$= 46.215\ \text{ft.}$$

$$\text{Area} = \sqrt{\text{s}(\text{s} - \text{a})(\text{s} - \text{b})(\text{s} - \text{c})}$$

$$= \sqrt{46.125\ \text{ft.}(46.125\ \text{ft.} - 27.42\ \text{ft.})(46.125\ \text{ft.} - 27.42\ \text{ft.})(46.125\ \text{ft.} - 37.59\ \text{ft.})}$$

$$= \sqrt{137738.4938\ \text{ft.}^4}$$

$$= 371.131\ldots \text{ or } 371.13\ \text{ft.}^2$$

If the height and the length of all three sides cannot be measured, the trig function equation is the only other option for determining the area of the

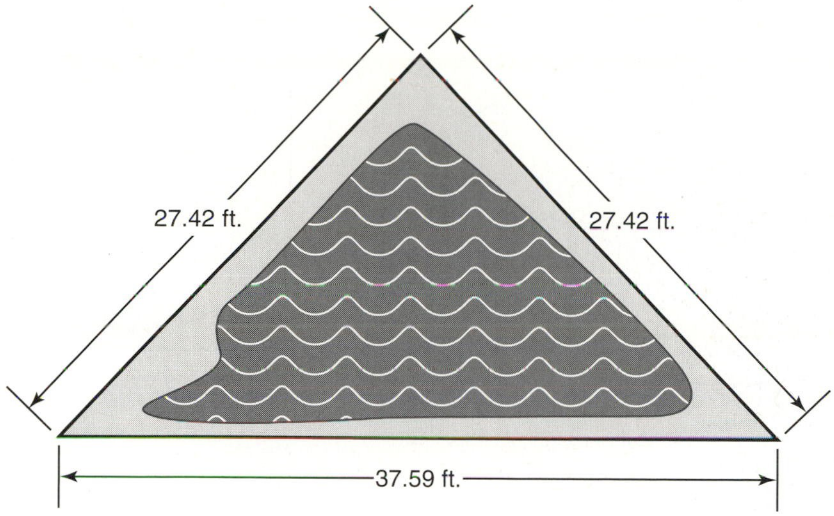

Figure 11-13 Area of isosceles triangle using Heron's equation.

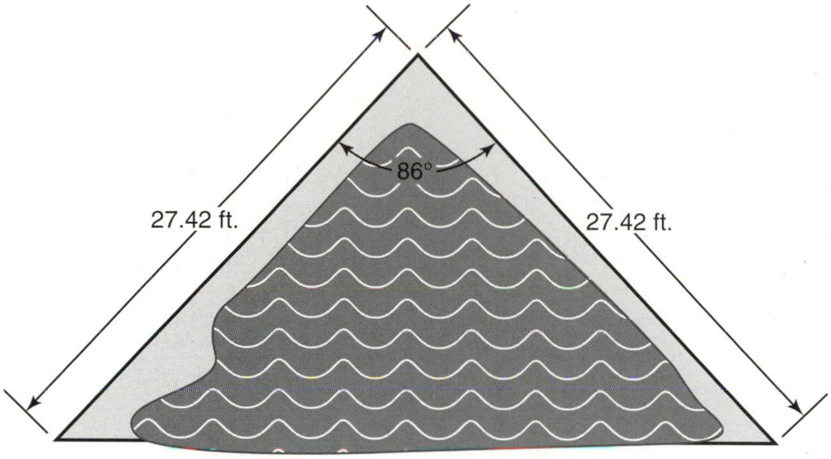

Figure 11-14 Area of an isosceles triangle using the trig function equation.

triangle. It can be used if an angle and the lengths of the adjoining sides can be measured, Figure 11-14.

The area of the triangle in Figure 11-14 is:

$$\text{Area} = \frac{a \times b \times \text{sine}\,\theta}{2}$$

$$= \frac{27.42\text{ ft.} \times 27.42\text{ ft.} \times 0.9976}{2}$$

$$= 375.0124\ldots \text{ or } 375.01\text{ ft.}^2$$

These examples illustrate that all three equations can be used to determine the area of an isosceles triangle. The "best" equation is the one that requires the least expenditure of resources to collect the necessary dimensions. Selecting the best equation can be influenced by factors such as the features of the site, the equipment that is available, and the expertise of the surveyor. Often the structures, trees and terrain of the site determine the best equation to use.

Scalene Triangle All three triangle equations can be used with scalene triangles, but the common area equation for triangles is more complicated because of the difficulty in measuring the height. Measuring a line from the apex to the midpoint of the base is not the height of the scalene triangle because the line does not form a 90 degree angle with the base, Figure 11-15. The height must be measured where a line perpendicular to the base passes through the apex of the triangle.

One suggested method when determining the height of a scalene triangle in the field is to walk along the baseline and sight toward the apex until you

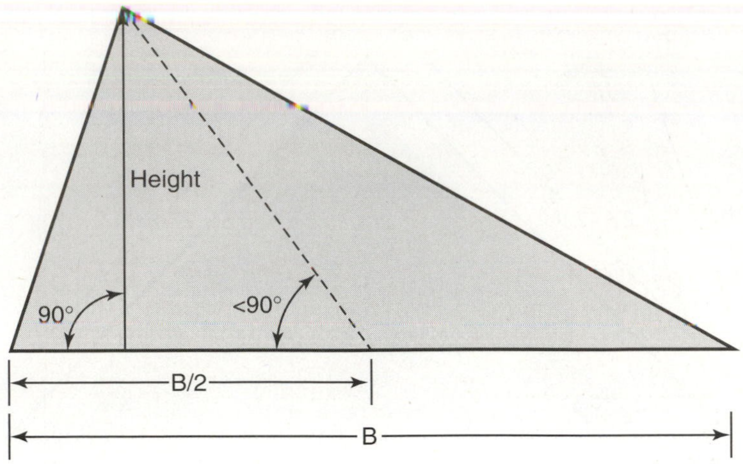

Figure 11-15 Height of scalene triangle.

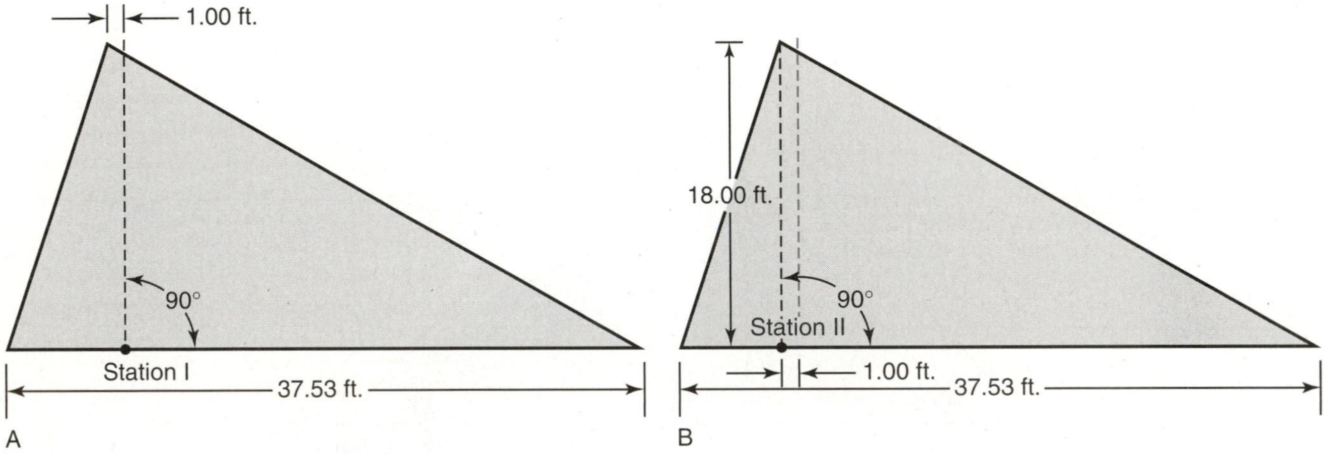

Figure 11-16 Determining the height of a scalene triangle.

estimate you are opposite the apex. Then use a direct or indirect method to lay out a 90 degree angle and extend the angle to the apex of the triangle, Figure 11-16 Illustration A. If you miss the apex, measure the distance from the line to the apex, shift the vertex of the angle the same distance along the baseline, reestablish the 90 degree angle and measure the distance from the baseline to the apex of the triangle, Figure 11-16 Illustration B.

For field measurements, this method is feasible for small, unobstructed triangles, but not very practical for large or obstructed areas. It works very well with maps because a right triangle can be slid along the base until it aligns with the apex, thereby locating the correct position of the line perpendicular with the baseline.

Although the math is more complicated, Heron's equation is a better fit than the standard area equation for scalene triangles.

Note: If Heron's equation is used frequently it is a simple task to set up a computer spreadsheet to do the calculations. The steps are the same as an equilateral triangle. The area of the scalene triangle in Figure 11-17 using Heron's equation is:

$$s = \frac{a + b + c}{2} \qquad A = \sqrt{s(s - a)(s - b)(s - c)}$$

$$s = \frac{18.97\text{ ft.} + 36.72\text{ ft.} + 38.00\text{ ft.}}{2}$$

$$= \frac{93.69}{2} = 46.845\text{ ft.}$$

$$A = \sqrt{46.845\text{ ft.}(46.845\text{ ft.} - 18.97\text{ ft.})(46.845\text{ ft.} - 36.72\text{ ft.})(46.845\text{ ft.} - 38.00\text{ ft.})}$$

$$= \sqrt{116942.1269\text{ ft.}^4}$$

$$= 341.968\ldots \text{ or } 341.97\text{ ft.}^2$$

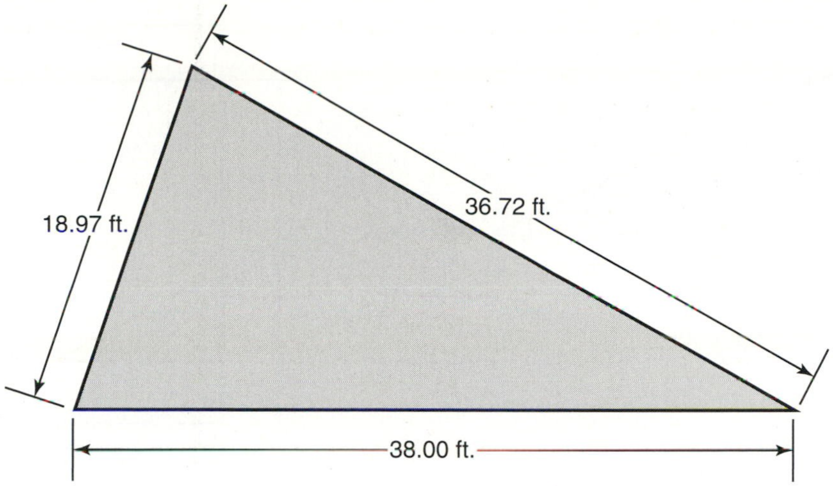

Figure 11-17 Area of scalene triangle using Heron's equation.

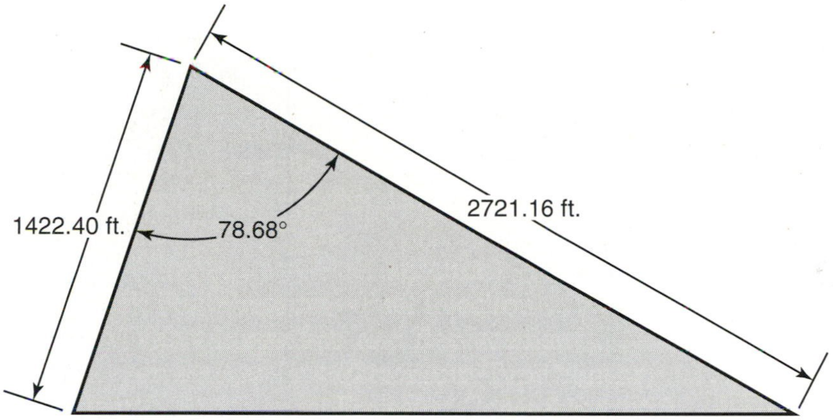

Figure 11-18 Area or scalene triangle using trig equation.

The trig function equation can also be used with a scalene triangle when an angle and the adjoining sides can be measured. The area for the scalene triangle in Figure 11-18 is:

$$\text{Area} = \frac{\text{a} \times \text{b} \times \text{sine } \theta}{2}$$

$$= \frac{1422.40 \text{ ft.} \times 2721.16 \text{ ft.} \times 0.9805}{2}$$

$$= 1{,}897{,}640.27 \text{ ft.}$$

The area for the triangle in Figure 11-18 is 1,897,640.27 square feet or 43.56 acres.

Right Triangle All three equations can be used to determine the area of a right triangle. Determining the height of the triangle is not a problem because it is the length of one of the sides. When the length of the base and the height can be measured the standard equation is used, Figure 11-19:

$$\text{Area} = \frac{\text{base} \times \text{height}}{2}$$

$$\frac{12.00 \text{ ft.} \times 25.20 \text{ ft.}}{2} = \frac{626.4 \text{ ft.}^2}{2} = 313.2 \text{ ft.}^2$$

The triangle in Figure 11-19 has an area of 313.2 ft^2.

Heron's equation is seldom used with right triangles because if the two sides adjoining the 90 degree angle can be measured, the standard equation can be used and the math is much less rigorous.

In situations in which either one of the adjoining sides to the right angle cannot be measured, the trig area equation is a possibility. To use the trig area equation, the side opposite the right angle and one adjoining side must be measurable. It also requires the measurement of the included angle. Determine

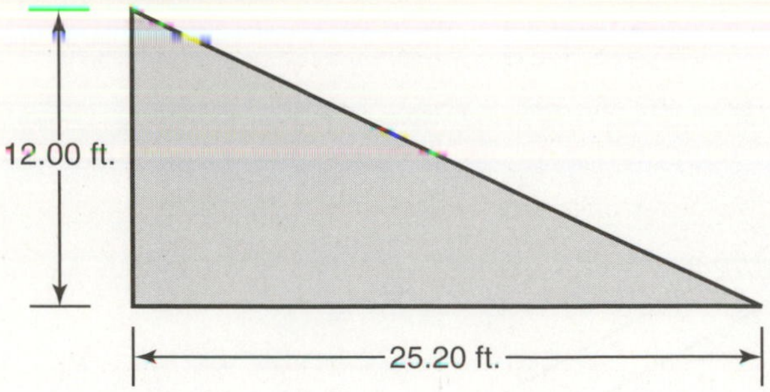

Figure 11-19 Using standard equation for right triangle.

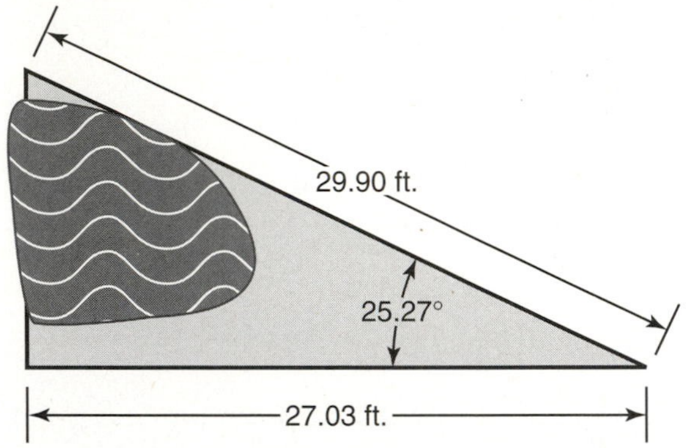

Figure 11-20 Determining the area of a right triangle using the trig function equation.

the area of the triangle in Figure 11-20 using the trig area equation:

$$\text{Area} = \frac{a \times b \times \text{sine } \theta}{2}$$

$$= \frac{29.90 \text{ ft.} \times 20.03 \text{ ft.} \times 0.42688}{2}$$

$$= 255.659\ldots \text{ or } 255.66 \text{ ft.}^2$$

The area of the triangle in Figure 11-20 is 255.66 square feet.

Obtuse Triangle A triangle with an obtuse angle will be either an isosceles or a scalene triangle. The area is determined using the same methods that were used for an acute triangle.

Square

A square is a four-sided polygon with same length sides and four 90 degree corners, Figure 11-21.

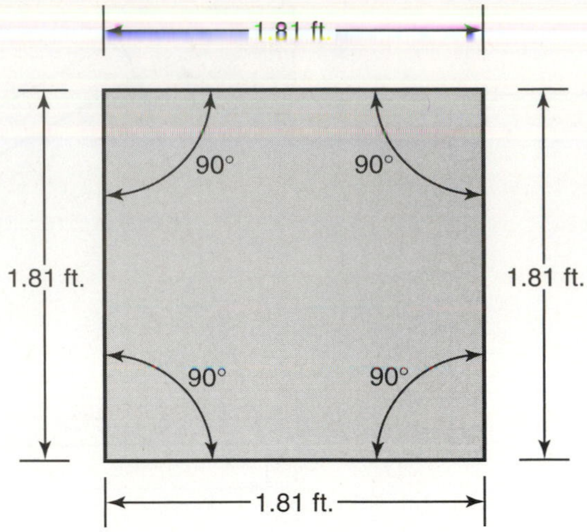

Figure 11-21 Square.

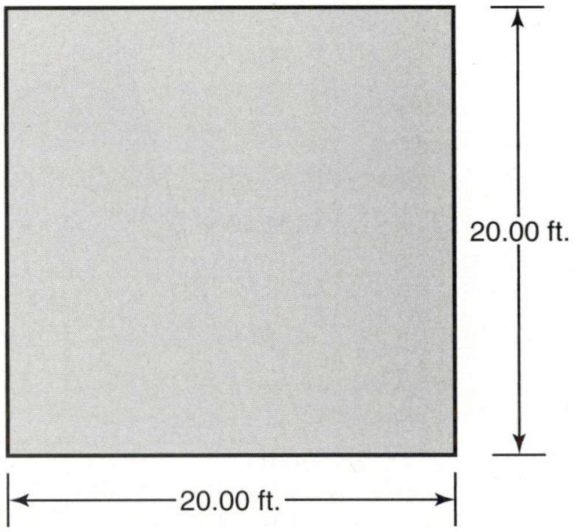

Figure 11-22 Determining the area of a square.

It would be a rare occasion for a parcel of land to have 90 degree corners and equal length sides, but if the shape of the land approximates a square it is appropriate for a low-precision survey to assume the unknown area is a square. Many man-made structures will have square components. The equation for determining the area of a square is:

$$\text{Area} = \text{base} \times \text{height}$$

The area of the square in Figure 11-22 is:

$$\text{Area} = \text{base} \times \text{height}$$

$$= 20 \text{ ft.} \times 20 \text{ ft.}$$

$$= 400 \text{ ft.}^2$$

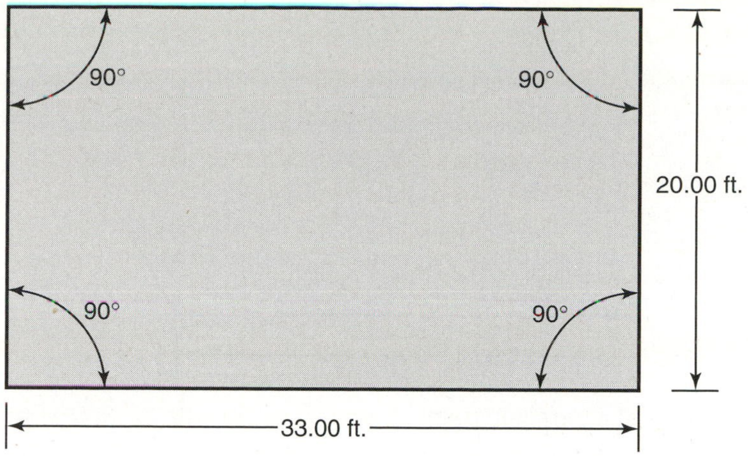

Figure 11-23 Rectangle.

Rectangle

A rectangle is a four-sided polygon where the sides are not equal length, but opposite sides are the same, Figure 11-23 and all four angles are 90 degrees. The area of a rectangle is determined using the same equation used to determine the area of a square:

$$\text{Area} = \text{base} \times \text{height}$$

The area of the rectangle in Figure 11-23 is:

$$\begin{aligned}\text{Area} &= \text{base} \times \text{height}\\ &= 33\text{ ft.} \times 20\text{ ft.}\\ &= 600\text{ ft.}^2\end{aligned}$$

Very few parcels of land will be a rectangle, but as with a square if the land shape approximates a rectangle, the rectangle equation can be used to for a low-precision survey. Many structures and man-made landscape features will be rectangular.

Parallelogram

Parallelograms are four-sided polygons with opposite sides that are the same length, two opposite angles measure greater than 90 degrees, and two opposite angles less than 90 degrees, Figure 11-24. They will also have two pairs of parallel sides.

The area of a parallelogram is determined using the same equation as for squares and rectangles, except the height must be measured perpendicular to a parallel side. This requires selecting a station along one side in such a position that a perpendicular line will intersect with the other parallel side, Figure 11-25.

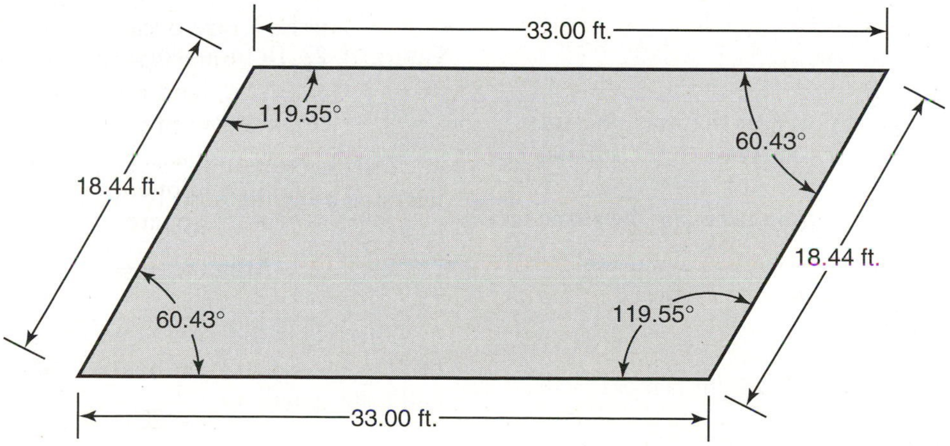

Figure 11-24 Parallelogram.

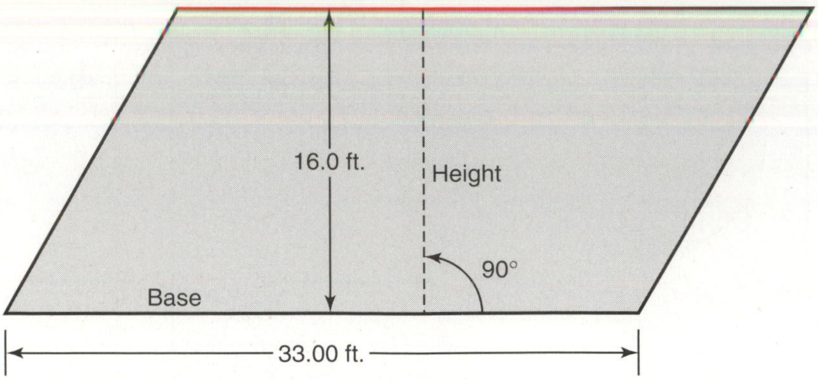

Figure 11-25 Determining the area of a parallelogram.

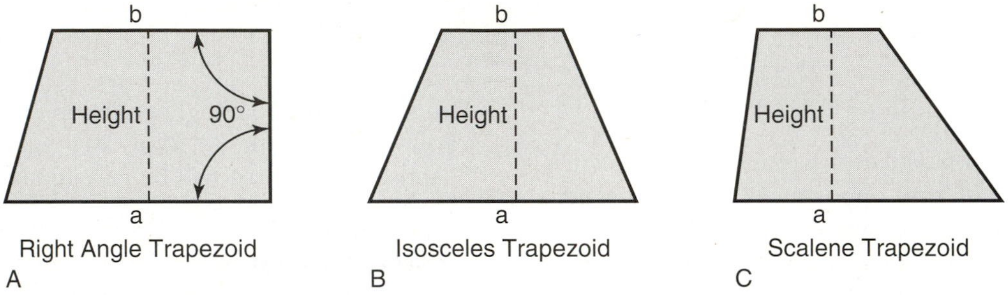

Figure 11-26 Trapezoids.

The area of the parallelogram in Figure 11-25 is:

$$\text{Area} = \text{base} \times \text{height}$$
$$= 33.00 \text{ ft.} \times 16.00 \text{ ft.}$$
$$= 528.00 \text{ ft.}^2$$

Parallelograms do not occur very frequently when determining the area of land or man-made structures, but the areas can be determined as long as the height is measured correctly.

Trapezoid

A trapezoid is a four-sided polygon that has one set of parallel sides. Trapezoids can take three forms, Figure 11-26.

The same area equation can be used for the three different trapezoids.

$$\text{Area} = \frac{A + B}{2} \times \text{height}$$

The only difference is how the height is measured. The height of the right angle trapezoid is the length of the side perpendicular to the parallel sides, Figure 11-26 Illustration A. The height of the isosceles and scalene trapezoid must be measured perpendicular to the parallel sides.

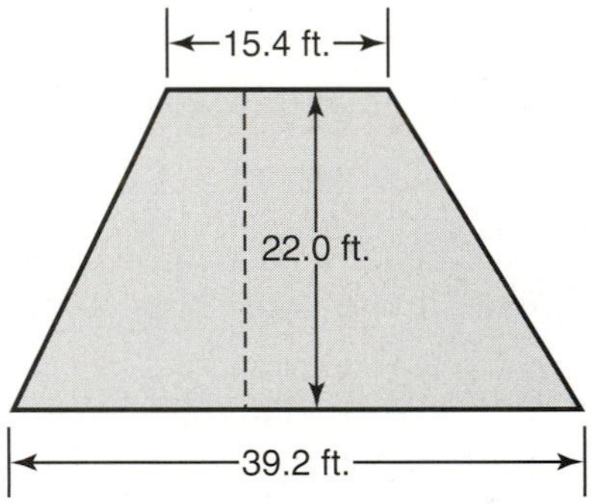

Figure 11-27 Determining the area of a trapezoid.

The use of the trapezoid area equation that will be illustrated using the isosceles trapezoid in Figure 11-27 is:

$$\text{Area} = \frac{A + B}{2} \times \text{height}$$

The trapezoid area equation is useful for determining the area of parcels of land because many lots and other small areas have two parallel sides.

Circle

A circle is a line that closes on itself, Illustration A Figure 11-28.

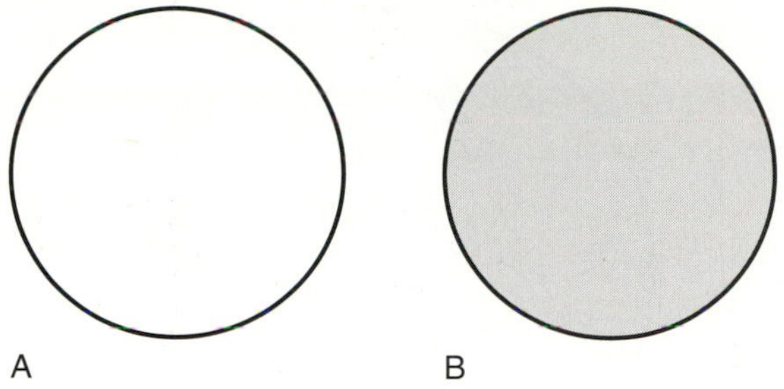

Figure 11-28 Circle.

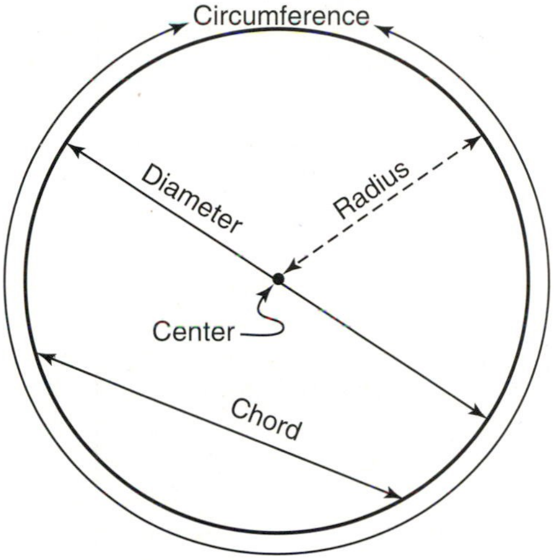

Figure 11-29 Circle parts.

Because a circle is a line it has no area, Illustration A Figure 11-28, but the disc inside the circle does, Illustration B Figure 11-28, and this is the area that is used when reference is made to the area of a circle. A circle consists of five parts: the center, radius, diameter, chord and perimeter (circumference), Figure 11-29.

Three equations are available for determining the area of a circle. The best equation is determined by which dimensions of the circle are known.

When the radius is known, the area can be determined by:

$$\text{Area} = \pi r^2$$

When the diameter is known, the equation is

$$\text{Area} = \pi \times \frac{D^2}{4}$$

When the circumference is known, the equation is

$$\text{Area} = \frac{C^2}{4 \times \pi}$$

Circle Area Using Radius

The radius equation is listed frequently as the method for determining the area of a circle, and

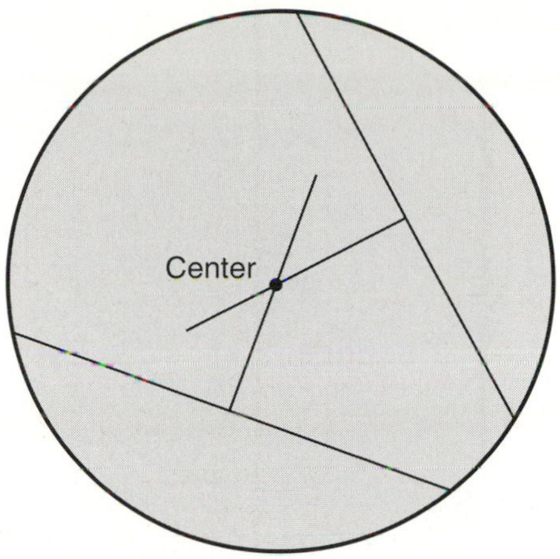

Figure 11-30 Locating the center of a circle.

it may be very useful and appropriate for math classes, but it is difficult to use in the field or with maps because to measure the radius of a circle, the center of the circle must be located. There are several math techniques for accomplishing this, but they would have a high probability of error when applied to field measurements because of the difficulty in making accurate measurements. For small, unobstructed circles, the center of a circle can be located with the chord or right angle techniques can be used. The chord method requires the establishment of two chords and laying out a perpendicular line at the midpoint of each cord. The intersection of the perpendicular lines is the center of the circle, Figure 11-30.

When completing field measurements for small, unobstructed circles this technique is useful, but as the circle size increases and as the number of surface obstructions increases the center becomes more difficult to locate and the opportunity for error increases. This method is more useable with maps.

Another technique is based on the principle that when the vertex of a right triangle is superimposed on the circumference of a circle, the two adjoining sides will intersect the circumference at two separate points. A line connecting these two intersects is the diameter of the circle, and the midpoint of the diameter is the center of the circle, Figure 11-31.

This technique has the same limitations as the chord technique. It could be used to find the center of a circle in a small or unobstructed field, but it is probably more useful for finding the center of a circle on a map.

In many landscapes a fountain, flagpole or some other structure may identify the center of the circle. When the center of the circle can be located, the area

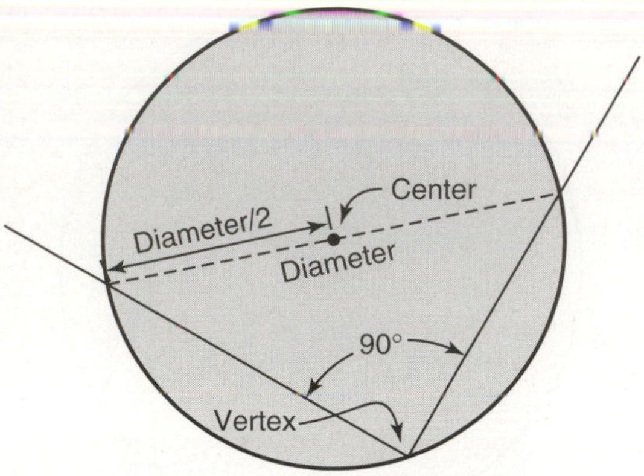

Figure 11-31 Locating circle center using 90 degree angle.

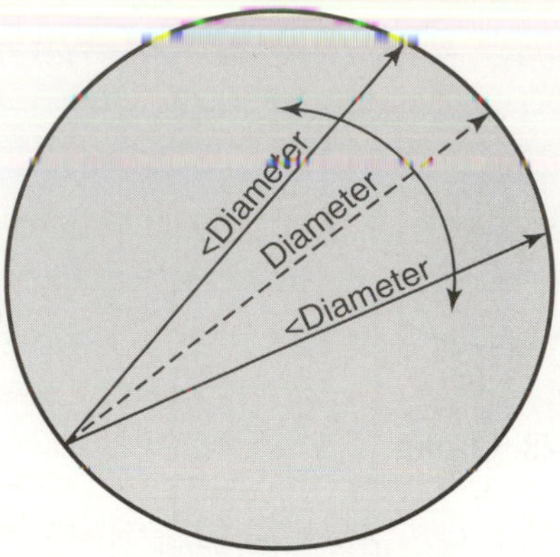

Figure 11-33 Determining circle diameter.

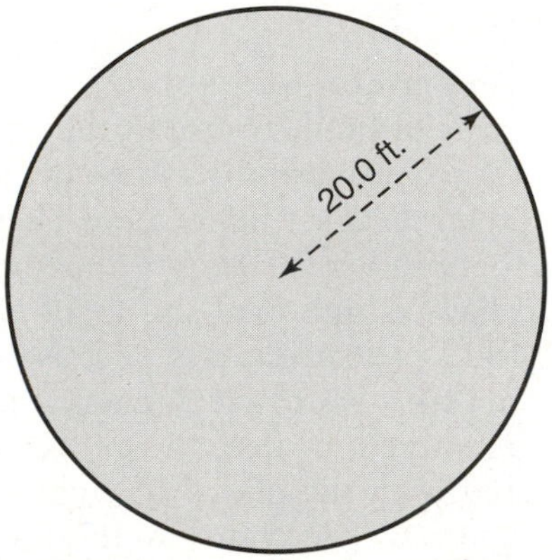

Figure 11-32 Determining area of a circle using the radius.

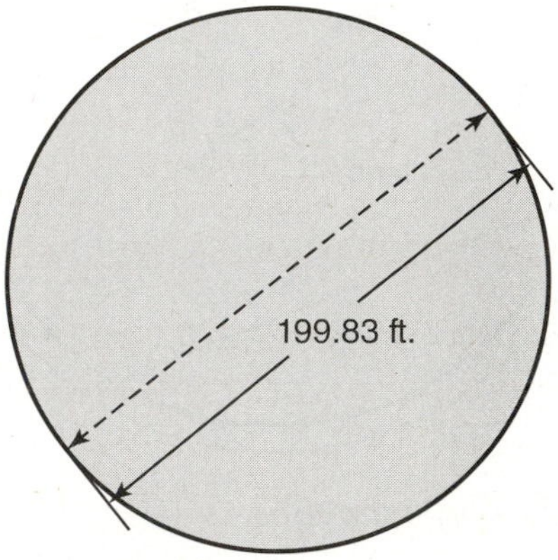

Figure 11-34 Determining the area of a circle using diameter.

by radius equation can be used. The area of the circle in Figure 11-32 is:

$$\text{Area} = \pi \times r^2$$
$$= 3.14 \times 20.0^2 \text{ ft.}$$
$$= 1256 \text{ ft.}^2$$

Circle Area Using Diameter

The area of a circle can also be determined using the diameter. Any straight line that intersects the perimeter of the circle at two points and passes through the center will be the diameter of the circle. The difficulty is finding the center of the circle. In the previous section two techniques were illustrated for locating the center of a circle.

If the circle is free of obstructions and not too large, the diameter can be determined by the sweep method. In this method one end of the measuring tape is fixed at the perimeter of the circle and the second end is stretched across the circle at the estimated position of the diameter. Multiple measurements are recorded along the perimeter of the circle. The longest measurement is the diameter, Figure 11-33.

When the diameter can be measured, the diameter equation can be used to determine the area of the circle as shown in Figure 11-34.

$$\text{Area} = \frac{\pi \times D^2}{4}$$
$$= 3.14 \times \frac{199.83^2 \text{ ft.}}{4}$$
$$= 31346.642\ldots \text{ or } 31346.64 \text{ ft.}^2$$

Circle Area Using Circumference

The third equation for determining the area of a circle uses the circumference of the circle. The circumference of a circle is a continuous arc, which makes it difficult to measure with chain because when the chain is pulled tight it will form a straight line. A large number of stakes or pins must be used to

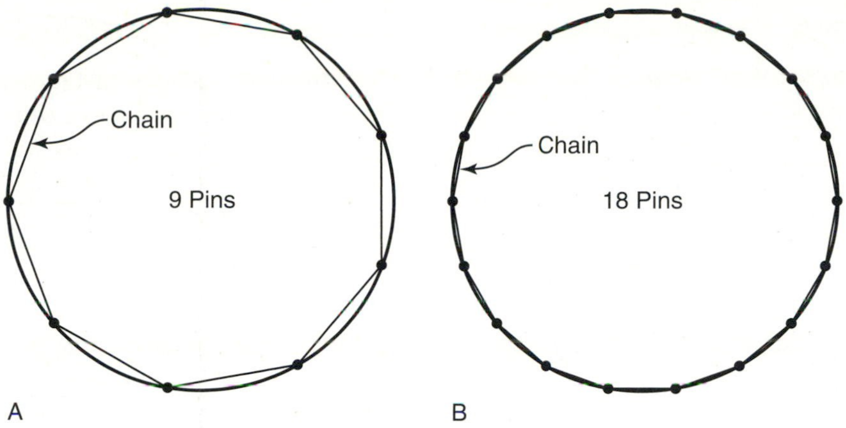

Figure 11-35 Measuring circle circumference.

insure the chain approximates the circumference as close as possible, Figure 11-35.

The need for a large number of pins along the circumference of the circle can be demonstrated with a graphical solution. Both circles in Figure 11-35 were drawn the same size. The circumference of the circles is 6.283 inches and the area is 3.14 square inches.

In Illustration A, 9 pins were used to measure the circumference. Using 9 pins the circumference of the polygon formed by the chain stretched around the pins is measured as 6.164 inches and the area is 2.90 square inches. In Illustration B using 18 pins the circumference of the polygon formed by the chain and pins is measured as 6.242 inches and the area is 3.07 square inches. This clearly shows that when measuring the circumference of a circle a large number of pins are required to achieve a high level of accuracy.

In many situations it may require fewer resources to measure the circumference than to measure the radius or diameter. When this is true, the area of a circle can be determined using the area by circumference equation. The area of the circle in Figure 11-36 is:

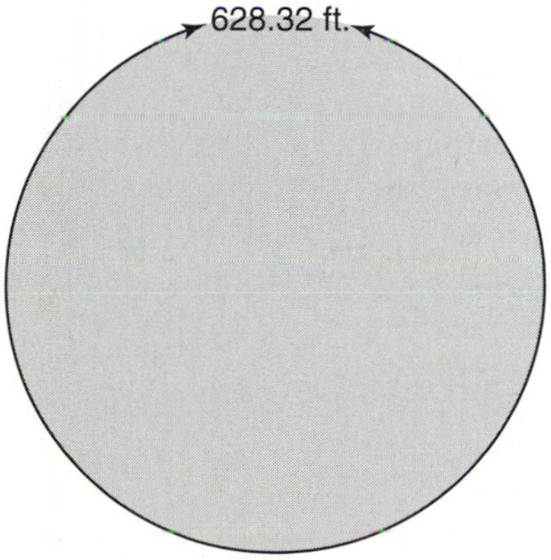

Figure 11-36 Determining circle area using circumference.

$$\text{Area} = \frac{C^2}{4 \times \pi}$$

$$= \frac{(628.32\text{ ft.})^2}{4 \times 3.14}$$

$$= 31432.008\ldots \text{ or } 31431.00\text{ ft.}^2$$

Sector

A sector is a part or segment of a circle. It would be rare for a parcel of land to form a sector. They are more common when reducing complex, irregular shapes into standard shapes. A sector has three parts: the radius, angle, and arc length. Two different equations are used depending on which measurements are known, Figure 11-37.

Measuring the arc length has the same concerns as measuring the circumference of a circle. A large number of stations must be established along the arc. The radius can be difficult to determine when the sector is part of a complex shape. For small, unobstructed sectors, the center of the arc can be determined using three chains, Figure 11-38.

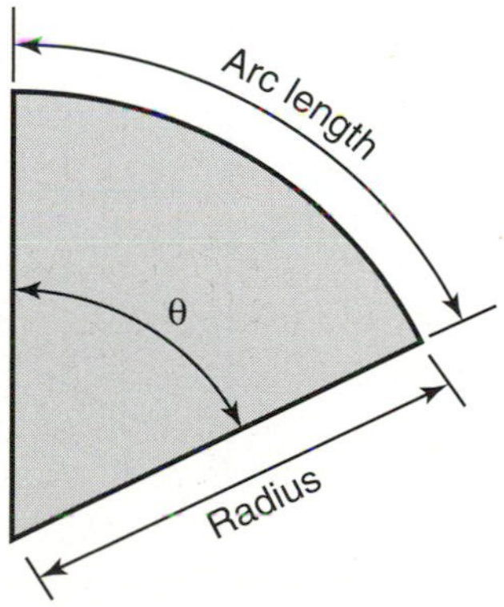

Figure 11-37 Sector parts.

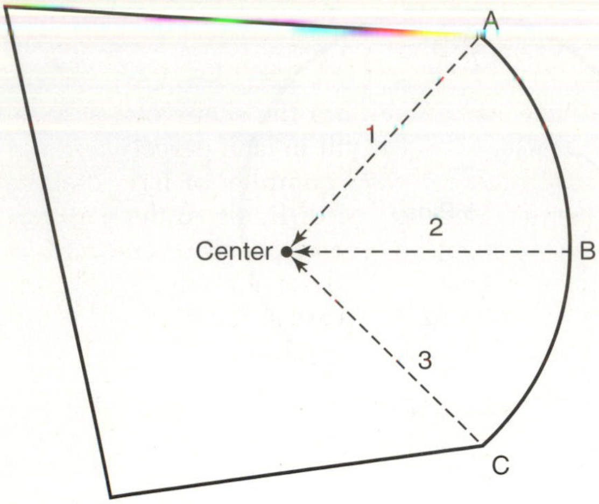

Figure 11-38 Locating the center of a sector.

Procedure: Attach chains to stations A, B, and C to the perimeter of the arc. Insure they are spaced as far apart as possible and insure the chain B is at the approximate center of the arc. Walk toward the estimated center of the sector and move around until all three chains read the same distance. If the arc is uniform, this is the center of the sector. The measurement on the chains is the radius for the sector.

When the length of the arc and the radius can be measured, the following equation can be used.

$$\text{Area} = \frac{\text{radius} \times \text{arc length}}{2}$$

When the length of the radius and the included angle can be measured, this equation can be used. This method requires measuring the angles with one of the direct or indirect methods discussed in Chapter 7.

$$\text{Area} = \pi \times r^2 \times \frac{\theta}{360}$$

Determine the areas for the sectors in the following illustrations.

In Illustration A of Figure 11-39 the radius and arc length are the known dimensions; therefore, the area is:

$$\text{Area} = \frac{\text{radius} \times \text{arc length}}{2}$$

$$= \frac{143.76 \text{ ft.} \times 157.72 \text{ ft.}}{2}$$

$$= 1336.913\ldots \text{ or } 1336.91 \text{ ft.}^2$$

In Illustration B of Figure 11-39 the radius and the angle are the known dimensions; therefore, the area is:

$$\text{Area} = \pi \times r^2 \times \frac{\theta}{360}$$

$$= 3.14 \times (143.76 \text{ ft.})^2 \times \frac{62.89°}{360°}$$

$$= 11336.653\ldots \text{ or } 11336.65 \text{ ft.}^2$$

Segment

A segment is the area formed between a chord and the arc of the circle. Segments are classified as either major, greater than half of a circle, or minor, less than half a circle, Figure 11-40.

Figure 11-40 shows that a segment is part of a sector. Therefore, the most accurate method for determining the area of a minor segment is to determine the area of the sector and subtract the triangle that is not part of the segment. This requires knowing the radius and the angle of the sector.

$$\text{Area} = \left(\pi \times r^2 \times \frac{\theta}{360}\right) - \frac{r^2}{2}$$

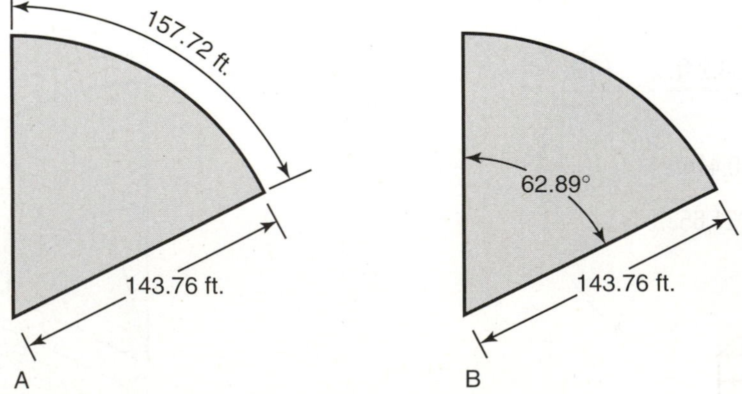

Figure 11-39 Determining the area of sectors.

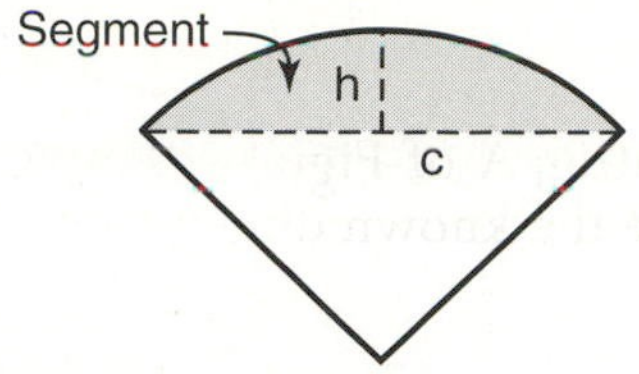

Figure 11-40 Segment.

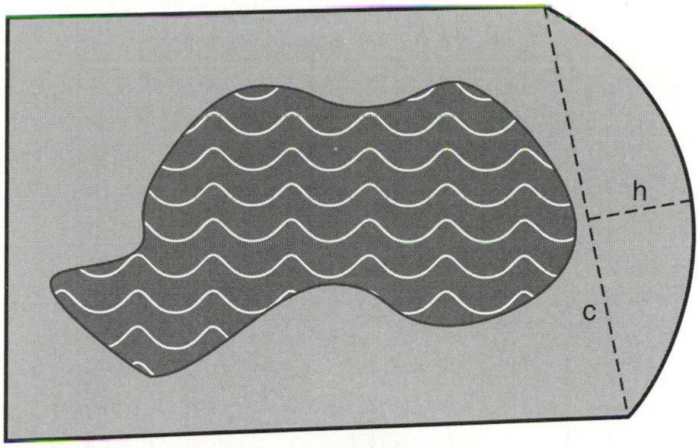

Figure 11-41 Segment example.

It is not always possible to measure the radius and the angle of the sector when determining the area of land segments. For example, the center of the arc in Figure 11-41 is in the water, but as long as the chord and the height can be measured, equations can be used to estimate the area of the segment.

Note: The height is measured by establishing a perpendicular line at the midpoint of the chord.

One equation used to estimate the area for a minor segment using only the chord length and height is:

$$\text{Area} = \frac{2hc}{3} + \frac{h^2}{2c}$$

Determine the area for the segment in Figure 11-42.

$$\text{Area} = \frac{2hc}{3} + \frac{h^2}{2c}$$

$$= \frac{2 \times 195.3\text{ ft.} \times 934.2\text{ ft.}}{3} + \frac{(195.3\text{ ft.})^2}{2 \times 934.2\text{ ft.}}$$

$$= 121632.84\text{ ft.}^2 + 20.414\ldots$$

$$= 121653.254\ldots \text{ or } 121{,}653.2\text{ ft.}^2$$

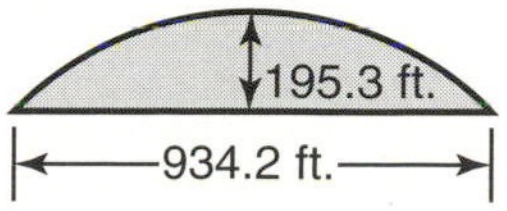

Figure 11-42 Determining the area of segment.

DETERMINING THE AREA OF COMPLEX SHAPES

Land parcels are seldom the shape of standard figures. Many factors result in land parcels with irregular shapes. The area of complex or irregular shaped parcels of land can be difficult to determine. The desired accuracy of the calculations and the terrain of the land parcel will determine the "best" method. Four common methods are:

- Standard geometric figures
- Offsets from a line
- Coordinate squares
- Coordinates

Area of Complex Shapes Using Standard Geometric Figures

Determining the area of a complex shape using standard geometric figures works best with maps or parcels of land that are small and unobstructed by trees, buildings, and other features. The method divides the complex shape into standard shapes, records the necessary dimensions, and calculates the area for each shape. The total area is the sum of the individual areas.

Many complex shapes can be divided up into more than one set of standard figures. The different divisions are not right or wrong; some will take less time and resources to measure than others. Obstructions to collecting measurements will also dictate one method over another. Study the example in Figure 11-43.

It should be clear that a portion is a sector, and it should also be clear that with no 90 degree corners the remaining area cannot be divided into squares or rectangles, but trapezoids and triangles are possible. Figure 11-44 illustrates three possible ways to divide the parcel into standard figures. The "best" method is the one that requires the least amount of resources to collect the necessary dimensions.

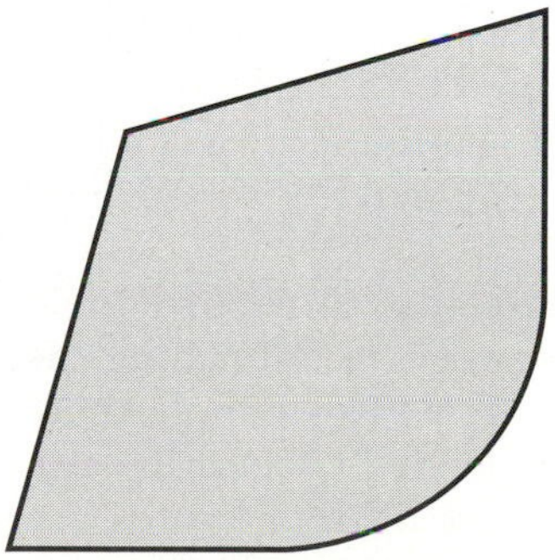

Figure 11-43 Parcel with irregular shape.

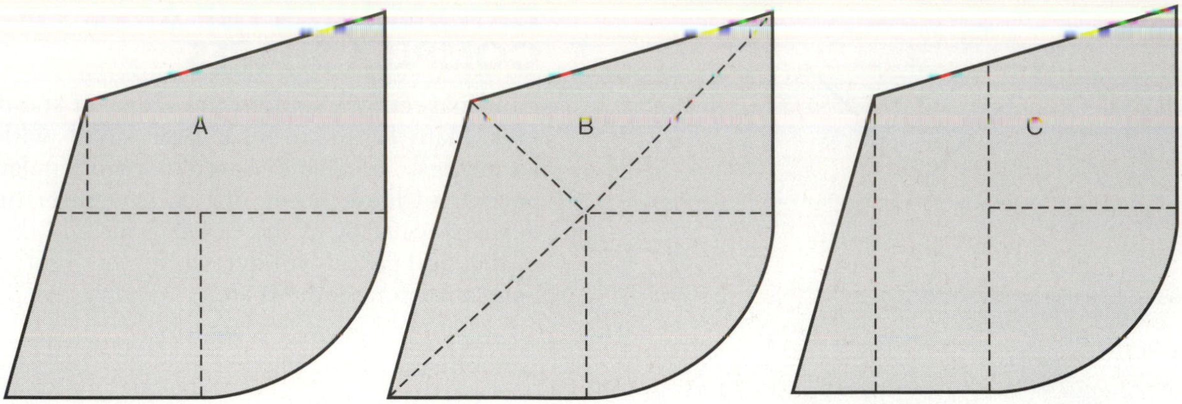

Figure 11-44 Three possible solutions.

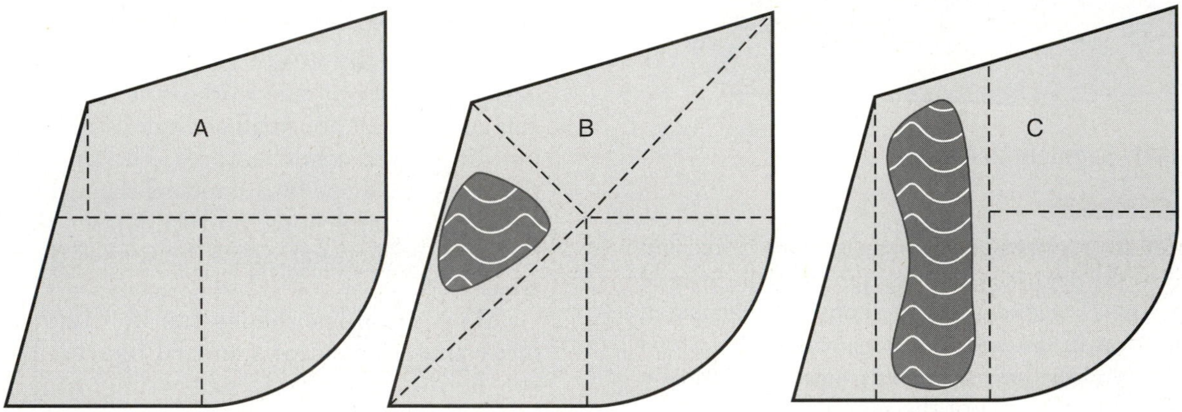

Figure 11-45 Division determined by water.

It is also possible that man-made and natural features will influence the division of the complex shape, Figure 11-45. In this illustration the presence of water is used to illustrate the influence of natural features even though it would not pose a problem if EDM, stadia, or GPS were used to measure the distances.

Illustration A of Figure 11-45 will be used as an example of determining area by the division into standard figures method. Note the parcel has been partitioned into four figures and each one has been labeled, Figure 11-46. This will reduce the opportunity for making errors when calculating the area.

The first step is to record the necessary distances. Step two is to determine the area for each of the figures. For this example there are two trapezoids, a triangle, and a sector. The last step is to add the area of each of the four figures.

Figure A is a right triangle. The area of A is:

$$\text{Area} = \frac{\text{base} \times \text{height}}{2}$$

$$= \frac{49.63 \text{ ft.} \times 176.0 \text{ ft.}}{2}$$

$$= 4367.44 \text{ ft.}^2$$

Figure B is a trapezoid. The area of B is:

$$\text{Area} = \frac{\text{A} + \text{B}}{2} \times \text{height}$$

$$= \frac{176.00 \text{ ft.} + 315.83 \text{ ft.}}{2} \times 470.41 \text{ ft.}$$

$$= 245.915 \text{ ft.} \times 470 \text{ ft.}$$

$$= 115580.88 \text{ ft.}^2$$

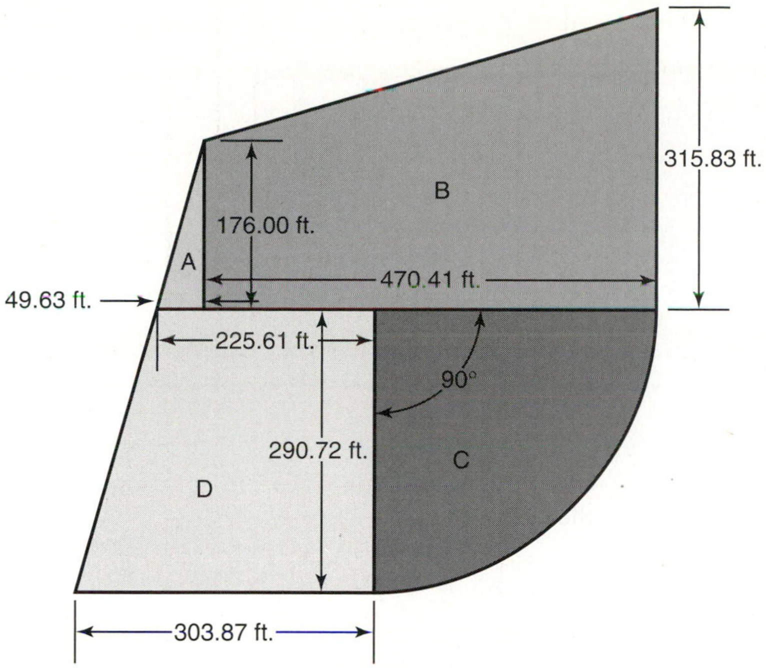

Figure 11-46 Example of division into standard shapes.

Figure C is a Sector. The area of C is:

$$\text{Area} = \pi \times r^2 \times \frac{\theta}{360°}$$

$$= 3.14 \times 290.72^2 \text{ ft.} \times \frac{90°}{360°}$$

$$= 265386.8918 \text{ ft.}^2 \times 0.25$$

$$= 66346.72 \text{ ft.}^2$$

Figure D is a trapezoid. The area of D is:

$$\text{Area} = \frac{A + B}{2} \times \text{height}$$

$$= \frac{225.61 \text{ ft.} + 303.87 \text{ ft.}}{2} \times 290.72 \text{ ft.}$$

$$= 264.74 \text{ ft.} \times 290.72 \text{ ft.}$$

$$= 76965.21 \text{ ft.}^2$$

The area for the complex shape is the sum of the area for the four figures; A, B, C, and D. The total area is:

$$\begin{array}{r} 4367.44 \text{ ft.}^2 \\ 115580.88 \text{ ft.}^2 \\ 66346.73 \text{ ft.}^2 \\ +\ 76965.21 \text{ ft.}^2 \\ \hline 263260.26 \text{ ft.}^2 \end{array}$$

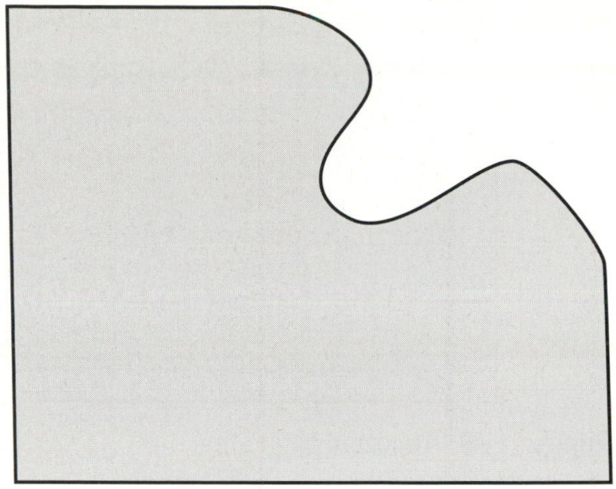

Figure 11-47 Land with stream boundary.

The total area of the irregular shaped parcel of land is 263,260.26 square feet.

Determining the area of a large complex shape by division into standard figures can be completed in the field, but a common problem is being able to visualize the shape of the parcel so that the best divisions can be established. Completing a boundary survey and sketching the boundary before attempting to divide up the parcel would reduce the amount of time and the number of mistakes that could occur. This method can also be used to determine the area of a land parcel from a map. The advantages of using a map include being able to see the shape and not being required to take the measurements in the field.

Area Using Offsets from a Line

For many parcels of land one or more boundaries are defined by a river or stream, Figure 11-47.

The meandering of a waterway makes area calculations difficult. One method that can be used is offsets from a line, Figure 11-50. In this method a series of trapezoids are used and if the same distance, d in Figure 11-48, is used the area of all of the trapezoids can be determined using one equation. Note this method estimates the area because the conversion to trapezoids does not follow the meandering of the stream exactly, Figure 11-48.

When all of the trapezoids have the same distance value, d in Figure 11-48, the summation trapezoidal equation can be used. This equation is:

$$\text{area} = d \times \left(\frac{h_o}{2} + \sum h + \frac{h_n}{2}\right)$$

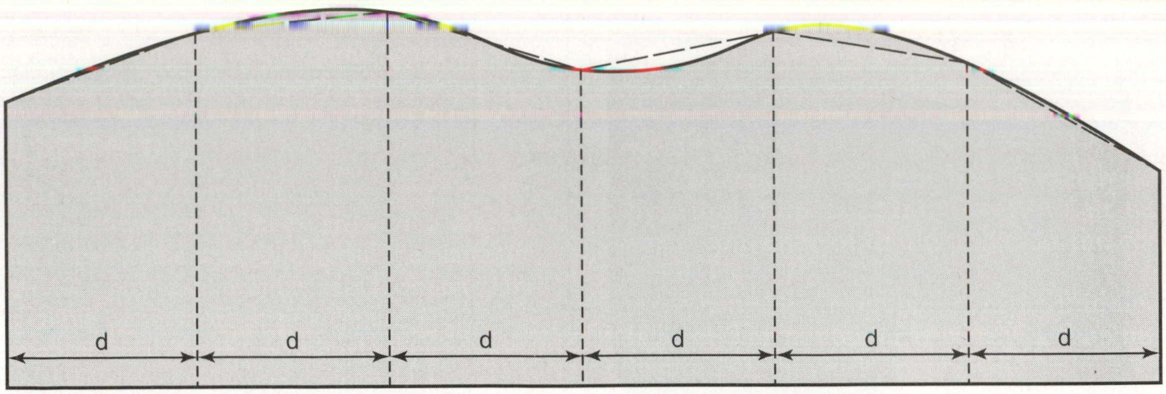

Figure 11-48 Trapezoid estimation.

TIP

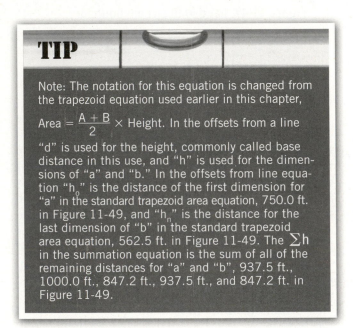

Note: The notation for this equation is changed from the trapezoid equation used earlier in this chapter, $\text{Area} = \frac{A + B}{2} \times \text{Height}$. In the offsets from a line "d" is used for the height, commonly called base distance in this use, and "h" is used for the dimensions of "a" and "b." In the offsets from line equation "h_o" is the distance of the first dimension for "a" in the standard trapezoid area equation, 750.0 ft. in Figure 11-49, and "h_n" is the distance for the last dimension of "b" in the standard trapezoid area equation, 562.5 ft. in Figure 11-49. The $\sum h$ in the summation equation is the sum of all of the remaining distances for "a" and "b", 937.5 ft., 1000.0 ft., 847.2 ft., 937.5 ft., and 847.2 ft. in Figure 11-49.

Using the summation trapezoid equation, the area for the land parcel in Figure 11-49 is:

$$\text{area} = d \times \left(\frac{h_o}{2} + \sum h + \frac{h_n}{2}\right)$$

$$= 500\text{ ft.} \times \left(\frac{750.0\text{ ft.}}{2} + (937.5\text{ ft.} + 1000.0\text{ ft.} + 847.2\text{ ft.} + 937.5\text{ ft.} + 847.2\text{ ft.})\frac{562.5}{2}\right)$$

$$= 500\text{ ft.} \times 5225.65\text{ ft.}$$

$$= 2{,}612{,}825\text{ ft.}^2$$

The offsets from a line method can be used by collecting measurements in the field or by determining distances on a map.

In the field, the meandering of the stream will not be as easy as the example in Figure 11-49. Figure 11-50 is one solution for the parcel of land with a meandering boundary illustrated in Figure 11-47.

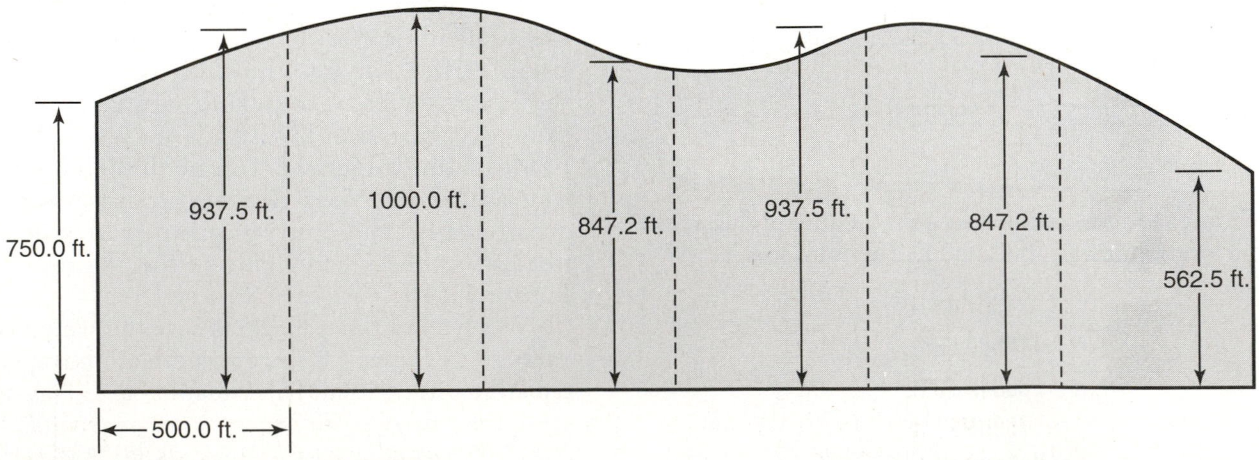

Figure 11-49 Summation trapezoid equation example.

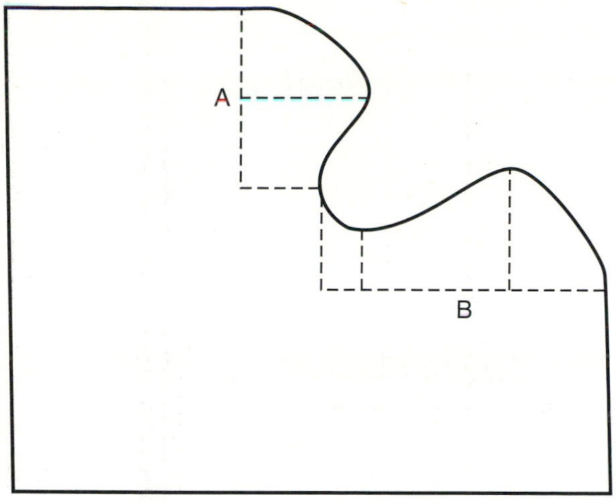

Figure 11-50 Meandering boundary divided into offsets.

In the example illustrated in Figure 11-50, offset A is divided into two trapezoids with the same base distance because dividing this portion of the meander in half is a good estimation of the area and having equal distances reduces the math requirements, Figure 11-52. Offset B does not use equal base distance trapezoids because the meandering of the stream would require a large number of divisions along the baseline to have equal distance trapezoids that match the stream boundary with the desired accuracy, Figure 11-51. Given the choice of spending more resources measuring in the field or completing a few more calculations, most surveyors would choose the additional calculations.

The next step is to measure the distances for meander A and B and calculate the areas.

Offset A in Figure 11-52 uses equal height trapezoids. In this situation the offsets form a line equation that can be used. Solving for the area A in Figure 11-52 results in:

$$\text{area} = d \times \left(\frac{h_o}{2} + \sum h + \frac{h_n}{2}\right)$$

$$= 229.13\text{ ft.} \times \left(\frac{88.11\text{ ft.}}{2} + 315.97\text{ ft.} + \frac{190.97\text{ ft.}}{2}\right)$$

$$= 229.13\text{ ft.} \times 455.51\text{ ft.}$$

$$= 104371.006\ldots\text{ or }104371.01\text{ ft.}^2$$

In this example only one dimension is used for the $\sum h$ because there were only three heights.

The area for offset B must be determined using the trapezoid equation three times because the base distance is not the same. The area of offset B is:

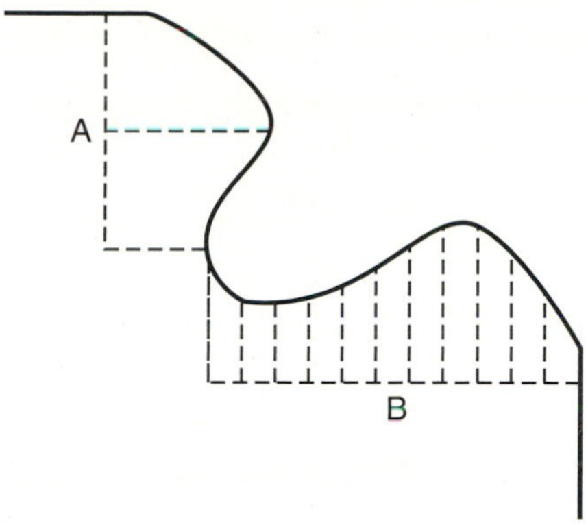

Figure 11-51 Equal distance trapezoids for meander B.

$$\text{Area} = \frac{a + b}{2} \times h$$

$$= \frac{247.42\text{ ft.} + 150.68\text{ ft.}}{2} \times 116.32\text{ ft.}$$

$$= 23{,}153.50\text{ ft.}^2$$

$$\text{Area} = \frac{a + b}{2} \times h$$

$$= \frac{150.68\text{ ft.} + 308.08\text{ ft.}}{2} \times 368.06\text{ ft.}$$

$$= 84{,}425.60\text{ ft.}^2$$

$$\text{Area} = \frac{a + b}{2} \times h$$

$$= \frac{308.08\text{ ft.} + 65.54\text{ ft.}}{2} \times 223.98\text{ ft.}$$

$$= 41{,}841.70\text{ ft.}^2$$

The total area for offset B is:

$$\begin{array}{r} 23{,}153.50\text{ ft.}^2 \\ 84{,}425.60\text{ ft.}^2 \\ +\ 41{,}841.70\text{ ft.}^2 \\ \hline 149{,}420.80\text{ ft.}^2 \end{array}$$

The total area of the meandering portion, A + B, is:

$$\begin{array}{r} 104{,}371.01\text{ ft.}^2 \\ +\ 149{,}420.80\text{ ft.}^2 \\ \hline 253{,}791.81\text{ ft.}^2 \end{array}$$

The remainder of the area in Figure 11-50 would be determined by dividing the area into squares, rectangles, or additional trapezoids, collecting the required distances, and calculating the area.

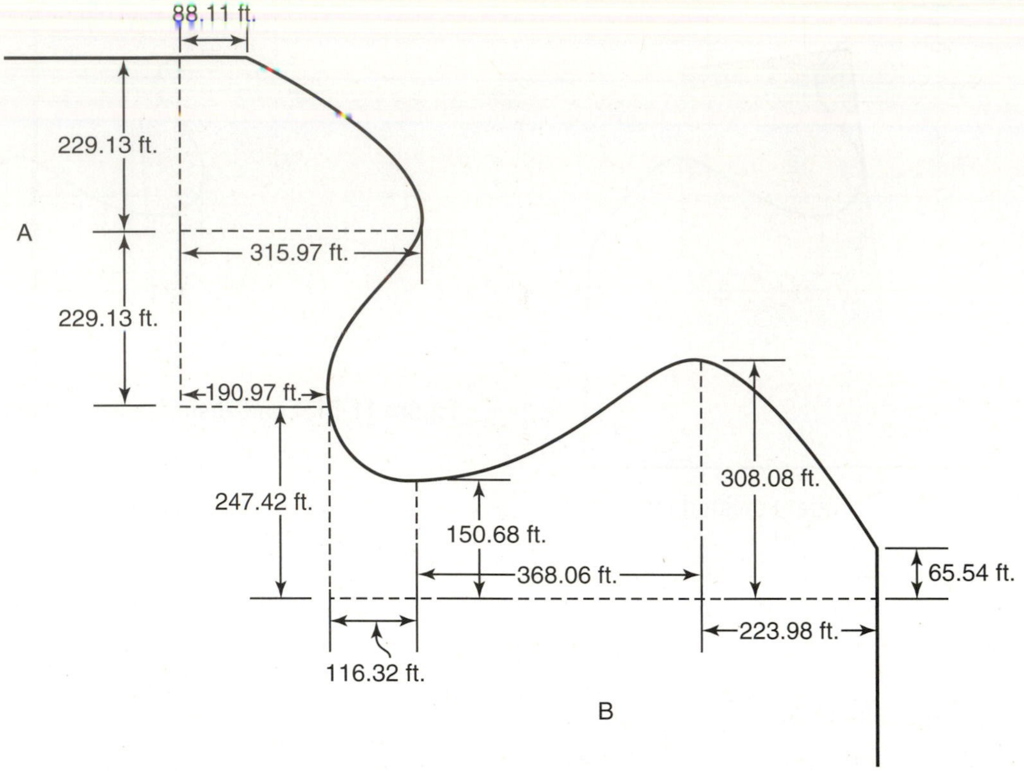

Figure 11-52 Measurements for offset from a line.

This example demonstrates the two common methods for determining the area when a boundary meanders. Determining the area by offsets methods can be used in the field or with maps.

COORDINATE SQUARES

The coordinate squares method overlays a map with a grid with a known size. It is a common practice to print the grid on transparent material so the map will show through. Knowing the size of the grid and the scale of the map, the area can be determined by counting squares. When the map scale is expressed as a ratio, the area for each grid is determined by:

$$\text{Area (ft.}^2) = \left(\text{grid size (in.)} \times \frac{\text{Map scale}}{12}\right)^2$$

When a 0.25 inch grid is used and the map scale is 1:1,000, then each square would be equivalent to:

$$\text{Area (ft.}^2) = \left(\text{grid size (in.)} \times \frac{\text{Map scale (in.)}}{12}\right)^2$$

$$= \left(0.25 \text{ in.} \times \frac{1000}{12}\right)^2$$

$$= 434.03 \text{ ft.}^2$$

When the map scale is expressed in units such as ft/in each grid area is:

$$\text{Area (ft.}^2) = (\text{grid size (in.)} \times \text{Map scale (ft./in)})^2$$

When a 1/2 inch grid is overlaid on a map with a scale of 500 feet per 1 inch, the area of each grid is:

$$\text{Area (ft.}^2) = (\text{grid size (in.)} \times \text{Map scale (ft./in)})^2$$

$$= (0.5 \times 500)^2$$

$$= 62500.0 \text{ ft.}^2$$

Determining the area of the polygon in Figure 11-53 requires three steps: counting the number of squares, determining the area per square, and then multiplying these two numbers.

When counting squares, estimate the partial squares as accurately as possible.

Figure 11-54 shows that the parcel of land in question has forty 0.5 inch squares. One through eleven are estimated by combining parts of squares. The next step is to determine the land area per square. The map scale is shown as 1:2000, therefore the area per square is:

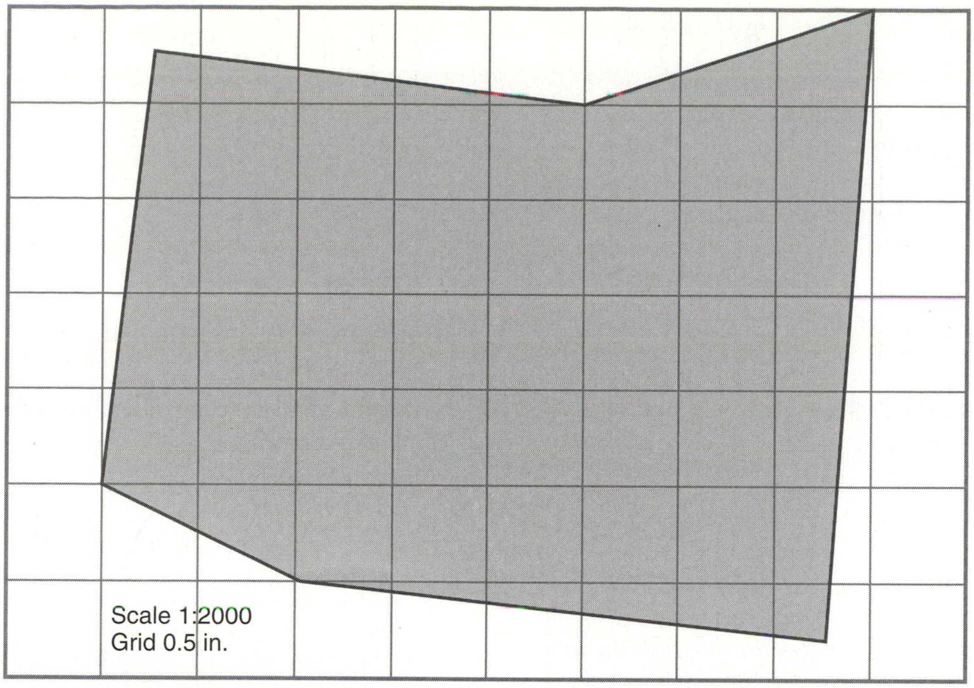

Figure 11-53 Area by coordinate squares.

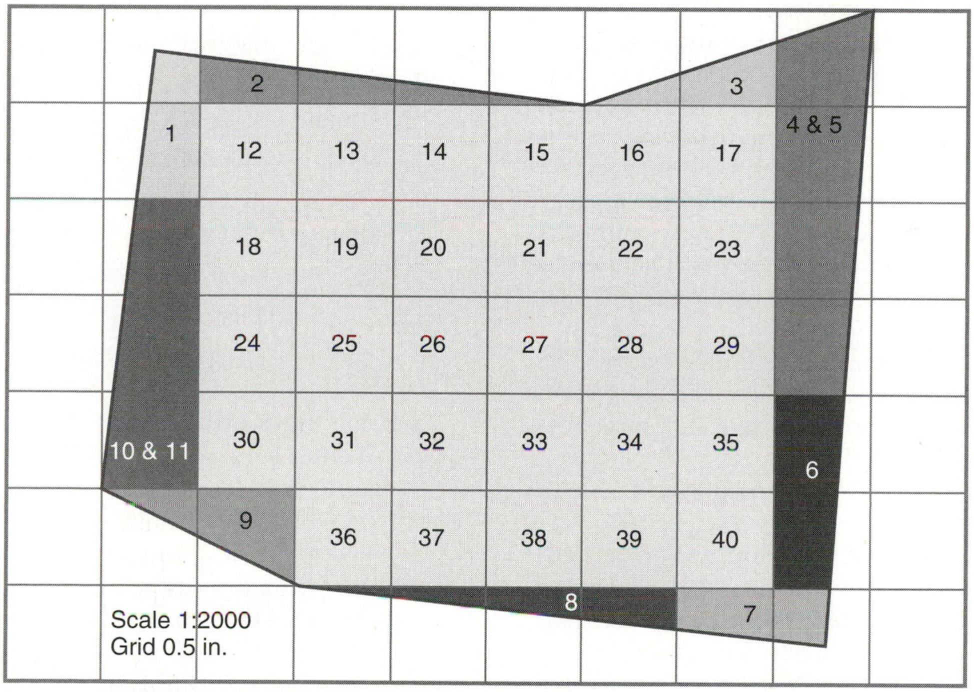

Figure 11-54 Estimated whole squares numbered for area by coordinate squares example.

$$\text{Area (ft.}^2) = \left(\text{grid size (in.)} \times \frac{\text{Map scale}}{12}\right)^2$$

$$= \left(0.5 \times \frac{2000}{12}\right)^2$$

$$= 6{,}944 \text{ ft.}^2$$

The area is:

$$\text{area} = \frac{6{,}944 \text{ ft.}^2}{\text{square}} \times 40 \text{ squares}$$

$$= 277{,}777.8 \text{ ft.}^2$$

COORDINATES

The coordinate method for determining area has been available for many years and was more popular when determining area from maps. When using a map, a grid with the correct scale can be superimposed on the map and the desired coordinates determined from the grid. Collecting field data requires knowing the X and Y coordinates for each station around the boundary of the parcel of land. Before the adoption of GPS, surveying this required completing a traverse survey and balancing the traverse before the area could be calculated. GPS surveying equipment can be set to record the coordinates for each station. For precise calculations the coordinates will still need to be balanced since all GPS coordinates have some error. For low-precision calculations the GPS coordinates can be used without balancing if sufficient care is taken to insure GPS errors are managed.

An example of the type of data used for area by coordinates is found in Table 11-1. These numbers were collected by a GPS unit and they will be used to complete a sample problem. The assumption is that the data has sufficient accuracy for the intended use.

Note: When coordinates are collected by GPS surveying instruments, the UTM coordinate system is used. The UTM coordinate system measures distances in meters only, see Chapter 10.

The coordinate method uses a matrix and table to organize the calculations.

The X and Y coordinates are multiplied according to the matrix, Figure 11-55, and sorted into the correct columns in the table, Table 11-2. Then the columns are summed. The sum of the + column is subtracted from the sum of the − column. The difference is twice the area of the parcel so it is divided by 2 to arrive at the area.

Table 11.1 GPS data

STA	N	E
A	404945.8	673157.5
B	404195.7	678155.3
C	400945.7	677919.2
D	400692.0	674406.9
E	401948.9	672900.0

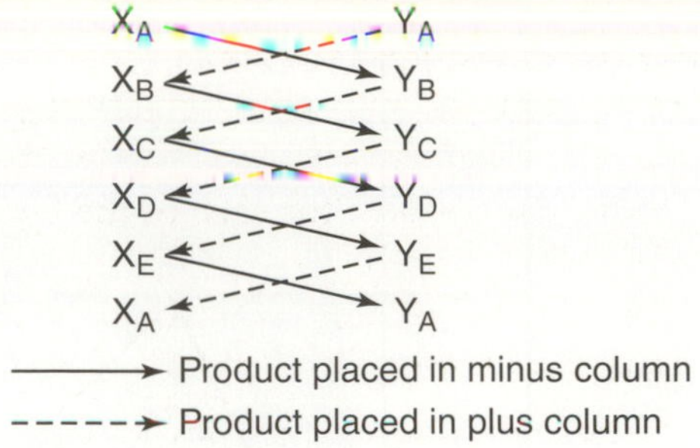

Figure 11-55 Coordinate matrix.

TIP

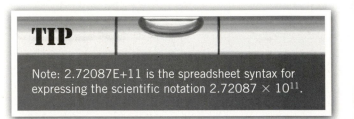

Note: 2.72087E+11 is the spreadsheet syntax for expressing the scientific notation 2.72087×10^{11}.

Table 11.2 Solution for area by coordinates

	X	Y	−	+
A	673157.5	404945.8		
B	678155.3	404195.7	2.72087E+11	2.74616E+11
C	677919.2	400945.7	2.71903E+11	2.74012E+11
D	674406.9	400692.0	2.71637E+11	2.70401E+11
E	672900.0	401948.9	2.71077E+11	2.69626E+11
A	673157.5	404945.8	2.72488E+11	2.70575E+11
		Sum	1.35919E+12	1.35923E+12
		Difference	36,517,223.49	
		Difference/2	18,258,611.74	Square meters

Using the coordinate method, the area of the parcel in Table 11-1 is 18,258,611.74 square meters.

SUMMARY

When determining areas, there are many different methods to choose from. The selection process can be daunting. It is important to remember that the "best" method is the one that produces the desired data, with the limitations of the site, with the least expenditure of resources.

Student Activities

1. Determine the area for each of the figures.

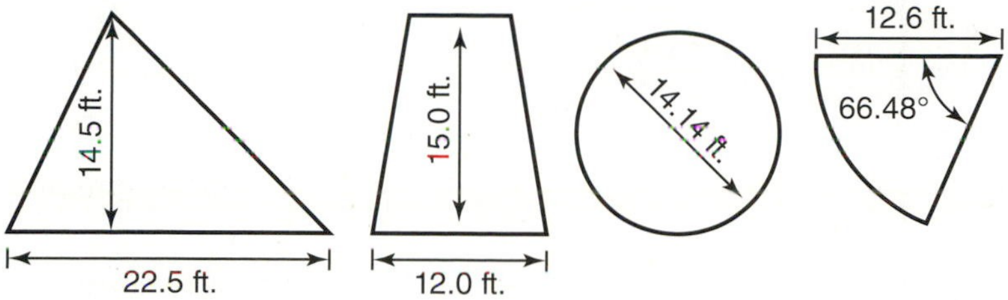

2. Determine the area for the following illustration using the coordinate squares method.

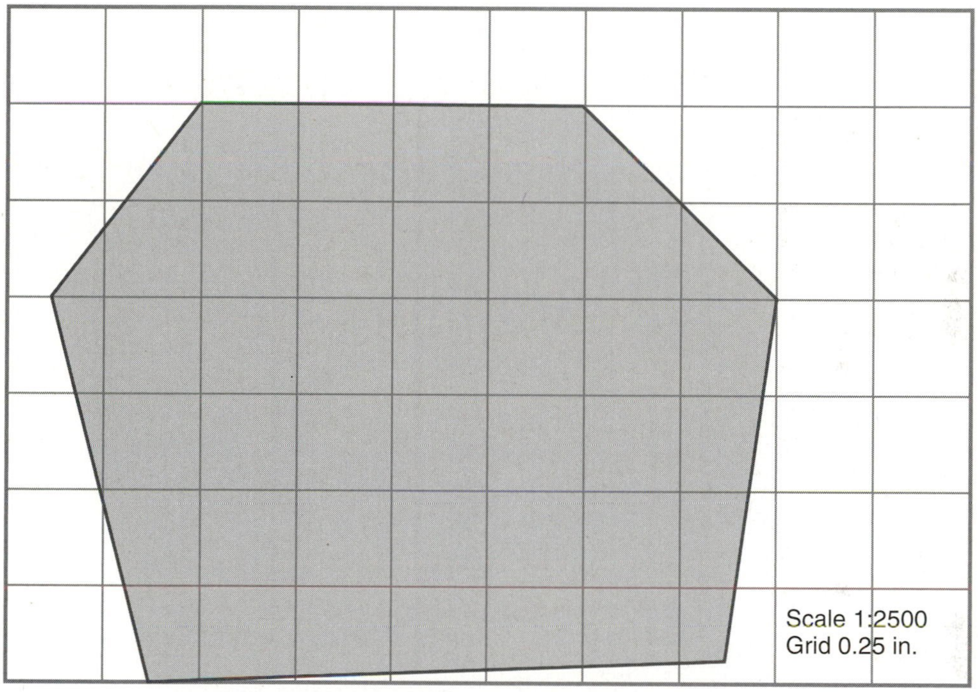

3. Determine the area of the following shape by division into standard figures.

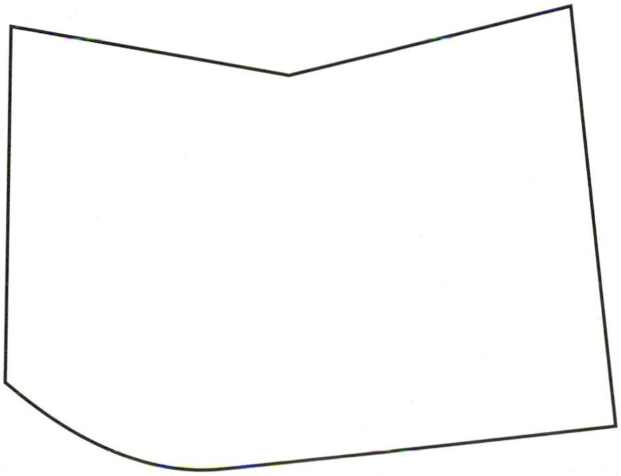

Note: Divide shape into standard figures and then use a ruler and a scale of 1 inch equals 100 feet to determine the area.

APPENDIX

Graphing Tips

INTRODUCTION

A graph (plot) is a very useful tool for producing visual representations of numbers. It is much easier to understand the relationship between variables when they are in visual form. Graphs can be two-dimensional, as shown in Figure 1, three-dimensional, as shown in Figure 2, or more. A two-dimensional graph gets its name because it shows the effect of two variables, one on the X-axis and one on the Y-axis. For example, the graph in Figure 1 shows how the elevation changes between stations zero and 49. The line in Figure 1 is connected from point to point; it could also be a smooth curve. The type of line is determined by the use of the data and the preferences for the discipline.

A good graph is scaled so that it uses as much of the page as possible, and has a title, labels, scales, units, and legends.

STEPS IN DRAWING A GRAPH

1. Select the appropriate style of graph paper for the data. When distances are being plotted with a precision of 1⁄10 or 1⁄100 of a foot, then paper with 10 squares per inch is easy to use.
2. Analyze the data to determine the maximum and minimum values for the X- and Y-axis. It is common practice when plotting profile data to place the distance along the X-axis and the elevation along the Y-axis.
3. Determine the orientation of the graph and the paper. It can be portrait, as shown in Figure 3, or landscape, as shown in Figure 4.

The data with the greatest range should be along the longest edge of the paper. For profile data, the distance data usually has the largest range. Therefore, the graph should be drawn in a landscape orientation.

4. Determine if both axes should have the same scale. The preferred technique for profile data is to use scales that are not equal. If both axes do not have the same scale, the relationship between the two sets of data will be distorted. This is the preferred method for profile data so that changes in evaluation are emphasized.

The largest range of numbers should be used on the longest edge of the paper when

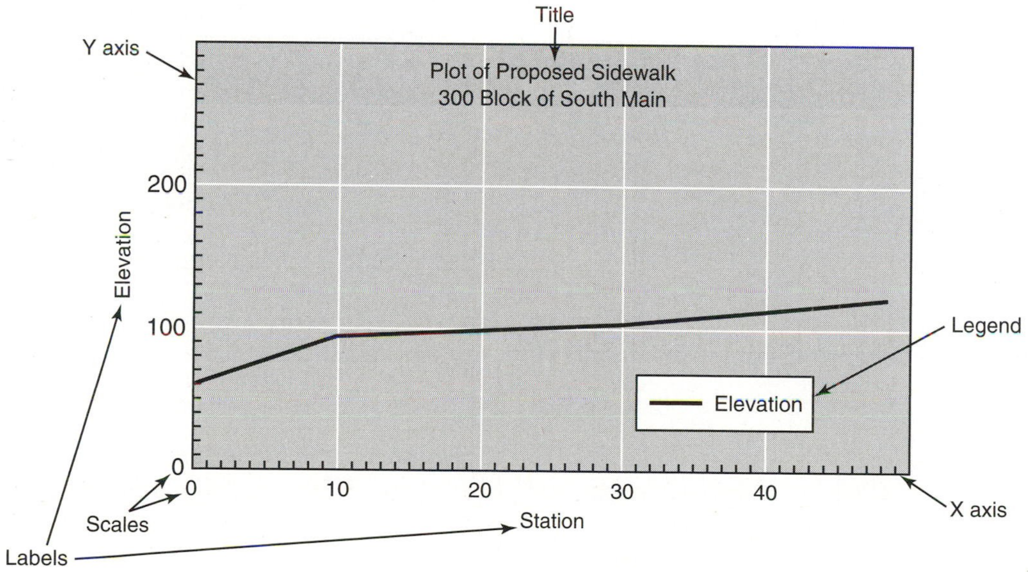

Figure 1 Parts of a graph.

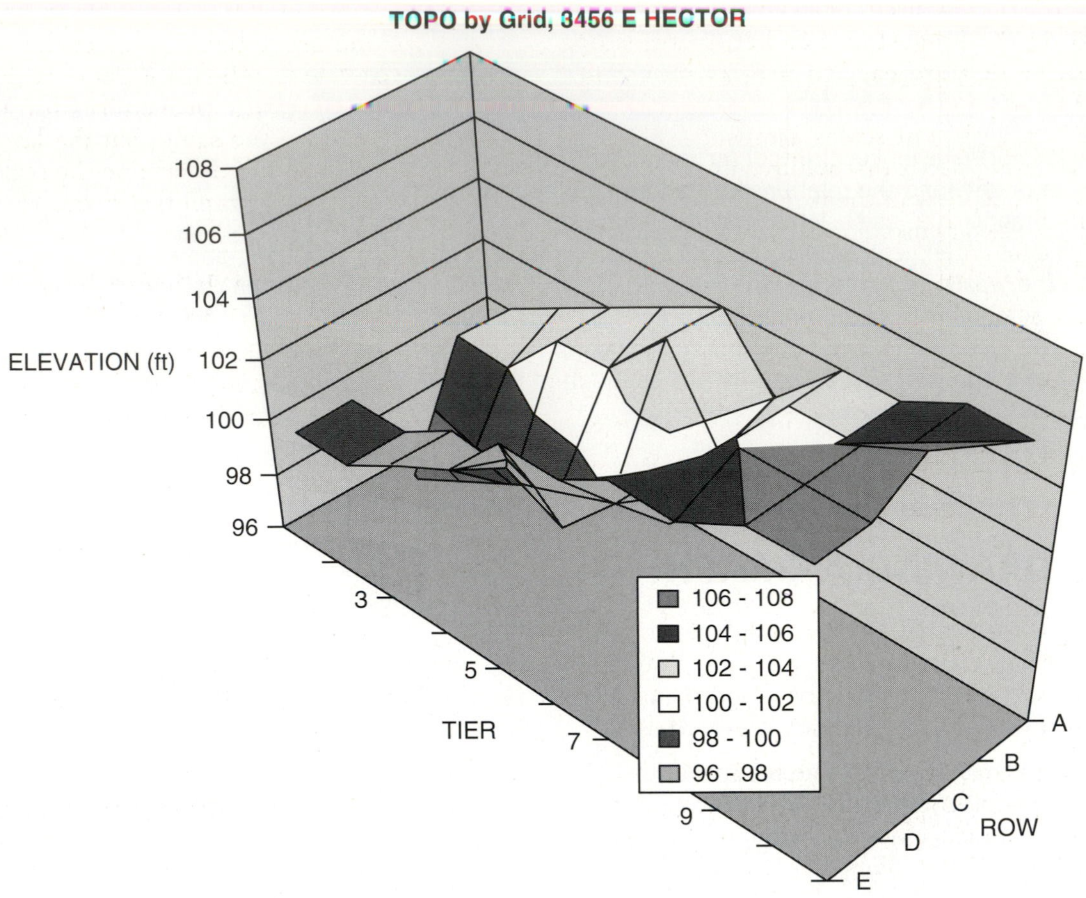

Figure 2 Three-dimensional graph.

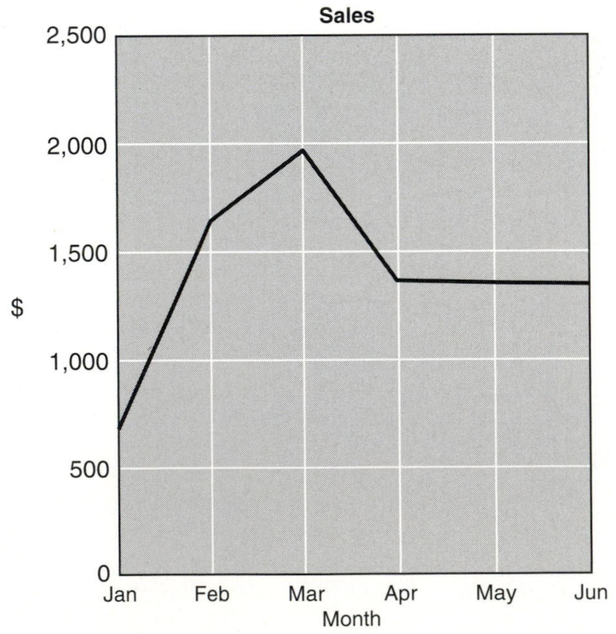

Figure 3 Portrait style of graph.

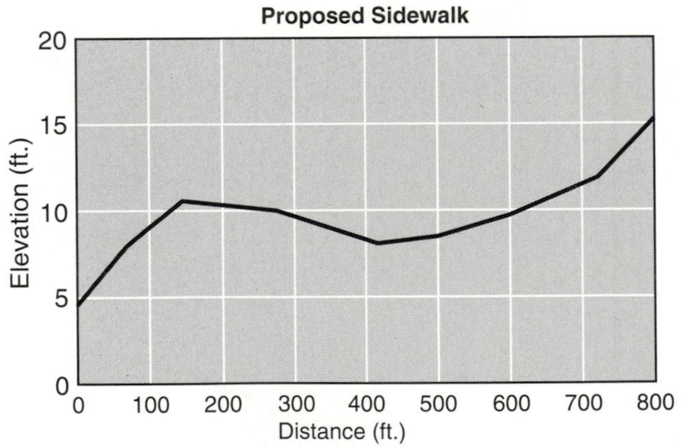

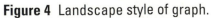

Figure 4 Landscape style of graph.

the same scale is used. Select a scale that will result in the longest axis and still have subdivisions that are easy to use, as shown in Figure 5.

When different scales are used, select a scale for the X-axis that will result in the line using the majority of the page, so the subdivisions are easy to use. For example: If the data ranges from 0.0 to 400.0 feet and the graph paper has 10 squares, then a good scale is 1 inch = 50 feet. With this scale, 8 of 10 squares will be used and each line on the paper will equal 5 feet. If all of the squares are used the value of each square will be 40 (400/10) and each line will have a value of 0.4.

When different scales are used, use the same procedures to select the best scale for the Y-axis. When the same scale is used for both axes the procedures for selecting the scales are the same, but the largest range of numbers will determine the scale. Often this results in a graph that is not very useable, as shown in Figure 6.

5. Draw the axis lines, mark the graduations and units on each scale, and plot the points.
6. Draw the appropriate line between the points. For profile data, the preferred method is to connect the points with a straight line.
7. Complete all the labels, legends, and titles.

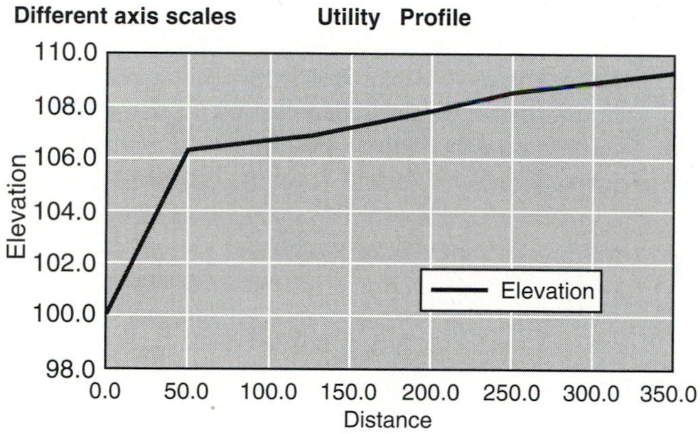

Figure 5 Profile plot with different scales.

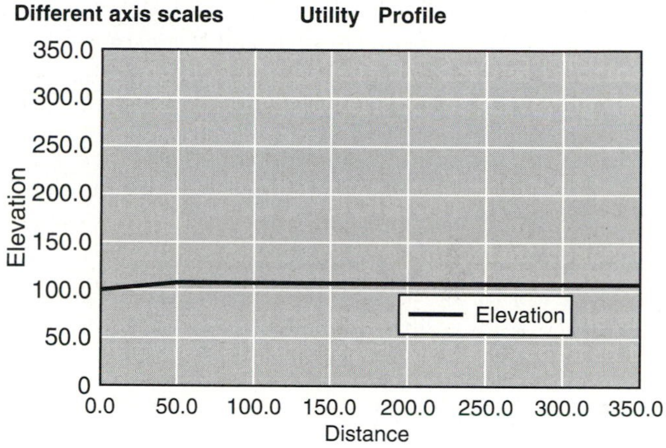

Figure 6 Profile plot with the same scale.

GLOSSARY

A

3-4-5 method — One of the indirect methods for laying out a 90 degree angle. Uses the principles of a right triangle.

Abney levels — A type of hand level that includes scales for direct reading percent slope and vertical angle.

accuracy — Mathematically, the number of decimal places included in a measurement.

add chain — Style of chain similar to a first foot chain, except that an additional foot has been added to the zero end for the graduated foot.

allowable error of closure — A system used to set standards for the amount of error that can occur in a survey. Several different mathematical equations are used based on the class order of the survey.

angle — Formed by the intersection of two lines and having three parts: a baseline, vertex, and second line. The measure of the rate of divergence for two lines.

angle of declination — The angular difference between true north and magnetic north. Available on USGS and other maps.

automatic level — A type of surveying instrument that has a built-in compensator to keep the instrument line-of-sight horizontal once the body of the instrument is close to level. It usually has three screws for leveling the instrument.

azimuth — The number of degrees from the north point. Angles to the right are + angles, and angles to the left are – angles.

B

backsight — A rod reading taken on a point of known or assumed elevation. Used to establish the height of the instrument.

balancing a traverse — The process of determining the amount of error that occurs during a traverse survey. The error is used to correct the angles and distances so they form a closed polygon.

balancing the sights — Setting the instrument halfway between the two stations being measured. Reduces the effect of instrument errors.

baseline — A reference line or line used as a basis for other measurements. The reference point for north and south tier identification in the rectangular system of land measurement.

bearing — A way of expressing a direction relative to a point.

benchmark — Point of known or assumed elevation.

breaking chain — Using less than a full chain, 100 feet, between pins when chaining. Used when horizontal chaining on slopes greater than 5 percent.

C

calibrating — A process that compares the performance of a measuring tool against a standard.

chaining — The process of measuring distance using a chain or tape.

chord — A line connecting two points on the perimeter of a circle.

clockwise — To rotate or move in the direction of the hands on a clock.

closed traverse — A series of angles and distances forming a polygon.

closing the loop — Surveying or measuring back to the starting point.

compass rule — One of the methods used to balance a traverse. In the compass rule method, the traverse must be oriented to north.

construction survey — Used to lay out a proposed construction project, or to evaluate a project after it is built.

construction transit — A term used to describe a group of instruments that are primarily a dumpy or automatic level that has been designed with a few degrees of vertical movement of the telescope. Sometimes called a transit level.

contour lines — Imaginary lines connecting points of equal elevation on a topographic map, and used to produce a three-dimensional map on a two-dimensional surface.

convergence/definition agreement — Coming together at one point. Principle meridian lines all converge at the North and South Poles.

cosine — One of the six trigonometric functions. It is the ratio of the adjacent side divided by the hypotenuse.

counterclockwise —To rotate or move in the opposite direction of the hands on a clock.

cut chain — Style of chain which is used by the head person holding the tape on the nearest foot mark and the rear person takes out the slack and reads the partial foot. To calculate distances, partial foot distances are subtracted from the measurement at the head of the chain.

D

decimal degrees — A method of recording angles that expresses fractions of a degree as a decimal.

deflection angle —The amount of deviation that has occurred from the direction of travel. Usually associated with surveys used for routes, such as utility lines or roads.

degrees-minutes-seconds —The sexagesimal system of recording angles. Each degree includes 60 minutes, and each minute includes 60 seconds.

departures — Used in the compass method for balancing a traverse to indicate the amount of movement to the east or west.

differential leveling — A surveying method used to determine the difference in elevation between two or more points, and commonly used to establish the elevation of a new benchmark.

differential surveys — Used to determine or establish the difference in elevation between two or more points.

distance —The amount of separation between two points, lines, surfaces, or objects measured along the shortest path joining them.

distance by stadia — A method of measuring distance using an instrument equipped with stadia crosshairs. Mathematically:

$$(\text{TSR} - \text{BSR}) \times \text{SF} = \text{Distance}$$

double reading — A method used to reduce rod-reading errors. The rod is read, the instrument is rotated off the rod and back, and then read the second time. Both readings are recorded.

dumpy level —Traditional four-legged level. Consisting of a telescope with crosshairs, a spirit level, a horizontal angle scale, and four leveling screws.

E

electronic distance measuring (EDM) — Used to describe a group of instruments that measure distance using microwave or laser signals.

electronic transit — A transit that includes a digital readout for all measurements.

elevation —The distance above or below a reference level surface or plane.

ellipse — A closed curve with two foci. The sum of the distances from any point on the perimeter to both foci is the same for all points.

ellipsoid — An ellipsoid is formed by rotating an ellipse. An ellipse rotated on its short axis, an ellipsoid is the preferred term used to describe the shape of the earth. It is the elevation formed by mean sea level.

extended foot tape — Surveyor chains having an extra foot added to the 0 end or both the 0 end and 100-foot end of the tape. It is used to determine partial feet.

F

field book — A small book with a specific style of lines on each page. The standard for recording surveying information.

first foot graduated chain — Surveyors chains having graduations only within the first (0 and 1) foot or first and last foot (0 and 1 and 99 and 100) of the chain. The graduated foot is used to determine partial feet.

foresight — A rod reading taken on a point of unknown elevation. The height of the instrument minus the foresight on the station equals the elevation of the station.

fully graduated chain — A surveying chain that has graduations between each foot mark on the chain.

G

geodetic surveying — A method of surveying that accounts for the curvature of the earth and that is used when surveying large areas and long distances.

geodetic surveys — Surveys that follow the principles of geodetic surveying.

Global Positioning System (GPS) — A system of satellites that provides signals, which a receiver can use to locate itself anywhere on the earth.

graduation —The subdivisions of a whole unit on a measuring scale.

guide meridians — A term used to describe a line used in the rectangular system of land measurement. A principle meridian is a continuous line between the North and South Poles. Guide meridians are in 24-mile sections between standard parrellels. They are not continuous lines because of convergence. The distance from the principle meridian is remeasured every 24 miles north and south of the baseline.

Gunter's chain — Early style of chain used in surveying. It was 66 feet long and consisted of 100 links. It has been replaced by the 100 foot chain.

H

hachure marks — Short lines, perpendicular to the contour line on topographic maps. They are used to indicate declining elevation.

half-stadia — A technique used when measuring distance by stadia when either the top or bottom stadia rod reading is obscured. Mathematically:

$$(\text{TSR} - \text{Elevation}) \times 2 \times \text{SF} = \text{Distance}$$

or

$$(\text{Elevation} - \text{BSR}) \times 2 \times \text{SF} = \text{Distance}$$

hand level — The simplest style of level, with a spirit level and a single crosshair.

height of instrument — Same as instrument height.

horizontal angle — An angle measured on a horizontal plane.

horizontal distance — A distance measured on a horizontal line or plane.

horizontal line — Formed when a line is established perpendicular to a vertical line, or when a line is established parallel with the horizon.

horizontal plane — A plane that is perpendicular to a vertical line or parallel with the horizon.

horizontal zero — One method for measuring vertical angles. Horizontal is 0 degrees. Angles measured up from horizontal are written as + angles, and angles measured down from horizontal are written as – angles.

I

inclinometer — Built-in feature available on laser rangefinders, which allows the user to determine vertical angles (object height).

initial point — The starting point. Commonly used to describe the junction of the baseline and the principle meridian in the rectangular system of land measurement. May also be used to describe the starting point of a route or other type of survey.

instrument height — The height of the instrument line-of-sight above the benchmark. It is determined by measuring the difference in elevation between the line-of-sight and the elevation of a benchmark.

interior angle — An angle that is always less than 180 degrees. Used to measure the deflection between adjoining sides of a polygon.

intermediate foresights — A foresight used to establish the elevation of a point that is not a turning point or benchmark. It is used to establish the elevation of a point important for the route or area being surveyed.

interpolating — The process of locating the position of a contour line between the elevations of two stations.

L

laser level — A type of surveying instrument that uses a laser beam to establish the measuring plane.

latitude — A measurement, given in degrees, of a position north or south of the equator.

latitudes — Used in the compass rule method for balancing a traverse to indicate the amount of movement to the north or south.

law of parallel lines — When a line connects two parallel lines, the opposite angles are equal.

least count — The smallest unit of measure for an angle scale or vernier scale.

level — The condition of an object when it is parallel with the horizon or perpendicular to a vertical line.

level surface — A continuous surface that is, at all points, perpendicular to the direction of gravity.

leveling — The process of determining if an object is parallel with the horizon, or if two or more objects are at the same elevation.

longitude — A measure of how far a position is east or west of zero longitude at Greenwich, England. It is measured in degrees.

LORAN — Long Range Navigation. An early location system that used radio waves.

lots — Parcels of land of regular or irregular size containing less than an acre or up to a few acres, identified by number.

M

metes and bounds — A system of land identification that uses natural and man-made structures as corner and boundary property markers. Distances may be estimated or given in pacing or other low-precision methods.

N

National Geodetic Vertical Datum of 1929 — The zero elevation reference point used for surveying. It was established by connecting 26 tidal benchmarks along the Atlantic, Gulf of Mexico, and Pacific Coasts.

Navigation Signal Timing and Ranging Global Positioning System — The full name for the Global Positioning System. See Global Positioning System.

note check — Used to determine the presence of mathematical errors in the data table. Mathematically:

$$|\Sigma \text{BS} - \Sigma \text{FS}| = |\text{BM1}_{\text{init}} - \text{BM1}_{\text{closing}}|$$

If the equation is true, the notes are okay. If not true, the notes contain a math error.

O

oblate spheroid — An older term used to describe the shape of the earth; a solid obtained by rotating an ellipse on its shorter axis.

odometer wheel — A mechanical revolution counter consisting of an odometer and a wheel.

open traverse — A series of angles and distances that define a line or route, but do not meet to form a complete geometric shape.

P

pacing — A technique for measuring distance that uses the length of a person's step, or pace. Distance equals number of paces times the length of pace.

parallax — An error that can occur when the viewer's line-of-sight through the telescope is not aligned with the optical line-of-sight of the telescope. It is eliminated by adjusting the eyepiece so that the crosshairs are at their darkest.

plane — A flat surface.

plane surveying — A method of surveying that assumes that the earth is flat and all elevations are recorded from a horizontal plane. Its use is appropriate for areas of a few acres and for short distances.

plane surveys — A survey that follows the principles of plane surveying.

plumb bob — A cylindrical object that tapers to a point on one end and a means for attaching a string at the centerline of the cone. It is used to transfer measurements in a vertical line.

plunging the telescope — Rotating the telescope on a transit 180 degrees vertically and also rotating the telescope 180 degrees horizontally. It is used to improve the accuracy of rod readings.

precision — Mathematically, the unit of measure. The smaller the unit of measure, the greater the precision.

principle meridian — A north-south line that passes through the initial point, and if extended far enough, through the North and South Poles.

profile — A side view of an object.

profile leveling — The process of determining the elevations of stations along a line to establish a profile.

profile survey — Used to establish a side view of the earth's surface. Commonly used for route surveys.

property survey — A survey that is used to establish the boundaries of a property.

Pythagorean theorem — Establishes the relationship between the lengths of the sides of a right triangle. Mathematically:

$$a^2 = b^2 + c^2$$

Q

quadrangle — A rectangular shape measuring 24 miles on each side used in the rectangular system of land measurement.

R

radio direction finding — One of the early attempts to use signals to locate positions. One type, LORAN, is still used today.

random error — An error that does not occur with a predictable pattern, and is difficult to control because the cause may not be known.

ranges — Used in the Rectangular System of Land Identification to identify a column of townships east and west of the principle meridian. Because each range is 24 miles wide, the range number indicates the distance east and west of the principle meridian.

range poles — Five- to six-foot tubes or rods with a solid, sharp point on one end, and painted in wide, alternating red and white stripes. They provide a visual reference for a line of travel or station location.

rangefinder — A device used to determine distances. Two types of instruments are used: optical and electronic.

Rectangular System of Land Description — A system developed by the U.S. government to identify parcels of land. The system results in a unique description of each parcel of land in the United States.

reference line — A line used as the reference or the basis for measurements.

reference plane — A plane used as a reference or the basis for measurements.

rod — A distance-measuring device used to determine the distance from the line-of-sight of the instrument to the ground.

rocking the rod — A technique used to eliminate the error that occurs when the rod is not held vertically. The rod is slowly rocked toward and away from the instrument. The shortest rod reading occurs when the rod is vertical.

rod level — A bull's-eye type of level attached to or held against the rod to ensure that it is vertical.

S

section — An area of land measuring 1 mile by 1 mile, containing 640 acres. Identified by number in the rectangular system of land description.

sine — One of the six trigonometric functions. The ratio of the opposite side divided by the hypotenuse.

slope — The rate of change in elevation.

slope distance — A distance measured while following the surface of the earth. May also be called surface distance.

spirit level — A tube or short cylinder filled with liquid and a bubble of air. The bubble is centered when the tube or cylinder is horizontal.

spot stations — Stations where elevations and locations are recorded because they are important to the survey, but are not parts of the route or the other stations being recorded.

standard parallel — A continuous line parallel to the baseline. Used in the Rectangular System of Land Description.

surveying — The art and science of measuring and locating points and angles on, above, and below the surface of the earth.

surveying pin — A metal pin 12 to 18 inches long that has traditionally been used to mark stations when chaining distances. Used in sets of 11.

surveying nail — A special nail with a cupped face that is used to locate the positions on top of a wooden stake.

systematic errors — Predictable errors, often due to limitations or problems associated with measuring instruments, and for which adjustments can be made.

T

tangent — One of the six trigonometric functions. The ratio of the opposite side divided by the adjacent side.

tape-sine method — An indirect method of measuring angles.

target — A disk divided into alternating red and white quadrants. Used when reading a rod indirectly or when recording rod measurements to 1/1000s of a foot. It contains a vernier, which provides an addition level of precision.

theodolite — A surveying instrument that looks like a transit, but has higher precision scales for measuring angles. Least counts may be as low as 1 second. Newer instruments are electronic.

three-wire leveling — A method used to reduce rod-reading errors. The rod reading for the top and bottom stadia and the elevation is averaged. Also called three-wire reading.

tiers — Used in the rectangular system of land measurement to describe a row of townships north and south of the baseline. Because each tier is 24 miles wide, the tier number indicates the distance north and south of the baseline.

topographic map — A map drawn with distances to scale and includes information about the surface of the earth in the map area, including relief, elevation, and the position of natural and man-made features.

topographic surveying — Used to collect data required to draw a topographic map, or three-dimensional drawing of the earth's surface.

total station — An instrument that is a combination of an electronic transit and an EDM.

township — An area measuring 6 miles by 6 miles, and divided into 36 sections.

transit — Traditional mechanical instrument of choice for professional surveyors. Includes a telescope that can be rotated 360 degrees horizontally and vertically, horizontal and vertical angle scales, a spirit level, compass and four leveling screws. Has been replaced by electronic equipment for professional surveys.

transit level — A term used to describe a group of instruments that are primarily a dumpy or automatic level that has been designed with a few degrees of vertical movements of the telescope. Same as construction transit.

traverse — A sequence of distances and angles.

triangulation — The process of locating a position by using at least one angle and distance from known points.

trilateration — A system used to locate a point by knowing the position of at least two reference points and three distances.

true foresight — A rod reading on a point with unknown elevation that will be used for a turning point or for a benchmark.

turning point — A station along a survey established as a temporary benchmark or reference point. Usually used to extend the survey. A stable structure such as a stake, curb, or sidewalk must be used.

U

Universal Transverse Mercator — A mapping projection of the earth's surface onto a flat surface.

United States Geological Service (USGS) — Government agency responsible for establishing and maintaining the elevation and location of a series of benchmarks across the United States.

V

verniers — A mechanical means of increasing the physical size of the last unit on a scale to provide an additional level of precision; read by identifying the line of coincidence.

vertex — The point where two lines meet.

vertical angle — An angle measured on a vertical plane, using either horizontal zero or zenith zero.

vertical line — A line that follows the direction of gravity. Can be established using a plumb bob.

vertical plane — A flat surface that is vertical.

W

Wide Area Augmentation System — A modification of the GPS established by the Department of Defense. Used by the Federal Aeronautics Association.

Z

zenith zero — One method for measuring vertical angles. Zero degrees are vertically overhead of the instrument.

INDEX

Page numbers followed by "f" indicate figure.

U

V

W

Z

Notes

Notes

Notes

Notes

Notes

Notes

Notes

Notes

Notes

Notes